METHODS IN PLANT ELECTRON MICROSCOPY AND CYTOCHEMISTRY

METHODS IN PLANT ELECTRON MICROSCOPY AND CYTOCHEMISTRY

Edited by

WILLIAM V. DASHEK

Mary Baldwin College, Richmond, VA

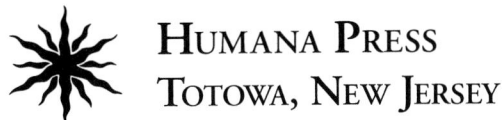

HUMANA PRESS
TOTOWA, NEW JERSEY

© 2000 Humana Press Inc.
999 Riverview Drive, Suite 208
Totowa, New Jersey 07512

For additional copies, pricing for bulk purchases, and/or information about other Humana titles, contact Humana at the above address or at any of the following numbers: Tel.: 973-256-1699; Fax: 973-256-8341; E-mail: humana@humanapr.com, or visit our Website: http://humanapress.com.

Cover design by Patricia F. Cleary.

This publication is printed on acid-free paper. ∞
ANSI Z39.48-1984 (American National Standards Institute) Permanence of Paper for Printed Library Materials).

Photocopy Authorization Policy:
Authorization to photocopy items for internal or personal use, or the internal or personal use of specific clients, is granted by Humana Press Inc., provided that the base fee of US $10.00 per copy, plus US $00.25 per page, is paid directly to the Copyright Clearance Center at 222 Rosewood Drive, Danvers, MA 01923. For those organizations that have been granted a photocopy license from the CCC, a separate system of payment has been arranged and is acceptable to Humana Press Inc. The fee code for users of the Transactional Reporting Service is: [0-89603-589-1/99 $10.00 + $00.25].

Printed in the United States of America.   10 9 8 7 6 5 4 3 2 1

Library of Congress Cataloging-in-Publication Data

Methods in plant electron microscopy and cytochemistry / edited by William V. Dashek.
        p. cm.
    Includes bibliographical references (p.).
    ISBN 0-89603-589-1 (alk. paper)
    1. Botanical microscopy--Technique. 2. Plant cytochemistry--Technique. 3. Electron microscopy--Technique. I. Dashek, William V.

    QK673 .M48 2000
    580'.28'25--dc21

                                                        00-022021

PREFACE

Although there is a copious supply of electron microscopy text-books and monographs concerned with animal cells and tissues, there are few whose thrust is plant cell biology. Because of the unique characteristics of the higher plant cell, i.e., the cell wall and often a large central vacuole, the techniques employed for animal cells are often not directly applicable to plant cells. Because many of the available methods for the optical microscopy of plant cells/tissues are now quite dated, *Plant Electron Microscopy and Cytochemistry* includes chapters regarding light microscope cytochemistry, autoradiography, and immunocytochemistry. Recent developments in fluorescence, confocal, and dark-field microscopies are highlighted. Light microscopy is often employed in conjunction with electron microscopy as correlative microscopy. With regard to electron microscopy, recent advances in conventional transmission and scanning electron microscopies are presented together with highly contemporary ancillary techniques. The latter include: high-resolution radioautography, immunoelectron microscopy, x-ray microanalysis, and electron systems imaging, as well as atomic force and scanning tunneling microscopies. Prior to the summation, *Plant Electron Microscopy and Cytochemistry* concludes with a chapter centering about the uses of electron microscopy in molecular biology.

Although this manual is concerned with higher plants, some chapters present information relevant to lower plants. In this connection, the position of the fungi has been debated for years. Most taxonomists do not include fungi in the plant kingdom. However, this manual includes some fungal systems since they are involved in wood decay.

Finally, though Humana Press has published *Electron Microscopy Methods and Protocols* by M. A. Nasser Hajibagheri in 1999, *Plant Electron Microscopy and Cytochemistry* is dedicated to plant studies, and should be quite useful to professors, certain graduate and undergraduate students, and postdoctorals, as well as government and industrial scientists.

William V. Dashek

DEDICATION

This volume is dedicated to Kristin Dashek Simpson and Karin Dashek Waddill who patiently dealt with my life-long dedication to scholarship. I am grateful to Drs. W. G. Rosen, W. F. Millington, D. T. A. Lamport, and J. E. Varner for my scientific development. The financial support of the National Science Foundation, the National Institutes of Health, the Department of Energy, and the USDA-Forest Service for aspects of the reported research is acknowledged. Gratitude is extended to Ms. Rebecca Kittelberger, Adult Degree Program, Mary Baldwin College, for her intelligent assistance with administrative details. Lastly, I express my deepest appreciation to Ms. Kim Chancey, Department of Botany, University of Georgia, for her patience and attention to detail during her word-processing preparation of this volume. I am grateful to the librarians at the Academic Campus of Virginia Commonwealth University and the Henrico Tuckahoe Library. I am thankful, also, to the Humana Press staff, especially Tom Lanigan, Paul Petralia, and Elyse O'Grady for their work on the publication of *Methods in Plant Electron Microscopy and Cytochemistry*.

CONTENTS

CONTRIBUTORS

WILLIAM V. DASHEK • *Adult Degree Program, Mary Baldwin College, Richmond, VA*

SEIZO FUJIKAWA • *Environmental Cryobiology Group, Institute of Low Temperature Science, Hokkaido University, Sapporo, Japan*

ELIOT M. HERMAN • *Agricultural Research Service, Climate Stress Laboratory, United States Department of Agriculture, Beltsville, MD*

TETSUYA HIGASHIYAMA • *Department of Biological Sciences, Graduate School of Science, University of Tokyo, Tokyo, Japan*

MARTIN J. HODSON • *School of Biological and Molecular Sciences, Oxford Brookes University, Oxford, UK*

TAKAKO S. KANEKO • *Department of Chemical and Biological Sciences, Faculty of Science, Japan Women's University, Tokyo, Japan*

ALYCE LINTHURST • *Center for Biotechnology, Old Dominion University, Norfolk, VA*

ROSLYN A. MARCH-AMEGADZIE • *GESS Division, Ivy Tech State College, Indianapolis, IN*

JOHN E. MAYFIELD • *Department of Biology, North Carolina Central University, Durham, NC*

JOHN E. NIELSEN • *Danisco Biotechnology, Copenhagen, Denmark*

MASAKO OSUMI • *Department of Chemical and Biological Sciences, Faculty of Science, Japan Women's University, Tokyo, Japan*

MAMIKO SATO • *Department of Chemical and Biological Sciences, Faculty of Science, Japan Women's University, Tokyo, Japan*

DAVID W. SEABORN • *Department of Biological Sciences, Old Dominion University, Norfolk, VA*

VIRGINIA A. SHEPHERD • *Department of Biophysics, School of Physics, The University of New South Wales, Sydney, Australia*

RENATA ŚNIEŻKO • *Department of Cell Biology, Institute of Biology, Maria Curie-Sklodowska University, Lublin, Poland*

HOWARD J. SWATLAND, *Department of Food Science, University of Guelph, Guelph, Canada*

ROSANNAH TAYLOR • *Department of Biology, American University, Washington, DC*

NORIFUMI UKAJI • *Environmental Cryobiology Group, Institute of Low Temperature Science, Hokkaido University, Sapporo, Japan*

JENNIFER L. WOLNY • *Fish and Wildlife Conservation Commission, Florida Marine Research Institute, St. Petersburg, FL*

SHIZUO YOSHIDA • *Environmental Cryobiology Group, Institute of Low Temperature Science, Hokkaido University, Sapporo, Japan*

1

Plant Cells and Tissues: Structure–Function Relationships

William V. Dashek

1. INTRODUCTION

This chapter is an introduction to methods-oriented microscopy. Because the contributing authors present methods in relation to their researches, plant cell structure–function relationships as revealed by light and electron microscopies are reviewed. Much of this conceptual and terminological information is summarized in tables that are augmented with references to either photomicrographs or electron micrographs of cells and tissues.

2. STRUCTURE OF CELLULAR MEMBRANES

Transmission electron microscopy, together with biochemical methods, have unveiled a copious number of functional organelles comprising most higher plant cells. In addition, the marriage of these disparate techniques has yielded significant information regarding the structure–function relationships of the diverse cellular organelles. Furthermore,

From: *Methods in Plant Electron Microscopy and Cytochemistry*
Edited by: W. V. Dashek © Humana Press Inc., Totowa, NJ

1

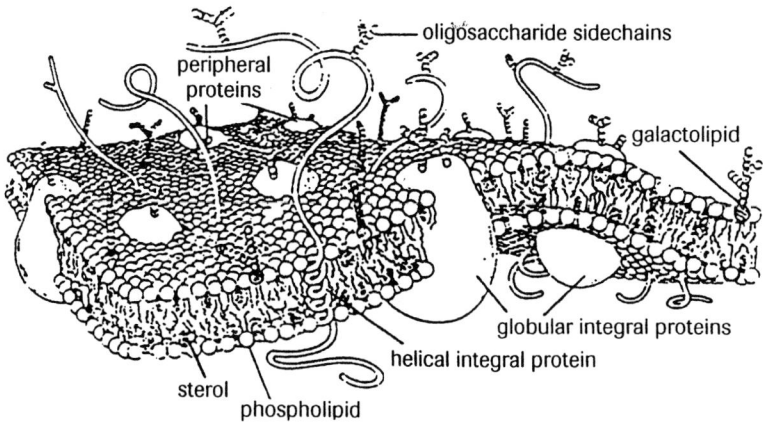

Fig. 1. Fluid mosaic model of membrane. From ref. *(5)* with kind permission of Kluwer Academic Publisher

applications of these techniques as well as ancillary technologies such as freeze fracturing and freeze etching have resulted in a critical rethinking of the Davson–Danelli model of cellular membranes. Today, most students of the life sciences learn the fluid-mosaic model *(1)* of the membrane (Fig. 1). According to this widely accepted model, cellular membranes are composed of a lipid bilayer in which globular proteins are embedded. Integral proteins often traverse the bilayer and protrude on either side of the membrane. Whereas the embedded protein is hydrophobic *(2)*, the exposed portion is hydrophilic. Short-chain carbohydrates attached to the protruding proteins are thought to function in cell-to-cell adhesion. Like most areas of contemporary life science research, both the structure and function of cellular membranes are continually being reexamined *(3)*. Table 1 summarizes the composition of cellular membranes. In addition, the reader is referred to Packer and Douce *(4)*, Leshem et al. *(5)*, Yeagle *(6)*, and Smallwood et al. *(7)* for in-depth recent reviews of membrane structure–function.

3. PLANT CELL ORGANELLES: STRUCTURE–FUNCTION RELATIONSHIPS

One striking advance in modern subcellular biology is a more complete understanding of the interrelationship of certain plant cell organelles (Fig. 2A–F). Rather than existing as discrete entities in physiological and biochemical isolation of one another, it is now generally accepted

Table 1
Composition of Certain Cellular Membranes

Chemical Composition	
Fatty acyl groups in membrane lipids	
16:0, 16:1, t-16:1, 16:3, 18:0, 18:1, 18:2, α 18:3, δ 18:3, 18:4, 22:0, 22:1, 24:0, 24:1	
Electroneutral phospholipids	Phosphatidylcholine, phosphatidyl-ethanol, phosphatidylethanolamine
Anionic phospholipids	Phosphatidylserine, phosphatidylglycerol, phosphatidylinositides
Lyo-phospholipids	
Sphingolipids	Cerebrosides
Chloroplast-specific glycerolipids	Galactolipids, sulpholipids
Mitochondrial phospholipids	Diphosphatidylglycerol and monophosphatidylglycerol
Sterols	
	Sitosterol
	Campesterol
	Stigmasterol
	Unusual sterols
	Cycloartenol
	Cholesterol, minute quantities
Sterolglycosides	
Lanosterol	Pathogenic fungal membranes
Water	
Extramembrane water	Membrane is a bilayer sandwiched between two layers of water
	Water located within the bilayer which is attached to or in approximate contact with the expanses of membrane constituents
Proteins	
Integral	May cross the membrane once or several times and are linked either electrostatically or by means of biophysical lipophilicity to the inner domains of the bilayer
Simple integral proteins	Classic α-helical structure that traverse the membrane only once
Complex integral proteins	Globular—comprised of several α-helical loops that may span the membrane several times
Peripheral proteins	Associated with only leaflet—easily isolated by altering ionic strength or pH of the encasing medium
Transport proteins	Pumps, carrier, and channels

Summarized from Leshem et al. *(5)*.

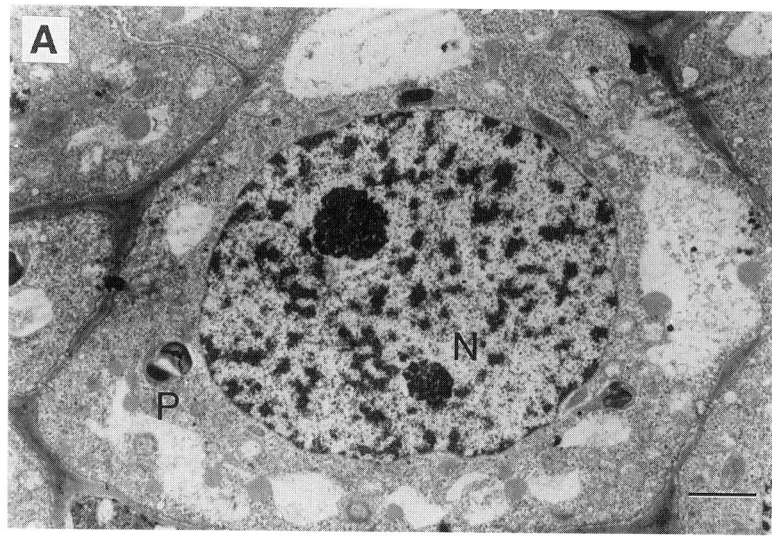

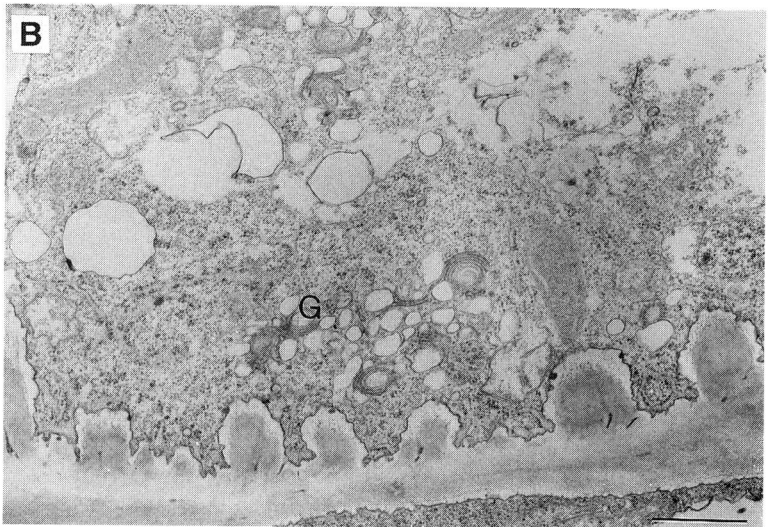

Fig. 2. Electron micrographs of a plant cell (**A**) nucleus (N) and plastids (P). (**B**) Golgi apparatus (G).

that the rough endoplasmic reticulum (rer), Golgi apparatus, and cell surface exist as a functional continuum, i.e., the endomembrane system *(26)*. Recently, Morré and Keenan *(68)* reviewed the current status of the system which, in their opinion, is "still a valid explanation for the transport of proteins and lipids from the er to the plasma membrane." The

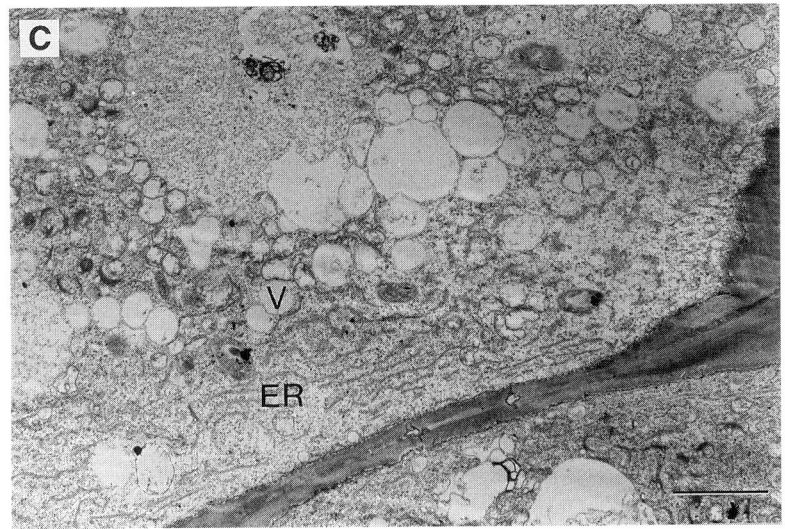

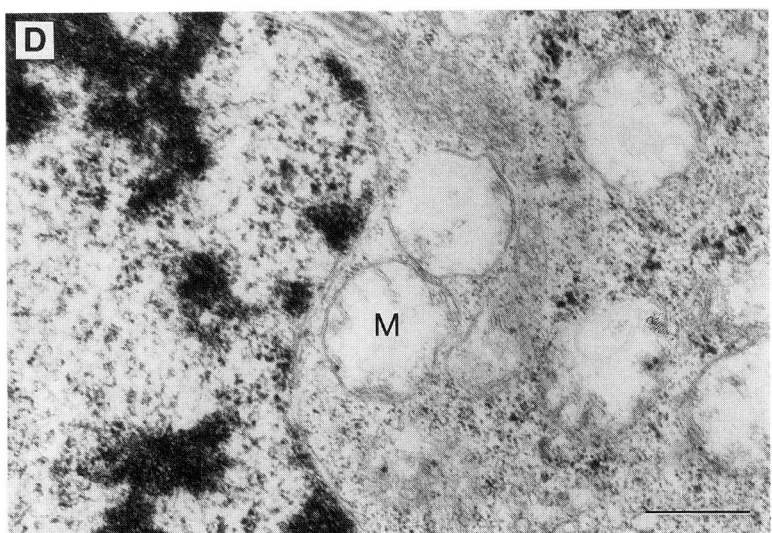

Fig. 2. (Continued). (**C**) Endoplasmic reticulum (ER). (**D**) Mitochondria (M).

hypothesis at the heart of the endomembrane concept is that membranes flow through an interconnected system of the er, Golgi apparatus, and plasma membrane (Fig. 3). Ultrastructural and biochemical evidence exists that the interconnections of these elements with one another involves "vesicular and transitional membrane forms." Morré and

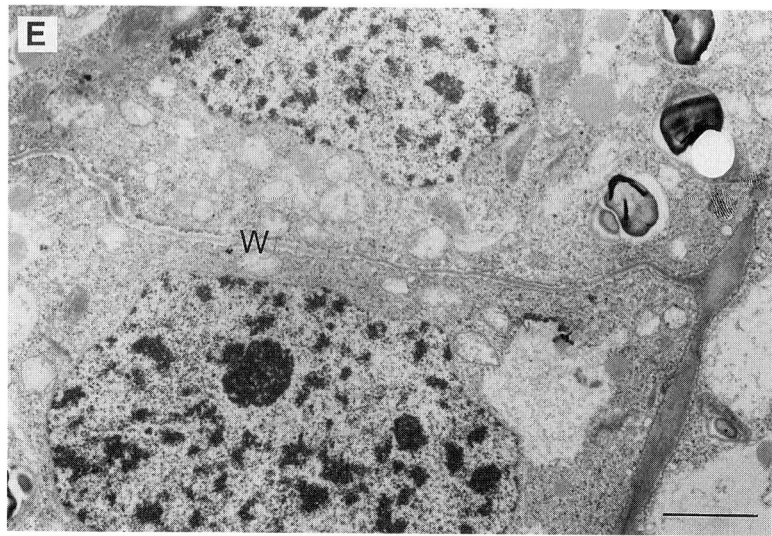

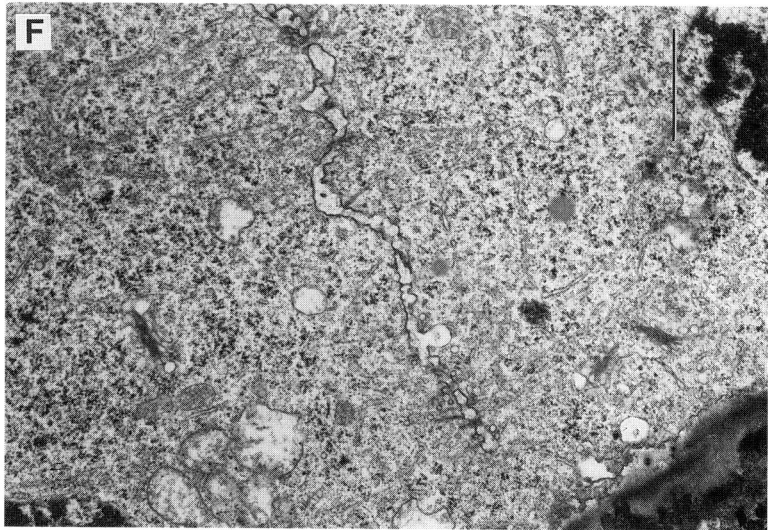

Fig. 2. (Continued). (**E**) Cell wall (W). (**F**) Cytoskeletal elements. From Mayfield JE, Dashek WV. *Methods in Plant Biochemistry and Molecular Biology* (Dashek WV, ed.), CRC Press, Boca Raton, FL, 1997. Bar = 2 µm (**A**), 1 µm (**B**), 2 µm (**C**), 0.5 µm (**D**), 0.5 µm (**E**), 1 µm (**F**).

Keenan *(68)* thoroughly discuss the nature of these vesicular and transitional membrane forms. In particular, these well-established, investigators evaluate er to Golgi apparatus transfer, Golgi apparatus to plasma membrane transfer, as well as models for conveying materials through

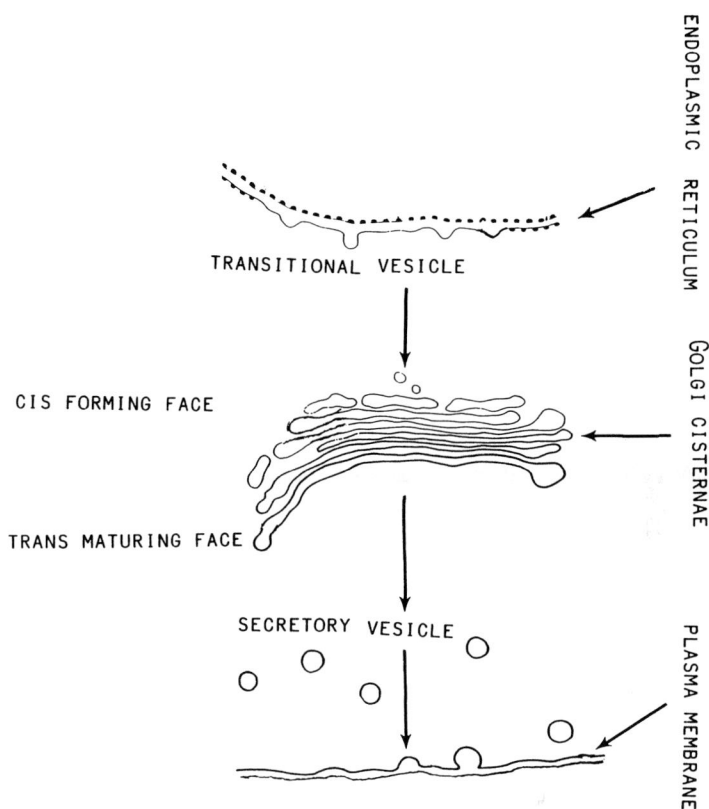

ENDOPLASMIC RETICULUM

GOLGI CISTERNAE

PLASMA MEMBRANE

TRANSITIONAL VESICLE

CIS FORMING FACE

TRANS MATURING FACE

SECRETORY VESICLE

Fig. 3. Endomembrane concept.

the Golgi apparatus. With regard to the models, Morré and Keenan *(68)* critically examine the evidence for the vesicle shuttle model (Fig. 4A) and the flow differentiation model (Fig. 4B). They conclude that the flow differentiation model of Golgi apparatus functioning remains as relevant as when it was first proposed. The reader is encouraged to read the thorough review of Morré and Keenan *(68)* to gain a historical perspective of current thinking regarding mechanisms of plant cell secretion.

3.1. The Plant Cell Wall

The components of the plant cell wall *(8–21)* are the middle lamella (intercellular substance), the primary wall, and the secondary wall. The middle lamella is the pectic layer between cells and holds adjoining cells together as do membrane carbohydrates. The primary wall is thin (1–3 μm) and flexible containing cellulose, hemicelluloses, pectins, and glycoproteins. This wall provides mechanical strength, maintains cell shape,

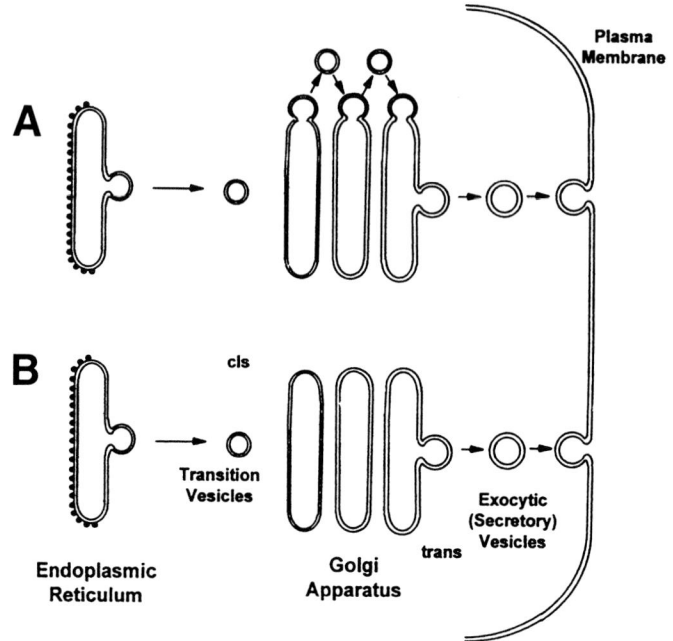

Fig. 4. Comparison of the vesicle shuttle (**A**) and flow differentiation (**B**) models of intra Golgi apparatus transport. From: Morré and Keenan *(68)* with permission.

controls cell expansion and intercellular transport, protects against certain organisms, and can function in cell signaling. The cellulose is a β1, 4D-glucan and the pectic substances consist of galacturonans, rhamnogalacturonans, arabnans, galactans, and arabinogalactans I. In contrast, the hemicelluloses are comprised of xylans, glucomannans, galactoglucomannans, xyloglucans, β-D-glucans, and other polysaccharides. The latter include β, 1-3-linked D-glucans (callose), arabinogalactans, and glucuronomannans. In addition to polysaccharides, primary walls contain proteins, e.g., hyp-containing glycoproteins and expansin. Expansin may increase plant cell wall plasticity in vitro and possibly tissue expansion in vivo.

The secondary wall is located inside the primary wall and can possess pits and sometimes three distinct layers, S_1, S_2, and S_3. The secondary wall can contain 25% lignin, and cellulose is more abundant than in primary walls. Support and resistance to decay are the main functions of the primary wall.

Plasmodesmata *(22–25)* may occur throughout the plant cell wall and provide a pathway for the transport of certain substances between cells.

The plasmodesmata may be aggregated in primary pit fields or in the pit membranes between pit pairs. The plasmodesmata appear as narrow canals (2 μm) lined by a plasma membrane and are traversed by a desmotubule, a tubule of endoplasmic reticulum. The plasmodemata are dynamic altering their dimensions and are functionally diverse. For example, whereas some transport endogenous plant transcription factors, others transport numerous proteins from companion cells to enucleated sieve elements.

3.2. The Plasmalemma

The plasmalemma (26–28) is the outer limiting membrane of the plant cell. An intervening space, the periplasmic space, can exist between the plasmalemma and the cell wall. Paramural bodies, plasmalemasomes, and multisvesicular bodies are derivatives of the plasmalemma and presumably function in solute transport, especially in transfer cells. The functions of the plasmalemma include mediating transport of substances, coordinating the synthesis of cell wall cellulosic microfilaments, and translating hormonal and environmental signals involved in the control of cell growth and differentiation.

3.3. Plastids

Plant cells abound with various types of plastids with each being bounded by an envelope consisting of two membranes. Plastids are usually classified on the basis of developmental stage and the pigments that they contain.

Chloroplasts (29–36) are the sites of photosynthesis and their ribosomes can carry out protein synthesis. Chloroplasts that contain chlorophylls and carotenoids, are disc shaped and 4–6 μm in diameter. These plastids are comprised of a ground substance (stroma) and are traversed by thylakoids (flattened membranous sacs). The thylakoids are stacked as grana. In addition, the chloroplasts of green algae and plants contain starch grains, small lipid oil droplets, and DNA.

Chromoplasts are pigmented plastids that contain carotenoids but lack chlorophyll. These plastids are attractants to insects and animals.

Leucoplasts are nonpigmented plastids that synthesize starch. Some leucoplasts, proteinoplasts, contain protein.

Proplastids are small colorless or pale green undifferentated plastids that occur in the meristematic cells of roots and stems. Proplastids are precursors of more highly differentiated plastids. Etioplasts are proplastids containing prolemellar bodies and are precursors of chloroplasts developmentally arrested by low light levels.

3.4. The Nucleus

The nucleus *(37–44)* is the most prominent structure within the cytoplasm. It is bounded by a nuclear envelope containing circular pores that are 30–100 nm in diameter. The outer nuclear envelope may be continuous with the er.

The nucleus contains chromatin within the nucleoplasm. The chromatin is comprised in part of DNA, the genetic information that controls cellular activities through determining which proteins are produced and when. Chapter 20 of this volume describes the electron microscopy of nucleic acids.

3.5. Mitochondria

Mitochondria *(45–56)* are organelles possessing a double membrane, the inner of which is invaginated as cristae. An intermembrane space exists between the inner and outer membranes. The inner membrane consists of an unusually high amount of protein and possesses spherically shaped particles approx 9 nm in diameter. These particles appear to be equivalent to F_0, F_1, and adenosine triphosphatase. In contrast to the inner membrane, the outer membrane is smooth and appears to be connected to the smooth er. This membrane is permeable to all molecules of 10,000 Dalton or less. A mitochondrial matrix is enclosed by the inner membrane and consists of a ground substance of particles, nucleoids, ribosomes, and electron-transparent regions containing DNA.

3.6. Protein Bodies and Endoplasmic Reticulum

These organelles occur in the endosperm of cereal grains and their structures are tissue specific. They are about 2–5 µm in diameter and often contain globoid and occasionally crystalloid inclusions. Prolamin accumulates in small or large spherical bodies. Crystalline protein bodies are the sites of accumulation of nonprolamin storage proteins.

The er is a three-dimensional membrane system *(57–62)*. As visualized in a transmission electron microscope, there are two parallel membranes with an intervening electron transparent space, the lumen. The form and abundance of the er vary. The rer are flattened sacs with numerous attached ribosomes (15–20 nm). In contrast, the smooth er (ser) lacks ribosomes. The er seems to function as a communication system within cells and can be continuous with the outer nuclear envelope. Although the rer is involved in protein synthesis, the ser functions in glycosylation.

3.7. Golgi Apparatus

The Golgi apparatus *(63–69)* or collection of all the dictyosomes is composed of flat-disc shaped sacs or cisternae. The Golgi apparatus appears to possess opposite poles, the *cis* (forming) and *trans* (maturing) faces. Whereas the *cis* membranes are similar to the er, the *trans* membranes resemble the plasmalemma. Direct evidence supports the existence of free vesicle intermediates between the er and the *cis*-Golgi apparatus and between the *trans*-Golgi apparatus and the plasma membrane. The question of whether vesicles convey materials through the Golgi apparatus remains controversial. Two models regarding the controversy are the *vesicle shuttle* model (two-compartment discontinuous model) and the *flow differentiation* model. Morré and Keenan *(68)* have reviewed the evidence for and against these theories. The functions of the Golgi apparatus are secretion, cell wall synthesis, transport of glycoproteins, and transformation of er-like membranes into plasma membrane-like membranes.

3.8. Vacuoles

Vacuoles *(70–78)* are membrane-bound regions of the cell filled with cell sap. Vacuoles are surrounded by a tonoplast (vacuolar membranes) and are diverse with distinct functions. Most investigators believe that lysosomes and the plant vacuoles are the same. Vacuoles develop turgor pressure and maintain tissue rigidity. They are storage components for various metabolites such as reserve proteins in seeds and malic acid in crassulacean acid metabolism (CAM) plants. Vacuoles can remove toxic secondary products and are the sites of pigment deposition.

3.9. Ribosomes

Ribosomes *(79–87)* are small organelles 17–23 nm in diameter. They can exist in clusters known as polysomes or be attached to the er where they bind to pores in the er membrane. A major constituent of the er pore is translocon, the heterotrimetric Sec 61 protein complex. Sec 61 binds to the 80s ribosomes *(86)*. Ribosomes consist of subunits, a 30s subunit (16srRNA and 21 proteins), and a 50s subunit (23s and 5s RNAs, > proteins and the catalytic site of peptidyl transferase). Ribosomes are the sites of protein synthesis.

3.10. The Cytoskeleton

The cytoskeleton *(88–96)* is a complex network of protein filaments. Microtubules are long cylindrical substances of approx 24 nm in diameter and varying lengths. Each microtubule (0.5–1.0 μm in diameter and 25 μm in internal diameter) is composed of tubulin subunits. The assembly of microtubules occurs at nucleating sites or a microtubule-organizing center. During the polymerization of tubulin, guanosine triphosphate is hydrolyzed to guanosine diphosphate.

With regard to microtubular ultrastructure, microfilaments (5–7 nm in diameter) are composed of filamentous actin. The tubule-like structures are formed by α, β-tubulin heterodimers. The wall is composed of 13 parallel protofilaments. Various microtubule-associated proteins and motor proteins (kinesin and dynein) are bound to the wall. The microtubule is a polar structure, i.e., plus and minus ends.

Among the functions of microtubules are the orderly growth of the cell wall through alignment of the cellulosic microfibrils, directing vesicles toward the developing wall, comprising spindle fibers, and cell plate formation. In addition, microtubules may be involved in cell wall deposition, tip growth of pollen tubes, nuclear migration, and cytoplasmic streaming.

3.11. Microbodies

Microbodies (97–101) are spherical organelles (0.1–2.0 µm in diameter) bounded by a single membrane. They possess a granular interior and sometimes crystalline protein body. A specialized type of microbody is the glyoxysome (0.5–1.5 µm) containing enzymes of the glyoxylate cycle. Glyoxysomes are found in the endosperm or cotyledons of oily or fatty seeds.

Finally, electron micrographs of these plant organelles have appeared many times in textbooks (106,107,109) and the original research literature (see references regarding organelles at the conclusion of this chapter).

4. PLANT CELLS AND TISSUES

A compilation of structure–function relationships for plant cells is depicted in Table 2 while Table 3 categorizes their locations within plant tissues. A color atlas of plant structure has been published by Bowles (102). The reader is referred to Esau (103), Fahn (105), Fosket (108), Maseuth (106), Moore and Clark (109), Steeves and Sussex (104), and Raven et al. (107) for in-depth discussions of both the morphology and physiology of stems, roots, leaves, and flowers, i. e., plant organs. Certain of these volumes consider the structural differences between monocot and dicot roots and stems.

5. CONCLUSIONS

Even the most cursory review of the past 50 years of research literature regarding plant organelles, cells, and tissues reveals that our current knowledge of their structure–function relationship has been dependent on advances in microscopic methodologies. Thus, the remainder of this

Table 2
Structure and Function of Plant Cells [a,b]

Cell type	Structure	Function
Parenchyma	Isodiametric, thin-walled primary cell wall; in some instances can have secondary cell walls; not highly differentiated	Photosynthesis, secretion, organic nutrient and water storage, regeneration as in would healing
Transfer cells	Specialized parenchyma cells, plasmalemma greatly extended, irregular extensions of cell wall into protoplasm	Transfer dissolved substances between adjacent cells, presence is correlated with internal solute flux
Chlorenchyma	Parenchyma cells containing chloroplasts	Photosynthesis
Collenchyma	Rectangular in longitudinal section, retain a protoplast at maturing, primary wall unevenly thickened, non-lignified, highly hydrated	Provide support for growing and mature organs, aerial portions of the plant only
Schlerenchyma	Thick cell walls containing lignin-nonextensible secondary walls, lack protoplasts at maturity, living or dead at functional maturity	Strength and supporting elements of plant parts, scattered throughout plant (also see conducting tissue)
Epidermal cells	Tabular; are layered sheets on surfaces of leaves and young roots, stems, flowers, fruits, seeds, ovules	Secrete the fatty substance, cutin, which forms a protective layer, the cuticle; cuticle covered by an epicuticular wax
Guard cells	Specialized epidermal cells, crescent-shaped, contain chloroplasts, form defines stomatal pore	Regulate stomatal aperture/pore for gas exchange
Subsidiary cells	Surround guard cells of stomata	Reservoirs for water and ions
Trichomes	Single-celled or multicellular outgrowths of epidermal cells	Produce volative oils for glandular trichomes; reflect bright light in some desert plants; increase boundary layer; mechanically discourage predators; salt secretion in some halophytes; digestions in sundews

(continued)

Table 2 (Continued)

Cell type	Structure	Function
Cork cells	Tabular with all walls suberized; occur in thick layers on the outer surfaces of older stems and roots	Secrete a fatty substance, suberin, into the walls, suberin renders cork cells waterproof and helps protect the tissues beneath
Lenticels	Loosely arranged group of cells in the periderm	Involved in gas exchange inward and outward
Cells of the vascular system		
Tracheids	Long tapering cell with lignified, secondary wall thickenings; or pitted secondary walls; lack protoplasts at maturity; more primitive (less specialized) type of cell than vessel members; widespread in vascular plants	Conduct water and mineral salts; water moves upward from tracheid to tracheid through pits
Vessel element of vessel member; arranged end to end	Perforated cell with lignified secondary wall thickenings or pitted secondary walls; protoplast is lacking at maturity; portion of end wall degraded to form performations; more evolutionarily advanced than tracheids; angio- sperms often contain both tracheids and vessels	Rapid transport of water and mineral solutes; more efficient than tracheids
Wood fiber (type of sclerenchyma cell)	Larger, more slender than the tracheid; tapering; heavy wall thickening and lignification; pits are reduced in size and number relative to tracheids; Protoplast lost as cell attains maturity	Strength
Sieve cells	More primitive than sieve tube members and occur in nonflowering plants; sieve cells contain clusters of pores, which are narrow and uniform in structure	Translocation of sugars and other organic nutrients

Table 2 (Continued)

Cell type	Structure	Function
Sieve tube elements (sieve tube is a vertical row of sieve tube elements)	Vertical row of elongated, specialized cells; possess multiperforate end walls or sieve plates; possess living protoplasts at maturity but lack nuclei; when young sieve element contains one or more vacuoles separated by a tonoplast; at maturity tonoplast disappears; protoplasts of sieve-tube members of dicots and some monocots contain P-protein	
Albuminous cells	Specialized parenchyma cells associated with sieve cells in gymnosperms	May function like companion cells
Companion cells	Specialized parenchyma cells; possess numerous plasmodesmatal connections with sieve tube members	May play a role in the delivery of substances to sieve tube members

[a] Summarized from refs. *(106)* and *(109)*.

[b] The anatomies of these cell types have been extensively reproduced *(102–109)*. Recently, Bowles *(102)* published a color atlas depicting many of the cell types.

volume presents certain methodological advances that should ultimately expand our knowledge base of plant cell biology. In this regard, contributing authors discuss novel, contemporary microscopical technologies in a user-friendly manner. The intent of the authors is to provide conceptual information and protocols for immediate microscopic application by the research scientist. The arrays of light and electron microscopic techniques available to the research scientist are displayed in Table 4. Finally, a variety of contemporary volumes is available providing methods for light microscope photomicrography *(160)* and video image analysis *(161–164)*.

ACKNOWLEDGMENT

This manuscript was prepared in part while W. V. Dashek was an adjunct faculty member in the Department of Biology at the University of Richmond. I express my appreciation to Dr. W. John Hayden for sharing his extensive knowledge of plant anatomy.

Table 3
Summary of Plant Tissues

Tissue type	Composition	Function
Meristematic	Meristematic cells	Cell division and growth
		Increase in length
		Increase in girth
Apical or primary	Meristematic cells	Produce primary body of root-and-shoot system
Lateral or secondary	Fusiform, ray initials	Produce secondary xylem and secondary phloem in woody plants
Protective or dermal		Mechanical protection
		Restriction of transpiration aeration
Epidermal tissue of leaves	Epidermal cells with and without modifications	Mechanical protection
		Restriction of transpiration aeration
Peridermal tissue of stems and leaves	Cork cells and cork cambium	Mechanical protection
		Restriction of transpiration aeration
Fundamental or ground		Production and storage of food; wound healing
Parenchyma	Parenchyma cells	
Chlorenchyma	Chlorenchyma cells	Photosynthesis
Collenchyma	Collenchyma cells	Flexible/extensible support for growing primary tissues
Sclerenchyma	Sclerenchyma cells	Mechanical support
Conductive tissue		
Xylem	Tracheids, vessel elements,	Conduction of water and minerals;
Phloem	sieve cells, sieve tubes	translocation of organic nutrients

See Table 2 for references.

Table 4
Summary of Light and Electron Microscopic Techniques

Light microscopy (see refs. 110–118)		Electron microscopy and ancillary models (see refs. 119–159)	
Technique	*Application*	*Technique*	*Application*
Bright field	Conventional microscopy	Atomic force	Mapping of surfaces to an atomic scale
Confocal scanning optical microscopy	Examination of cells in live tissue in bulk samples	Cryoelectron microscopy	Imaging of biological macromolecules in the absence of specimen dehydration and staining
Confocal fluorescence	DNA labeled with more than one fluorescent tag		
Dark field	Visualization technique for ashes produced by micro-incineration and fluorescence microscopy; useful for low-contrast subjects	Electron systems imaging	Detection, localization, and quantitation of light elements
		EM shadowing	Structural information from ordered arrays of macromolecules
Epi-illumination	Subcellular imaging structures	Freeze fracture	Preparation of cellular ultrastructures in frozen-hydrated and living state for electron microscopy; macromolecular organization of bilayer membranes
Reflection contrast	Quantification in gap between light and em microscopies	Immunoelectron	Localization of cellular antigens
Reflection-imaging microscopy	Useful for imaging highly reflective particles such as silver grains in autoradiographs		
Field ion microscopy	Atomic structure of crystals		
Nearfield scanning optical	Determine single molecules on surfaces	Negative staining	Useful for detergent-extracted cytoskeletons, membrane fractions, organelles

(continued)

Table 4 (Continued)

Light microscopy (see refs. 110–118)		Electron microscopy and ancillary models (see refs. 119–159)	
Technique	*Application*	*Technique*	*Application*
Nuclear magnetic resonance microscopy	High-resolution 3-D Imaging of living plants Forms images if H_2O in the body; water distribution and binding in transpiring plants and H_2O transport in plants with light-stressed foliage		
		Scanning electron microscopy	Surface topography
Nomarski differential interference contrast	Reveals edges in biological structures, e.g. organelle and nuclear boundaries, cell boundaries and cell walls; also images fibrous subcellular components, e.g., microtubules.	Scanning tunneling microscopy	Surface topography, surface spectroscopy, image internal structure of macromolecules such as proteins, liquid crystals, and DNA
Phase contrast microscopy	Produces visible differences in retardations of light waves, useful for examining biological material which possesses limited inherent direct contrast	Transmission Electron microscopy	Subcellular morphology
Polarization microscopy	Most useful for highly birefringent objects, e.g., cellulose microfibrils in cell walls and distinguishing crystalline and noncrystalline inclusions	X-ray microanalysis	Detection, localization, and quantitation of elements
Raman microscopy	Analysis of bioaccumulations in plant vacuoles		

18

REFERENCES

1. Singer SJ, Nicolson GL. The fluid mosaic model of the structure of cell membranes. *Science* 1972; 175: 720–731.
2. Engelman DM. Crossing the hydrophobic barrier. Insertion of membrane proteins. *Science* 1996; 274: 1850–1851.
3. Jacobson K, Sheets ED, Simson R. Revisiting the fluid mosaic model of membranes. *Science* 1995; 268: 1441–1442.
4. Packer L, Douce R. *Plant Cell Membranes*, Academic Press, San Diego, CA, 1987.
5. Leshem YY with the participation of Shewfelt RL, Willner CM, Pantoja O. *Plant Membranes: A Biological Approach to Structural Development and Science*, Elsevier, New York, 1991.
6. Yeagle PL. T*he Structure of Biological Membranes*, CRC Press, Boca Raton, FL, 1991.
7. Smallwood M, Knox JP, Bowles D. *Membrane Specialized Functions in Plants*, Bios Scientific Publishers, Herndon, VA, 1996.
8. Bacic A, Harris PJ, Stone BA. Structure and functions of plant cell walls, in *The Biochemistry of Plants,* Vol. 14 (Preiss J, ed.), Academic Press, New York, 1988, pp. 297–371.
9. Brett CT, Hillman JR. *Biochemistry of Plant Cell Walls*, Cambridge University Press, Cambridge, UK, 1985.
10. Brett G, Waldron K. *Physiology and Biochemistry of Plant Cell Walls*, Unwin Hyman, Boston, MA, 1990.
11. Carpita NC, Gibeaut DM. Structural models of primary cell walls in flowering plants: consistency of molecular structure with the physical properties of walls during growth. *Plant J* 1993; 3: 1–30.
12. Carpita NC, McCann M, Giffing LR. The plant cell extracellular matrix. News from the cell's frontier. *Plant Cell* 1996; 8: 1451–1463.
13. Cassab GJ. Plant cell wall proteins. *Ann Rev Plant Physiol Mol Biol* 1998; 49: 281–309.
14. Cosgrove DJ. Assembly and enlargement of the primary cell wall in plants. *Ann Rev Cell Dev Biol* 1997a; 13: 171–201.
15. Cosgrove DJ. Relaxation in a high-stress environment. The molecular bases of extensible cell walls and cell enlargement. *The Plant Cell* 1997b; 9: 1031–1041.
16. Fleming AJ, McQueen-Mason S, Mandel T, Kuhleimer C. Induction of leaf primordia by the cell wall protein expansin. *Science* 1997; 276: 1415–1417.
17. Fry SC, Miller JG. Toward a working model of the growing plant cell wall. Phenolic cross-linking reactions in the primary cell walls of dicotyledons. American Chemical Society, Washington, DC, 1989.
18. Fry S. Polysaccharide-modifying enzymes in the plant cell wall. *Ann Rev Plant Physiol Plant Mol Biol* 1995; 46: 497–520.
19. Lamport DTA. Structure and function of plant glycoproteins, in *The Biochemistry of Plants,* Vol. 3 (Preiss J, ed.), Academic Press, New York, 1980, pp. 501–541.
20. Lewis NG, Paice MG. *Plant Cell Wall Polymers, Biogenesis and Biodegradation*, American Chemical Society, Washington, DC, 1989.
21. Terrashima N, Fukushima K, He LF, Takabe K. Comprehensive model of the lignified plant cell wall, in *Forage Cell Wall Structure and Digestibility*, ASA-CSSA-SSSA, Madison, WI, Chapter 10, 1993.
22. Lucas WJ, Wolf, S. Plasmodesmata the intracellular organelles of green plants. *Trends Cell Biol* 1993; 3: 308–315.

23. McLean BG, Hempel FD, Zambryski PC. Plant intracellular communication via plasmodesmata. *The Plant Cell* 1997; 9: 1043–1054.
24. Robards AW, Lucas WJ. Plasmodesmata. *Ann Rev Plant Physiol Mol Biol* 1990; 41: 369–419.
25. Zambryski P. Plasmodesmata: Plant channels for molecules on the move. *Science* 1995; 270: 1943–1944.
26. Morré DJ. Endomembrane system of plants and fungi, in *Tip Growth in Plant and Fungal Systems* (Heath B, ed.), Academic Press, San Diego, CA, 1990.
27. Sussman MR, Harper JF. Molecular biology of the plasma membrane of higher plants. *Plant Cell* 1989; 1: 953–960.
28. Laisson C, Moller IM. *The Plant Plasma Membrane: Structure, Function and Molecular Biology*, Springer-Verlag, New York, 1990.
29. Edelman M, Hallick RP, Chua NH. *Methods in Chloroplast Molecular Biology*, Elsevier Biomedical Press, New York, 1982.
30. Ellis RJ. *Chloroplast Biogenesis*, Cambridge University Press, New York, 1984.
31. Halliwell B. *Chloroplast Metabolism. The Structure and Function of Chloroplasts in Green Leaf Cells*, Clarendon Press, New York, 1984.
32. Hoober JK. *Chloroplasts*, Plenum Press, New York, 1984.
33. Bogorad L, Vasil IK. *The Molecular Biology of Plastids*, Academic Press, San Diego, CA, 1991.
34. Hirsch S, Muckel E, Heemeyer F, vonHeijne G, Sool J. A receptor component of the chloroplast protein translocation machinery. *Science* 1994; 266: 1989–1992.
35. Kessler F, Blobel G, Patel A, Schnell DJ. Identification of the two GTP-binding proteins in the chloroplast import machinery. *Science* 1994; 266: 1035–1039.
36. Schnell DJ, Kessler F, Blobel G. Isolation of components of the chloroplast protein import machinery. *Science* 1994; 266: 1007–1012.
37. Jordan EG, Collins CA. *The Nucleolus*, Cambridge University Press, New York, 1982.
38. Hadjiolov AA. *The Nucleus and Ribosome Biogenesis*, Springer-Verlag, New York, 1985.
39. Adolph KW. *Molecular Biology of Chromosome Function*, Springer-Verlag, New York, 1989.
40. Adolph KW. *Advanced Technique in Chromosome Research*, Marcel Dekker, New York, 1991.
41. Adolph KW. *Chromosomes: Eukaryotic, Prokaryotic and Viral*, CRC Press, Boca Raton, FL, 1991.
42. Sobti RC, Obe G. *Eukaryotic Chromosomes: Structural and Functional Aspects*, Springer-Verlag, New York, 1991.
43. Wagern R, Magurire MP, Stallings RL. *Chromosomes: A Synthesis*, Wiley Liss, New York, 1993.
44. Aitchison, D, Blobel, G, Rout, MP. Kap 104p: a karyopherin involved in the nuclear transport of messenger RNA-DNA binding proteins. *Science* 1996; 274: 624–627.
45. Douce, R. *Mitochondria in Higher Plants*, American Society of Plant Physiologists monograph. Academic Press, New York, 1985.
46. Moore AL, Beechey RB. *Plant Mitochondria: Structural, Functional and Physiological Aspects*, Plenum Press, New York, 1987.
47. Hissel R, Wissinger B, Schuster W, Brennicke A. RNA editing in plant mitochondria. *Science* 1989; 246: 1632–1634.
48. Wolstenholme DR, Jeon KW. *Mitochondrial Genomes*, Academic Press, San Diego, CA, 1992.

49. Brennicke A, Kuck A. *Plant Mitochondria with Emphasis on RNA. Editing and Cytoplasmic Male Sterility*, VCH, Weinheim, Germany, 1993.
50. Thorsteinson KE, Woyln DJ. Simultaneous purification of chloroplast and mitochondrial DNA without isolation of intact organelles from green carrot tissue and suspension cultures. *Plant Mol Biol Report* 1994; 12: 26–36.
51. Ungerman C, Newpert W, Cyr DM. The role of HSP 70 in conferring unidirectionality on protein translocation into mitochondria. *Science* 1994; 266: 1250–1253.
52. Levings CS III, Vasil IK. *The Molecular Biology of Plant Mitochondria. Advances in Cellular and Molecular Biology of Plants,* Vol. 3, Kluwer Academic, Norwell, MA, 1995.
53. Binder S, Marchfelder A, Brennicke A. Regulation of gene expression in plant mitochondria. *Plant Mol Biol* 1996; 32: 303–314.
54. Thorsness PE, Weber ER. Escape and migration of nucleic acids between chloroplasts, mitochondria and the nucleus. *Int Rev Cytol* 1996; 165: 207–232.
55. Maier RM, Zeltz P, Kossel H. Bonnard G, Gualberto JM, Grienenberger JM. RNA editing in plant mitochondria and chloroplasts. *Plant Mol Biol* 1996; 32: 343–365.
56. Special Issue. Mitochondria. *Science* 1999; 283:1475–1497.
57. Koch GLE. Reticuloplasmins: a novel group of proteins in the endoplasmic reticulum. *J Cell Sci* 1987; 92: 491–492.
58. Lee C, Chen LB. Dynamic behavior of endoplasmic reticulum in living cells. *Cell* 1988; 57: 37–46.
59. Bednarek SV, Raikhel NV. Intracellular trafficking of secretory proteins. *Plant Mol Biol* 1992; 20: 133–150.
60. Sitia R, Meldolesi J. Endoplasmic reticulum: A dynamic patchwork of subregions. *Mol Biol Cell* 1992; 3: 1067–1072.
61. Okita TW, Rogers JC. Compartmentation of proteins in the endomembrane system of plant cells. *Ann Rev Plant Physiol Plant Mol Biol* 1996; 47: 327–350.
62. Etkin LD. A new face for the endoplasmic reticulum: RNA localization. *Science* 1997; 276: 1092–1093.
63. Whaley G. *The Golgi Apparatus*, Springer-Verlag, New York, 1975.
64. Pavelka M. *Functional Morphology of the Golgi Apparatus*, Springer-Verlag, New York, 1987.
65. Kumar J. Hanry Y, Sheetz M. Kinectin, an essential anchor for kinesin-driven vesicle mobility. *Science* 1995; 267: 1834–1837.
66. Staehelin LA, Moore I. The plant Golgi apparatus: structure, functional organization and trafficking mechanisms. *Ann Rev Plant Physiol Plant Mol Biol* 1995; 46: 261–288.
67 Berger EG, Roth J, eds. *The Golgi Apparatus*, Birkhauser Verlag, Basel, Switzerland, 1997.
68. Morré DJ, Keenan TW. Membrane flow revisited. *Bioscience* 1997; 47: 489–498.
69. Mazzarello P, Bentivoglio M. The centenarian Golgi apparatus. *Nature* 1998; 392: 543–544.
70. Marty F, Branton D, Leigh RA. Plant vacuoles, in *The Biochemistry of Plants. The Plant Cell* (Tobert NE, ed.), Academic Press, New York, 1980, pp. 625–658.
71. Mair B. Plant Vacuoles. *Their Importance in Solute Compartmentation in Cells and Their Applications in Plant Biotechnology. NATO ASI Series*, Plenum Press, New York, 1986.
72. Nakamura K, Matsuoka K. Protein targeting to the vacuole in plant cells. *Plant Physiol* 1993; 101: 1–6.
73. Wink M. The plant vacuole: a multifunctional compartment. *J Exp Bot* 1993; 44: 231–246.

74. Leigh RA, Sanders D, Callow JA. *The Plant Vacuole. Advances in Botanical Research*, Academic Press, San Diego, CA. 1997.

75. Matile P. *The Lytic Compartment of Plant Cells*, Springer-Verlag, New York, 1975.

76. Dean RT. *Lysosomes*, Arnold, London, UK, 1977.

77. Glauman H, Ballard J. *Lysosomes and Their Role in Protein Breakdown*, Academic Press, New York, 1987.

78. Holtzman E. *Lysosomes*, Plenum Press, New York, 1988.

79. Bielkor H. *The Eukaryotic Ribosome*, Springer-Verlag, New York, 1982.

80. Phelps CF, Arnstein HRV. *Messenger RNA and Ribosomes in Protein Synthesis*, Biochemical Society, London, UK, 1982.

81. Spirin AS. *Ribosomes Structure and Protein Biosynthesis*, Benjamin Cummings, Menlo Park, CA, 1986.

82. Spedding G. *Ribosomes and Protein Synthesis: A Practical Approach*, Oxford University Press, Oxford, UK, 1990.

83. Steitz JA, Tycowski KT. Small RNA chaperones from ribosome biogenesis. *Science* 1995; 270: 1626–1627.

84. Zimmerman RA. Ins and outs of the ribosome. *Nature* 1995; 376: 391–392.

85. Heilek GM, Noller HF. Site-directed hydroxyl radical probing of the rRNA neighborhood of ribosomal protein S5. *Science* 1996; 272: 1659–1662.

86. Powers T, Walter P. A ribosome at the end of the tunnel. *Science* 1997; 278: 2072–2073.

87. Simpson J, Weiser B. Mapping the inside of the ribosome with an RNA helical ruler. *Science* 1997; 278: 1093.

88. Roberts K, Hyans JS. *Microtubules*, Academic Press, New York, 1979.

89. Dustin P. *Microtubules*, Springer-Verlag, New York, 1981.

90. Saifer D. *Dynamic Aspects of Microtubules Biology*, New York Academy of Sciences, New York, 1986.

91. Vallee RB. *Molecular Motors and the Cytoskeleton*, Academic Press, San Diego, CA, 1991.

92. Fosket DE, Morejohn LC. Structure and function organization of tubulin. *Ann Rev Plant Physiol Plant Mol Biol* 1992; 43: 201–240.

93. Hyans JS, Lyod CW. *Microtubules*, Wiley-Liss, New York, 1994.

94. Shibaoka H. Plant hormone-induced changes in the orientation of cortical microtubules: alterations in the cross-linking between microtubules and the plasma membrane. *Ann Rev Plant Physiol Mol Biol* 1994; 45: 527–544.

95. Tabon, J. Morphological bifurcations involving reaction-diffusion processes during microtubule formation. *Science* 1994; 264: 245–248.

96. Deloof A, Broeck JV, Jensen I. *Hormones and the Cytoskeleton of Animals and Plants. International Review of Cytology* (Jeon KW, ed.), Academic Press, New York, 1996.

97. Hogg JF. *The Nature and Function of Peroxisomes (Microbodies, Glyoxysomes)*, New York Academy of Sciences, New York, 1969.

98. Kindl H, Lazarow PB. *Peroxisomes and Glyoxysomes*, New York Academy of Sciences, New York, 1982.

99. Huang AKC, Trelease RN, Moore TS. *Plant Peroxisomes. American Society of Plant Physiologists*, Academic Press, New York, 1983.

100. Angermuller S. *Peroxisomal Oxidases: Cytochemical Localization and Biological Relevance*, G. Fischer Stuttgart, New York, 1989

101. Latruffe N, Bugart M. *Peroxisomes*, Springer-Verlag, New York, 1994.

102. Bowles BG. *A Color Atlas of Plant Structure*, Iowa State University Press, Ames, IA, 1996.
103. Esau K. *Anatomy of Seed Plants*, Wiley, New York, 1997.
104. Steeves TA, Sussex IM. *Patterns in Plant Development. A Molecular Approach*, Academic Press, New York, 1989.
105. Fahn A. *Plant Anatomy*, Pergamon Press, Elmsford, New York, 1990.
106. Maseuth JD. *Botany. An Introduction to Plant Biology*, Saunders College Publishing, Philadelphia, PA, 1991.
107. Raven PH, Evert RF, Eichhorn SE. *Biology of Plants*, Fifth Edition, Worth Publishers, New York, 1982.
108. Fosket DE. *Plant Growth and Development. A Molecular Approach*, Academic Press, New York, 1994.
109. Moore R, Clark WD. *Botany. Plant Form and Function*, Wm. C. Brown Publishers, Dubuque, IA, 1995.
110. Burrells W. *Microscope Technique. A Comprehensive Handbook for General and Applied Microscopy*, A Halsted Press Book, Wiley, New York, 1977.
111. Marmasse C. *Microscopes and Their Uses*, Gordon and Breach, New York, 1980.
112. Herman B, LeMasters JJ. *Optical Microscopy*, Academic Press, San Diego, CA, 1993.
113. Shotton D. (ed.). *Electronic Light Microscopy. Modern Biomedical Microscopy*, Wiley-Liss, New York, 1993.
114. Jones C, Mulloy B, Thomas H. (eds.). *Microscopy, Optical Spectroscopy and Macroscopic Techniques*, Humana Press, Totowa, NJ, 1994.
115. Cherry RJ. *New Techniques of Optical Microscopy and Microspectroscopy*, CRC Press, Boca Raton, FL, 1991.
116. Shaw PJ, Rawlins DJ. An introduction to optical microscopy for plant cell biology, in *Plant Cell Biology A Practical Approach* (Harris N, Oparka KJ, eds.), Oxford University Press, Oxford, UK, 1994, pp 1–26.
117. Williams PM, Cheema MS, Davies MC, Jackson DE, Tedler SJB. *Methods in Molecular Biology*, Vol. 22, *Microscopy, Optical Spectroscopy and Microscopic Techniques* (Jones C, Mulloy B, Thomas AH, eds.), Humana Press, Totowa, NJ, 1994.
118. Turrell G, Corset J. *Raman Microscopy Developments and Applications*, Academic Press, New York, 1996.
119. Juniper BC, Cox CC, Gilchrist AJ, Williams, PR. *Techniques for Plant Electron Microscopy*, Blackwell Scientific Publications, Oxford, UK, 1970.
121. Tribe MA, Evant RM, Snook RK. *Electron Microscopy and Cell Structure*, Cambridge University Press, Cambridge, UK, 1975.
122. Hall JL. *Electron Microscopy and Cytochemistry of Plant Cells*, Elsevier North Holland Biomedical Press, Amsterdam, 1978.
123. Hayat MA. *X-ray Microanalysis in Biology*, University Park Press, Baltimore, MD, 1980.
124. Postek MT, Howard KS, Johnson AH, McMichael KL. *Scanning Electron Microscopy: A Student's Handbook*, Ladd Research Industries, Inc, Burlington, VT, 1980.
125. Goldstein JI. *Scanning Electron Microscopy and X-ray Microanalysis: A Text for Biologists, Material Scientists and Geologists*, Plenum Press, New York, 1981.
126. Gersh I. *Submicroscopic Cytochemistry*, Academic Press, New York, 1973.
127. Mohanty SB. *Electron Microscopy for Biologists*, Charles Thomas, Springfield, IL, 1982.

128. Murr LE. *Electron amd Ion Microscopy and Microanalysis Principles and Applications*, Marcel Dekker, New York, 1982.
129. Morgan JA. *X-ray Microanalysis in Electron Microscopy for Biologists*, Oxford University Press, Oxford, UK, 1985.
130. Egerton RF. *Electron Energy Loss Spectroscopy in the Electron Microscope*, Plenum Press, New York, 1986.
131. Hayat MA. *Basic Techniques for Transmission Electron Microscopy*, Academic Press, New York, 1986.
132. Lawes G. *Scanning Electron Microscopy and X-ray Microanalysis*, Wiley, New York, 1987.
133. Steinbrecht RA, Zierold K. *Cryotechniques in Biological Electron Microscopy*, Springer-Verlag, Berlin, Germany, 1987.
134. Plattner H. *Electron Microscopy of Subcellular Dynamics*, CRC Press, Boca Raton, FL, 1989.
135. Spence JCH. *Experimental High-Resolution Electron Microscopy*, Oxford University Press, New York, 1988.
136. Hayat MA. *Principles and Techniques of Electron Microscopy*, CRC Press, Boca Raton, FL, 1989.
137. Griffin RL. *Using the Transmission Electron Microscope in the Biological Sciences*, Ellis Horwood, New York, 1990.
138. Hawkes P, Valdre V. *Biophysical Electron Microscopy, Basic Concepts and Modern Techniques*, Academic Press, New York, 1990.
139. Hayat MA. *Negative Staining*, McGraw Hill, New York, 1990.
140. Hall JL, Hawes C. *Electron Microscopy of Plant Cells*, Academic Press, New York, 1991.
141. Harris R. *Electron Microscopy in Biology. A Practical Approach*, IRL Press, Oxford, UK, 1991.
142. Harris R, Horne R. Negative staining, in *Electron Microscopy in Biology—A Practical Approach* (Harris JR, ed.), IRL Press, Oxford, UK, 1991, pp. 203–228.
143. Hyatt AD. Immunogold labeling techniques, in *Electron Microscopy in Biology— A Practical Approach* (Harris JR, ed.), IRL Press, Oxford, UK, 1991, pp. 59–81.
144. Roos N. Freeze-substitution and other low temperature embedding methods, in *Electron Microscopy in Biology—A Practical Approach* (Harris JR, ed.), IRL Press, Oxford, UK, 1991, pp. 39–58.
145. Sexton R, Hall J. *Enzyme Cytochemistry. Electron Microscopy of Plant Cells* (Hall JL, Hawes C, eds.), Academic Press, San Diego, CA, 1991, pp. 105–180.
146. Bozzola JJ, Russell LD. *Electron Microscopy Techniques for Biologists*, Jones and Bartlett Publishers, Boston, MA, 1992.
147. Bonnell A. *Scanning Tunneling Microscopy and Spectroscopy Techniques and Applications*, VCH, New York, 1993.
148. Chen C. *Introduction to Scanning Tunneling Microscopy*, Oxford University Press, New York, 1993.
149. Dykstra MJ. *A Manual of Applied Techniques of Biological Electron Microscopy*, Plenum, New York, 1993.
150. Kaestner G. Many-beam electron diffraction related to electron microscope diffraction contrast, Akademie Verlag, Berlin, Germany, 1993.
151. Morel G. *Hybridization Techniques for Electron Microscopy*, CRC Press, Boca Raton, FL. 1993.
152. Neddermeyer H. *Scanning Tunneling Microscopy*, Kluwer Academic Publisher, Dordrecht, Boston, MA, 1993.

153. Othmar M, Amrein M. *STM and SRM in Biology*, Academic Press, San Diego, CA, 1993.
154. Reed SJB. *Electron Microprobe Analysis*, Cambridge University Press, Cambridge, MA, 1993.
155. Sigee DC. *X-ray Microanalysis in Biology: Experimental Techniques and Applications*, Cambridge University Press, Cambridge, UK, 1993.
156. Slayter EM. *Light and Electron Microscopy*, Cambridge University Press, Cambridge, UK, 1993.
157. Harris N. Immunocytochemistry for light and electron microscopy, in *Plant Cell Biology* (Harris N, Oparka KJ, eds.), IRL Press, Oxford, UK, 1994, pp. 157–176.
158. Harris N, Oparka KJ. (eds.). *Plant Cell Biology*, IRL Press, Oxford, UK, 1994.
159. Hawes C. Electron microscopy, in *Plant Cell Biology* (Harris N, Oparka KJ, eds.), IRL Press, Oxford, UK, 1994, pp. 69–52.
160. Smith RF. *Microscopy and Photomicrography*, CRC Press, Boca Raton, FL, 1994.
161. Hader DP. *Image Analysis in Biology*, CRC Press, Boca Raton, FL, 1992.
162. Russ JC. *The Image Processing Handbook*, CRC Press, Boca Raton, FL, 1995.
163. Jahne B. *Practical Handbook on Image Processing for Scientific Applications*, CRC Press, Boca Raton, FL, 1997.
164. Swatland HJ. *Computer Operation for Microscope Photometry*, CRC Press, Boca Raton, FL, 1997.

2

Methods for the Cytochemical/Histochemical Localization of Plant Cell/Tissue Chemicals

William V. Dashek

CONTENTS

1. INTRODUCTION

1.1. Value of Cytochemistry and Histochemistry

Although many published volumes exist regarding the cytochemical/histochemical localizations of cellular and tissue chemicals for animal systems *(1–10)*, there are only a few relatively recent monographs concerning plant cell/tissue cytochemistry and histochemistry *(11–15)*.

Perhaps the main reason for the rather numerous volumes centering about animal systems stems from the obvious importance of localizing cellular/tissue chemicals for clinical histopathology. For example, embryonic surface antigens appear during transformation of a healthy cell to a malignant one. Although there are plant cancers, e.g., crown galls, the development of cytochemical stains to reveal possible surface antigens in plant neoplasms has not been extensively explored. Nevertheless, plant cytochemistry has yielded a wealth of information regarding

From: *Methods in Plant Electron Microscopy and Cytochemistry*
Edited by: W. V. Dashek © Humana Press Inc., Totowa, NJ

the distribution of cellular and tissue chemicals in diverse systems. Furthermore, cytochemistry/histochemistry has provided significant details about the organization of the vascular system in monocot and dicot roots and stems.

1.2. Continued Evolution of Cytochemistry

Plant cytochemistry/histochemistry continues to evolve as fluorescence microscopy *(16–19)*, confocal fluorescence microscopy *(20,21)*, and microspectrophotometry *(22)* expand our quantitative knowledge of the distributions of chemical constituents in plant cells and tissues. With regard to microspectrophotometry, this is possible for single cells, as the Arcturus Corporation (Mountain View, CA) has developed an instrument capable of isolating single cells.

2. PURPOSE OF THE CHAPTER

Since the classic plant histochemistry volume of Jensen *(23)*, a few volumes regarding the topic (see opening paragraphs of the introduction) have appeared over the last four decades. Certain of these volumes contain updated methods for fixation, dehydration, and embedding of plant cells and tissues for the light microscopic localization of certain of their chemical constituents. Much of the older botanical microtechnique volumes, e.g., Sass *(24)* abound with paraffin embedding and sectioning methods. These volumes remain very useful, as they contain highly relevant information regarding microtomy and affixing sections to slides.

This chapter offers some select, recent developments regarding fixation, dehydration, and embedding. In addition, some tried-and-true procedures are described for the localizations of cellular and tissue chemicals in stems and roots of young *Zea mays* seedlings. Also provided are more recently developed fluorochromes for DNA and RNA localizations *(18,19)*.

Finally, the localizations of low-molecular-weight compounds requires special specimen preparation techniques, as these compounds are often diffusible, water- or organic-solvent soluble, and solubilized by conventional fixation and dehydration procedures. The reader is referred to ref. *(12)* for the processing of cells and tissues for the cytochemical and histochemical localization of these compounds.

3. PROTOCOLS

3.1. Preparation of Plant Cells and Tissues
for Light Microscope Cytochemistry/Histochemistry

1. Cut tissue block, with at least one dimension a maximum of 5 mm, into 1.5% (w/v) formaldehyde, 2.5% (v/v) glutaraldehyde in 0.05 M phos-

Table 1
Fixatives Employed for Light Microscopy

Acetic acid 45%
Acetic acid—alcohol
Acetic acid—alcohol—chloroform
Chromium tetroxide
Chromium—formal
Ethanol 50-70% aqueous
Formalin—calcium
Formalin 10%
Formalin—alcohol—acetic acid (FAA)
Glutaraldehyde 20%[a]

[a] Harris et al. *(13)* suggest using a combination
of glutaraldehyde and paraformaldehyde.

phate buffer, pH 7.0. Note that speed is important to prevent autolytic changes. *See* Table 1 for other fixatives employed for light microscopy.
2. Fix overnight at 4°C (volume of fixative >10X volume of sample).
3. Wash twice in phosphate buffer, 30 min each time.
4. Dehydrate through a graded alcohol series (10%, 25%, 40%, 60%, 75%, 95%) with two changes at each step (15 min each change) and three 15–30 min changes in 200-proof alcohol.
5. Embed in 1:1 ethanol:polyethylene glycol 1000 overnight at 40°C.
6. Infiltrate with polyethylene glycol for 48–72 h at 56°C with changes to fresh PEG each morning and evening.
7. Place in prewarmed embedding molds with fresh polyethylene glycol and cool on ice at 4°C.

3.1.1. WAX EMBEDDING PROCEDURE FOR SECTIONING *(11,12)*

1. Wash with 2:1 ethanol: Histo-Clear for 2 h at room temperature.
2. Repeat with 1:1 and 2:1 ethanol: Histo-Clear and leave in Histo-Clear overnight.
3. Infiltrate with Histo-Clear: wax (or paraplast) at 1:1 for 8 h at 56°C.
4. Infiltrate with wax (or paraplast) for 96 h at 56°C with changes every 24 h.
5. Place in prewarmed embedding molds with fresh wax (or paraplast) and cool on ice at 4°C. Adapted from ref. *(13)*. The reader is referred to Jensen *(23)* and Berlyn and Miksche *(11)* for older methods of clearing with xylene and subsequent progressive embedding in graded mixtures of paraplast and xylene or toluene with final embedding in pure paraplast. These methods have endured and are still widely used today. The reader is urged to examine the early papers of Rosen et al. *(25)* and Reynolds and Dashek *(26)* for celloidin-embedding procedures.

Table 2
Summary of the Specificity of Cytochemical Stains Available
for the Detection of Various Classes of Cellular Chemicals [a,b]

Compound	Stain
Carbohydrates	Periodic Acid—Schiff's
Callose	Aniline blue fluorescence
Cellulose	Zinc chlor-iodide
Pectin	Hydroxylamine—ferric chloride
	Ruthenium red
Nucleic acids	Calcofluro white M2R fluorescence
DNA/RNA	Methyl green—pyronin
	Azure B
DNA	Feulgen
	Acridine orange as a fluorochrome
	Ethidium bromide as a fluorochrome
	4', 6'- diamido-2-phenylindole
	as a fluorochrome
Lignin	Acidic phloroglucinol
Lipids	Nile blue
	Sudan black B
	Sudan IV
Phospholipids	Acid haematin
	Bromine—Sudan black
	Bromine—Acetone—Sudan black
Protein (total)	Fast green pH2
	Ninhydrin—Alloxan Schiff's
	Mercuric-bromphenol blue
Proteins	
Containing tyrosine	Million's diazotization
Containing arginine	Sakaguchi reaction
Containing tryptophan	N-(1-Naphthyl)-ethylenediamine
Containing sulfhydrils or	Rosindole
disulfide	
Tannins	Tetrazolium
	Mercaptide formation
	Ferric chloride—HC1

[a] Adapted from refs. *(23), (11), (12), (18)*.
[b] See ref. *26a*.

3.2. Protocol

3.2.1. CYTOCHEMICAL/HISTOCHEMICAL LOCALIZATIONS OF CHEMICALS IN STEMS AND ROOTS OF ZEA MAYS SEEDLINGS (SEE TABLE 2)

Chemicals	Plant material
Adhesive such as Haupt's	*Zea mays* seedlings
dH_2O	Prepare in advance

Chemicals	Plant material
Ethanol	Excise stem or roots
Glutaraldehyde	(see Introductory Material)
Fast green	*Equipment*
Paraformaldehyde	Analytical balance and weighing paper
Paraplast	Greenhouse or hood light banks (Grolux)
Periodic acid	Incubator
Permount or Polymount	Light microscope with or without camera
Phosphate buffer	Microtome with blade
Polyethylene Glycol 1000	Ocular micrometer
Safranin	Slide warmer
Schiff's reagent	Tissuetek or Paraplast dispenser
Sudan stains	(optional—embedding can be
Vermiculite or perlite	accomplished without them—see text)
Xylene (histological grade)	Top-loading balance and weighing boats
or Histo-Clear, a recent	Water bath
commercially available	
clearing agent	

Supplies
Aluminum foil
Camel's hair brush
Coplin jars or staining dishes
Coverslips 22 × 50 mm
Embedding molds
Embedding rings
Flats for growing corn seedlings
Graduated cylinder
Ice bucket
Kimwipes
Microscope slides—frosted end
Pasteur pipets
Permount
Pipets 1,5, and 10 mL
Probes
Pro-pipets
Pyrex bottles
Single-edge razor blades
Vials for fixation

3.2.2. Use Coplin Jars or a Rack of Staining Dishes

Carbohydrates—periodic acid— Schiff (use ref. 12 controls)	Nucleic acids—azure blue[a]
Deparaffinize with two xylene changes, 5 min each	Use freeze-dried or freeze-substituted tissue; can also use chemically fixed

(continued)

*Carbohydrates—periodic acid—
Schiff (use ref. 12 controls)*   *Nucleic acids—azure blue[a]*

Hydrate
    100% ethanol, 5 min
    95% ethanol, 5 min
    70% ethanol, 5 min
Staining
    Place sections in 0.4 g periodic
    acid, 35 mL; absolute ethanol,
    5 mL 0.2 *M*; sodium acetate,
    10 mL dH_2O for 10 min
Rinse the sections in 70%
    aqueous ethanol
Transfer the section to reducing
    bath for 3 min
Reducing bath = 1 g potassium
    iodide and 1 g sodium thiosulfate
    in 30 mL absolute ethanol and
    20 mL distilled H_2O; add
    0.5 mL 2NHCl (make reducing
    bath fresh daily)
Rinse the section in 70%
    aqueous ethanol
Stain the section in Schiff's
    reagent for 20 min
Schiff's—dissolve 1 g basic
    fuchsin in 200 mL boiling
    $dH_2O\bar{c}$ stirring; cool solution
    to 50°C and filter, add 30 mL
    NHCl and then 3 g $K_2S_2O_5$
Keep in dark for 25 h in a well-
    stoppered bottle; add 0.5 g
    charcoal and shake for 1 min;
    filter and store filtrate in dark
    in tightly stoppered bottle
Wash the sections in three changes
    of freshly prepared SO_2H_2O,
    each 10 min (INHCl, 5 ml.
    $K_2S_2O_5$, 5 mL dH_2O, 100 ml)
Dehydrate the sections through
    a graded ethanol series
Mount in Permount, Polymount,
    or Euparol
Aldehyde groups stain pink

Lipids and fatty acids—Sudan III
Unfixed or fixed frozen sections
Take sections to 50% aqueous
    ethanol

tissues if chromic acid or other heavy
    metals are absent
Deparaffinize with two changes of xylene
Hydrate through a decreasing alcohol
    series
Immerse slides in an 0.25-mg mL^{-1}
    solution of azure B in pH 4.0 citrate
    buffer at pH 4.0 for 2 h at 50°C
Wash in dH_2O
Place in tertiary butyl alcohol (TBA)
    for 30 min
Take through two changes in TBA
    for 30 min each time
If additional differentiation is required,
    can allow sections to remain overnight
    in TBA
Xylene two changes, 5 min each
Mount in Permount, Polymount,
    or Euparol
DNA = blue green
RNA = purple or dark blue

*Total protein—ninhydrin—alloxan—
    Schiff's reaction (see Jensen, 1952
    for Deamination and Acetylation
    Controls)*
Unfixed, freeze dried, freeze substituted,
    or chemically fixed
Chemically fixed, use 15–25 μm
    sections
Deparaffinize with two changes
    of xylene
Place sections in 0.5% ninhydrin or in
    1.0% alloxan in absolute alcohol at
    37°C, 20–24 h
Rinse in two changes of absolute
    ethanol
Rinse in dH_2O
Immerse in Schiff's reagent
    (see PAS method) for 10–30 min
Rinse in dH_2O
Place in 2% sodium bisulfite for
    1–2 min
Wash in running tap water 10–20 min.
Dehydrate through an increasing
    alcohol series

Lipids and fatty acids—Sudan III
Stain in Sudan III in 70% ethanol for 30 min
Rinse sections in 50% aqueous ethanol
Mount in glycerine
Avoid the use of absolute ethanol, as lipids will be soluble
Mount in Permount, Polymount, or Euparol xylene two times, 5 min each time
Neutral fats and fatty acids stain red

[a] The specificity of azure B for DNA and RNA must be verified in each system by DNase and RNase treatments as well as other cytochemical tests, The Feulgen reaction for DNA and acridine orange (DNA and RNA) coupled with fluorescence microscopy are particularly useful. Similarly, the specificity of fast green at pH2 for total protein must be verified by treating sections with proteases.

Table 3
Summary of Wood-Decay Fungal H_2O_2 Investigations

Tests employed	*References*
3' 3-Diaminobenzidine; horseradish peroxidase, and ABTS or o-diansidine; titanium reagent	Forney et al. *(27)*; Highley and Murmanis *(28)*; Illman and Highley (30); Micales and Highley *(31)*

3.3. Cytochemical/Histochemical Localizations of Low-Molecular-Weight Compounds—H_2O_2

Some of the most comprehensive investigations of H_2O_2 localizations in plant tissues have been those of Highley and his co-workers (Table 3). These investigators were concerned with localizing H_2O_2 in decaying wood and wood decay fungi as H_2O_2 is thought to function in proposed Fenton chemistry-mediated wood decay (*see* Chapter 12 and refs. *32*). Highley and his co-workers both present and cite methods for localizing H_2O_2. With modification for systems differences, the tests cited in Table 3 should be applicable to a wide variety of plant systems.

4. CONCLUSIONS

The future of cytochemistry resides in its usefulness as an adjunct to biochemistry. As mentioned, fluorescence *(33–35)* and confocal *(36–40)* microscopies have provided new dimensions to cytochemistry.

Finally, photomicrography is the culmination of the preparation of specimens for optical microscopy. This is a very technical area requiring proper illumination *(41–43)*, focusing, choice of films, as well as exposure and appropriate film development. This critical area of microscopy should see continued technological innovations as much of photomicrography is being computerized (44, 45). This effort is witnessing the concomitant improvement of basic measuring techniques for light microscopy and image analysis *(46–48)*.

REFERENCES

1. Galigher AE, Kosloff EN. *Essentials of Practical Microtechnique*, Lea and Febiger, Philadelphia, PA, 1964.
2. Hayat MA. *Stains and Cytochemical Methods*, Plenum Press, New York, 1993.
3. Horubin RW. *Understanding Histochemistry: Selection, Evaluation and Design of Biological Stains*, Hopwood, Chichester, UK, 1988.
4. Kiernan JA. *Histological and Histochemical Methods: Theory and Practice*, Pergamon Press, Oxford, UK, 1990.
5. Lillie RD, Fuller HM. *Histopathology Techniques and Practical Histochemistry*, McGraw-Hill, New York, 1976.
6. Pearse AGE. *Histochemistry Theoretical and Applied*, Little, Brown, Boston, MA, 1964.
7. Sannes PL. *The Histochemical and Cytochemical Localization of Proteases*, Deerfield BEA, Stuttgart, Germany, 1988.
8. Sheehan DC. *Histotechnology—Theory and Practice*, The CV. Mosby Co., St. Louis, MO, 1980.
9. Horobin RW. Histochemistry and the light microscope, in *Light Microscopy in Biology* (Lacey AJ, ed.), IRL Press, Oxford, UK, 1989, pp. 137–162.
10. Sumner BEH. *Basic Histochemistry*, Wiley, Chichester, New York, 1988.
11. Berlyn GP, Miksche JP. *Botanical Microtechnique and Cytochemistry*, Iowa State University Press, Ames, IA, 1976.
12. Gahan PB. *Plant Histochemistry and Cytochemistry: An Introduction*, Academic Press, London, UK, 1984.
13. Harris N, Spence J, Oparka KJ. General and enzyme histochemistry, in *Plant Cell Biology* (Harris N, Oparka KJ, eds.), IRL Press, Oxford University Press, Oxford, UK, 1994, pp. 51–68.
14. Vaughn KC. *Handbook of Plant Cytochemistry*, CRC Press, Boca Raton, FL, 1987.
15. Vigil EL, Hawes CR. *Cytochemical and Immunological Approaches to Plant Cell Biology*, Academic Press, London, UK, 1989.
16. Rost FWD. *Quantitative Fluorescence Microscopy*, Cambridge University Press, New York, 1991.
17. Rost FWD. *Fluorescence Microscopy*, Vols. I and II, Cambridge University Press, New York, 1992.
18. Oparka KJ, Read ND. The use of fluorescent probes for studies of living plant cells, in *Plant Cell Biology* (Harris N, Oparka KJ, eds.), IRL Press, Oxford University Press, Oxford, UK, 1994, pp. 27–50.
19. Oparka KJ, Roberts AG, Santa Cruz S, Bocvnik P, Prior D, Smallcombe A. Using GFP to study virus invasion and spread in plant tissues. *Nature* 1997; 388: 401–402.
20. Wilson T. *Confocal Microscopy*, Academic Press, New York, 1990.
21. Shuming N, Chiu DT. Probing molecules with confocal fluorescence microscopy. *Science* 1994; 266: 1018.
22. Cherry RJ. *New Techniques of Optical Microscopy and Microspectrophotometry*, CRC Press, Boca Raton, FL, 1990.
23. Jensen WA. *Botanical Histochemistry*, Freeman, San Francisco, CA, 1952.
24. Sass JE. *Botanical Microtechnique*, 3rd ed., Iowa State College Press, Ames, IA, 1958.
25. Rosen WG, Gawlik SR, Dashek WV, Siegesmund KA. Fine structure and cytochemistry of *Lilium* pollen tubes. *Am J Bot* 1964; 51: 61–67.

26. Reynolds JD, Dashek WV. Cytochemical analysis of callose distribution in *Lilium longiflorum* pollen tubes. *Ann Bot* 1976; 40: 409–416.

26a. Prasad B. *Staining Techniques in Botany*, State Mutual Book and Periodical Services, New York, 1986.

27. Forney LJ, Reddy CA, Pankvatz HS. Ultrastructural localization of hydrogen peroxide production in ligninolytic *Phanerochaete chrysosporium* cells. *Appl Environ Microbiol* 1982; 44: 732–736.

28. Highley T, Murmanis LL. Determination of hydrogen peroxide production in *Coriolus versicolor* and *Poria placenta* during wood degradation. *Mater Org* 1985; 29: 241–252.

29. Highley T. Effect of carbohydrate and nitrogen on hydrogen peroxide formation by wood decay fungi in solid medium. *FEMS Microbiol Lett* 1987; 48: 373–378.

30. Illman B, Highley TL. Hydrogen peroxide formation by wood decay fungi in liquid medium. *Phytopathology* 1988; 78: 1590.

31. Micales JA, Highley TL. *In vitro* production of hydrogen peroxide by degradative and non-degradative isolates of brown-rot wood decay fungi. *Phytopathology* 1989; 77: 988.

32. Highley TL, Dashek WV. Biotechnology in the study of white-rot and brown-rot decay, in *Forest Products Biotechnology* (Bruce A, Palfreyman JW, eds.), Taylor and Francis, London, UK. 1998, pp. 15–36.

33. Slavik J. *Fluorescence Microscopy and Fluorescent Probes, Proceedings of a Conference Held in Digre, Czech Republic*, Plenum Press, New York, 1996.

34. Tarke HJ. *Fluorescence Microscopy*, Bios Scientific UK, Cornett Books, Oxford, UK, 1996.

35. Wang XF, Herman B. *Fluorescence Imaging Spectroscopy and Microscopy*, Wiley, New York, 1996.

36. Van der Voort HTM, Valkenburg JAC, Van Spronsen EA, Woldringh CL, Brakenhoff GJ. Confocal microscopy in comparison with electron and conventional microscopy, in *Correlative Microscopy in Biology. Instrumentation and Methods* (Hayat MA, ed.), Academic Press, Orlando, FL, 1987, pp. 60–81.

37. Pauley B. *Handbook of Biological Confocal Microscopy*, Plenum Press, New York, 1990.

38. Corle TR, Kino GS. *Confocal Scanning Optical Microscopy and Related Imaging Systems*, Academic Press, New York, 1996.

39. Sheppard C, Shottan, D. *Confocal Laser Scanning Microscopy*, Springer-Verlag, Berlin, Germany, 1997.

40. Paddock SW. *Confocal Microscopy Methods and Protocols*, Humana Press, Totowa, NJ, 1998.

41. Smith RF. *Microscopy and Photomicrography. A Working Manual*, 2nd ed., CRC Press, Boca Raton, FL, 1994.

42. Thompson DJ, Bradbury S. *An Introduction to Photomicrography* (Book 13), Oxford University-Royal Microscopical Society, Oxford, UK, 1991.

43. Weiss DG, Marle W, Wick RA. Video microscopy, in *Light Microscopy in Biology: A Practical Approach* (Lacey AJ, ed.), IRL Press, Oxford, UK, 1989, pp. 221–278.

44. Bradbury S. *Basic Measurement Techniques for Light Microscopy*, Oxford University- Royal Microscopical Society, Oxford, UK, 1991.

45. Swatland HJ. *Computer Operation for Microscopy Photometry*, CRC Press, Boca Raton, FL, 1997.

46. Hader DP. *Image Analysis in Biology*, CRC Press, Boca Raton, FL, 1992.

47. Russ JC. *The Image Processing Handbook*, CRC Press, Boca Raton, FL, 1995.

48. Jahne B. *Practical Handbook on Image Processing for Scientific Applications*, CRC Press, Boca Raton, FL, 1997.

3

Methods in Light Microscope Radioautography

Alyce Linthurst and William V. Dashek

CONTENTS

1. INTRODUCTION

Light microscope radioautography *(1–12)* is a technique that involves the administration of a radiolabeled compound of suitable specific activity (see radioisotopes dose and route of administration) to cells, tissues in culture, or even whole tissues followed by the preparation of sections from paraffin/paraplast or resin-embedded specimens (*see* Chapter 2). The sections are coated with the appropriate fine grain, photographic emulsion (*see* Table 1) and the section–emulsion sandwich is incubated in a light-tight container for various times. Then, the section–emulsion sandwich is developed with photographic solutions appropriate for the selected photographic emulsion. Subsequent to development, the task of grain quantitation can be laborious. Such methods as manual grain counting have been performed using a hemocytometer or observing the image

From: *Methods in Plant Electron Microscopy and Cytochemistry*
Edited by: W. V. Dashek © Humana Press Inc., Totowa, NJ

Table 1
Commercially Available Emulsions and Their Applications[a]

Emulsion type	Required safelight	Development[b]
Kodak NTB-2 and NTB-3 nuclear track emulsion (store up to 2 mo at 4°C)	Kodak safelight filter no. 2 using 15-W bulb 4 ft from working area	Kodak Dektol developer or Kodak developer 19 followed by washing and fixing in Kodak fixer— followed by washing
NTB-2 (less sensitive than NTB-3 and useful for [^{3}H], [^{125}I], [^{14}C], and [^{35}S]		
Amersham LM-1	See Amersham Technical Bulletin	
Kodak AR-10 stripping film (store up to 6 mo at 4°C in its shipping container)	Kodak safelight filter no. 1 (red)	Kodak Dektol developer or D-19
Ilford G5, K0, K1, K2, K5 and L4, G5, and K5 differ in crystal size but both have a high level of sensitivity; K2 is useful for [^{3}H] and [^{125}I]		See Chapter 17, electron microscope D-19, Microdol X

[a] The amount of time the emulsion is exposed to the radiolabeled specimen is dependent on the thickness of the histological section, uniformity of radioisotope labeling, crossfire effects resulting from choice of radioisotope, emulsion type, emulsion thickness, storage conditions during exposure, development methodology, and developer solution activity.

[b] See extensive references in ref. *12*.

with a dark-field microscope while illuminating the grains as white specks; the remainder of the tissue is black. Advances in image analysis software have eliminated human counting error; however, the task is still rather labor intensive.

The value of light microscope autoradiography is the cellular/tissue locations of either the radiolabeled compound or its metabolic products are easily ascertained.

A useful adjunct is the well-known pulse–chase technique, in which exposure to a radiolabeled compound for a specified period of time (pulse) is followed by the administration of the same compound that is not labeled (chase).

An obvious limitation of autoradiography is the necessity to couple it to the biochemical reisolation of the exogenously supplied radiolabeled

Table 2
Radiation Terminology

Term	Symbol	Definition
Becquerel	Bq	SI quantity of radioactivity $Bq = 1\,dps$ or $2.7 \times 10^{-11}\,Ci$
Curie	Ci	Quantity of radioactivity $1\,Ci = 3.7 \times 10^{10}\,dps$ $1\,Ci = 3.7 \times 10^{10}\,Bq$
Microcurie	μCi	Quantity of radioactivity $1\,\mu Ci = 2.2 \times 10^{6}\,dpm$
Electron volt	eV	Unit of energy $1\,eV = 1.6 \times 10^{-12}\,erg$ $1\,eV = 1.6 \times 10^{-19}\,J$

compound to rule out its metabolite. Other drawbacks are the safety precautions and training that are necessary to work intelligently and safely with radioisotopes.

Let us now examine the methods of light microscope radioautography for organic compounds insoluble in a wide range of polar and nonpolar solvents (13). The reader is referred to the reviews of Roth and Stumpf (14), Williams (15), Eschrich and Fritz (16), Stumpf (17), and Inson and Sheridan (18) for microautoradiography of diffusible or watersoluble inorganic compounds.

2. RADIOISOTOPES
AND THEIR ADMINISTRATION

2.1. Routes and Dose

Whereas Table 2 presents some radiation terms that are pertinent to the following discussion, Table 3 lists the radioisotopes and their half-lives that are suitable to microautoradiography. Of the two types of nuclear radiation, i.e., wavelike electromagnetic radiation (X-ray and gamma ray) and "corpuscular" radiation (electrons, beta particles, and alpha particles) the latter are the most employable for microautoradiography. The reader is referred to Slater (20) for in-depth discussion of all aspects of radiobiology.

The dose of radioisotope is also dependent on the β-emitting strength of the isotope. For instance, tritiated ^3H samples will require a higher dose than ^{32}P samples for similar incubation times prior to development. However, the isotope chosen also is reflective of what the aim of

Table 3
Nuclides Used in Microautoradiography

Suitability for micro-autoradiography	Radio-isotope	Radiation β %	Energy MeV		t1/2
autoradiography of	^3H	100	0.018		12.3yr
organic compounds	^{14}C	100	0.156		5730yr
	^{35}S	100	0.167		87 d
autoradiography of	^{33}P	100	0.248		25.2d
inorganic anions {	^{32}P	100	1.71		14.3d
	^{36}Cl	98.3	0.71	1.7	3×10^5yr
		100	0.256		165 d
autoradiography of	^{45}Ca	46	0.27		45 d
inorganic cations {	^{59}Fe	53	0.46		
		plus			
		2			
		other	1.77		

Adapted from refs. *(6)* and *(7)*.

the research is, whether it is amino acid incorporation, monitoring of DNA replication, drug or metabolite localization, or cellular viability. Typically, in mammalian systems dosage is given as µCi of isotope/ µmol of investigated substance/mL of reaction mixture, µCi/µmol/mL. A typical dose for an experiment involving cellular viability is 1 µCi/1 µmol/1 mL and allowing the tissue to incubate 24 h with antibiotic supplementation. It should be pointed out that the administered dose can be influenced by the specific activity of the radionuclide. The reader is referred to refs. *(5), (7),* and *(21)* for discussions of dose.

3. SPECIMEN PREPARATION

3.1. Fixation, Washing, Dehydration

The choice of fixative for the preparation of one's specimen for light microscope radioautography depends on the macromolecules that you want to preserve. Table 4 summarizes the molecules that are likely to be retained by commonly utilized fixatives. In addition, the table presents some quantitative data for the retention of radioactivity by various fixatives subsequent to the administration of labeled precursors to biological systems. More recently, Morel *(21)* catalogued the current information regarding the molecules that are retained by various fixatives and embedding procedures. Although this information pertains mainly to electron microscope (EM) radioautographs, Morel does provide a light microscope radioautographic procedure for EM resin-embedded tissues *(21)*.

3.2. Embedding

Although the traditional dehydrating and subsequent paraffin/paraplast embedding methods for light microscopy (*see* Chapter 2) are applicable, many current investigators embed in electron microscope resins and cut 1-μm sections for light microscope radioautography. The main difference between embedding a routine histology and autoradiographic histology section is being attentive to properly disposing of waste solutions. Generally, the tissue begins dehydration with two incubations in 70% ethanol for 30 min, each followed by two 30-min incubations in 95% ethanol, then by two 30-min incubations in 95% ethanol and finally three 30-min incubations in 100% ethanol. All of these solutions are radioactive according to liquid scintillation quantitation in the author's laboratory and should be disposed of in an appropriately labeled container in accordance with the investigators' institutional Radiation Safety Officer's radioactive waste disposal plan. After dehydration in ethanol the infiltration process follows. Many protocols utilize xylene; however, the hazards of working with xylene are well documented and an excellent replacement (Hemo-De) has been developed by Fisher and has been used with great success in the author's laboratory. Typically, the tissue is first incubated in a Hemo-De/paraffin solution in the ratio of 2:1 for 1 h, followed by a ratio of 1:1 for 1 h and ending overnight in 100% paraffin. All of these incubations occur at 65°C, or above the melting temperature of the paraffin being used. As tested in the author's laboratory, these incubations yield waste that is not radioactive.

Sections for light microscope radioautography originate either from paraffin/paraplast or, more recently, resin-embedded specimens as mentioned.

Table 4
Retention of Macromolecules and Exogenously Supplied Radioactivity Following Fixation and Washing Procedures[a]

Fixative	Precursor	Molecules most likely retained	% Radioactivity retained	Investigator(s)
Formaldehyde or glutaraldehyde followed by buffer wash and postfixation in O_sO_4 and dehydration	[3H]-Leucine	Proteins and polypeptides, although investigators did not rigorously establish chemical form of radioactivity; compared results to those obtained by TCA precipitation	Formalin—all but 14.5% Glutaraldehyde —all but 3.5%	Vanha-Pertulla and Grimley (22)
Formaldehyde or glutaraldehyde Formaldehyde or Carnoy's	[3]-Thymidine	DNA 4SRNA		Vanha-Pertulla and Grimley (22) Sirlin and Leoning (23)
Formaldehyde or glutaraldehyde	[3]-Uridine	RNA	Formaldehyde —25% loss Glutaraldehyde —22% loss	Vanha-Pertulla and Grimley (22)
Osmium tetroxide	Precursors of RNA	RNA	Retention of radioactivity	Monneron and Moule (24)
Formaldehyde or glutaraldehyde	Labeled mannose	Polysaccharide	Considerable 20% loss of radioactivity with either fixative	Vanha-Pertulla and Grimley (22)

[a]See review of Williams (25) for retention of lipids with osmium tetroxide; generally high rates of phospholipids are retained.

42

In either case, the problem is to obtain sections of uniform thickness on a routine basis *(7)*. This problem appears to be more significant with light microscope radioautographs than with electron microscope autoradiographs, as estimates of EM section thickness are based on interference color charts. For light microscope radioautographs, sections of 4–5 μm are desirable for paraffin/paraplast-embedded specimens. However, 1–2 μm sections of plastic-embedded materials are employable (*see* protocol in ref. *21*). These are "thick" sections cut from blocks for EM radioautography.

Subsequent to sectioning, it is necessary to mount the sections on glass microscope slides that require meticulous cleaning even if the slides were precleaned by the manufacturer. For example, slides should be soaked overnight in a chromate solution (dangerous) and subsequently washed for several hours in cold, running tapwater followed by two changes of dH_2O (30 min each). After drying, the cleaned slides are dipped, at room temperature, into a freshly prepared and filtered solution consisting of 5.0 g gelatin, 0.5 g chrome alum, and dH_2O to yield 1 L (subbing). Next, the slides are allowed to drain and dry in a dust-free atmosphere. The resultant gelatin layer provides sufficient adhesion for both the section and emulsion. However, Rogers *(7)* reported that gelatin possesses different properties from other types of adhesives commonly employed in light microscopy. If the investigation involves paraffin-embedded specimens, the paraffin must be removed (*see* Chapter 2).

4. EMULSIONS

4.1. Types of Emulsions

The emulsions that are routinely employed in light microscope radioautography are summarized in Table 1. The authors are most familiar with Kodak products and the reader is referred to technical bulletins from Amersham and Ilford for the details of how to properly use these emulsions. Kodak NTB2 is most often used in biological systems that utilize tritium as the isotope. Briefly, it requires Kodak Safelight Filter Type 2, a darkroom which is absolutely light tight, water bath, Copland jar, and slide-drying rack. The emulsion is a gel until melted at 45°C in a water bath. The emulsion container should never be opened until it is fully ascertained that there are absolutely no light leaks or fluorescence in the dark room in which the slide coating and developing are to occur. It will require approximately 10–15 min for the eye to fully adjust to the darkroom whose only light source is the safelight filter. Common sources of leaks are around doors and fluorescent timers. Firmly fitting door trims will solve most leaks associated with doors and removing or

Table 5
Effect of Emulsion Thickness on Resolution

Isotope	Section thickness (μ)	Emulsion	Developer	HD (μm)
[³H]	0.5	AR-10	D-19	0.35
	0.4	L4	D-19	0.3
	1.0	L4	D-19	0.35
	0.5	NTB-2	Dektol	0.38
[¹⁴C]	0.5	L4	D-19	0.8
	0.5	AR-10	D-19	2.0

From ref. *(7)*, with permission.

placing fluorescent-faced timers in a cabinet will eliminate this other significant problem. By taking the time to ensure that no light is leaking into the darkroom, there is the benefit of minimal background on the slide and thereby decreased interference with quantitation of the silver grains on the slide.

Table 5 illustrates the effects of emulsion thickness on resolution as indicated by *HD* (distance from the line within which 50% of all the silver grains from it lie). The data indicate that resolution is improved as emulsion thickness is decreased. If the emulsion thickness exceeds 2 μm, an insignificant quantity of β-particles (*see* radioisotopes) will reach the emulsion from most biological specimens. Some factors which influence emulsion thickness include dilution of the emulsion, temperature of the emulsion, temperature and wetness of the slide as well as the temperature and humidity during drying. The reader is referred to Rogers *(7)* for thorough discussions of each.

4.2. Emulsion Application

The conditions for handling emulsions as well as the required equipment (*see* Table 6) have been detailed by Kodak Scientific Imaging Systems in their technical bulletin regarding microradioautography *(12)*.

With regard to emulsion coating, there are various ways by which this can be accomplished. When dealing with diffusible or water-soluble substances, one can either employ dry mounting or apposition procedures *(12)*. For the former, the specimen can be "press"-fixed to a previously coated emulsion layer. In the case of apposition, an emulsion coated slide is brought into contact with a specimen slide during exposure of the specimen to the emulsion. Subsequent to the appropriate exposure, the two slides are separated for both photographic processing and specimen staining. Finally, the two slides are reunited for silver grain analysis (*see* Subheading 6.).

Table 6
Conditions for Handling Emulsions

Emulsion handling	Darkroom equipment
Have emulsions delivered directly to your laboratory and not to a central facility. Perform a routine pretest for fog level immediately on receipt *(12)*	An area dedicated to handling and processing photographic materials
Store emulsions in the original light-tight container in a chemical and radiation-free refrigerator (4°C)	Source of light leaks and chemical contamination should be eliminated to prevent background fog, which may require special door stripping
Do not store bottled autoradiographic emulsions in a freezer	Darkroom should have a light-lock entry or a light-tight entry door that can be locked from the inside
Do not keep emulsions longer than 1 or 2 d at room temperature	Perform safelight sensitivity testing *(12)* if concerned about safelight contributing to background fog.
Use meticulously cleaned glassware; emulsion coats on glassware and are difficult to remove	Do not use fluorescent lights; if you cannot remove fluorescent lights, wait 30 min after they have been turned off
	Whenever possible use a darkroom with temperature and humidity controls as well as filtered air conditioning

Summarized from Kodak Scientific Imaging Systems *(12)*, with permission.

The hand dipping technique is most often used for mammalian tissue sections. The tissues are prepared as described in Subheading 3.2. and sectioned into 1–5 µm thick sections and affixed to the slides using standard mounting procedures that are briefly described. The sections are placed in a 45°C waterbath that contains Haupt's gelatin in a ratio of 1 mL gelatin per 500 mL of deionized water. From this point, a slide is carefully placed under the target section and is carefully lifted out of the water bath onto the slide. The excess water is carefully removed and the slide placed on a 40°C slide warmer for 40 min to dry. The slides are then ready for dipping. The NTB2 emulsion is liquefied for 1 h at 45°C and the appropriate amount is aliquoted into a preheated Copland jar, which remains in the water bath during the procedure. From the time the emulsion is poured out into the Copland jar until the dried slides are placed in a light-tight slide box, all procedures are performed with only the assistance of the Kodak Type 2 Safetylight. Once the emulsion is liquefied, dip the slide to a depth that surpasses the specimen by 0.67–1.27 cm. The speed with which the emulsion is entered should equal the speed by which it is removed. The object is to achieve a reproducibly thin layer of emulsion without introducing any bubbles into the emulsion. It is beneficial to use a simple trick to ensure the reproducibility of the emulsion thickness: count to three as the slide enters and exits the emulsion with no hesitation between entry and exit. The nonspecimen slide is wiped dry and placed on a dust-free rack to dry for 2 h. The slides are placed into a black slide box with dust-free desiccant, wrapped in aluminum foil, and stored at 4°C for the experimentally determined optimum incubation time usually 4–6 wk for tritium.

It is recommended that the investigator read ref. *(12)* concerning the meticulousness required to eliminate chemical contamination resulting from poorly cleaned glassware or equipment. In addition, the investigator is urged to perform chemography testing *(8,12)* to assess whether the selected emulsion will react to chemicals within the specimen. The Kodak technical bulletin provides a thorough discussion of latent image fading *(26–28)*.

In the case of nondiffusible substances, sections affixed to slides maybe coated by hand-dipping techniques (*see* Fig. 1, and refs. *28, 29, and 35*), automated dipping procedures *(30–32)*, or loop methods (*see* Fig. 2, and refs. *21, 33, and 34*). An alternative application technique employs stripping film (*see* Fig. 2, and refs. *21* and *36*). Tables 7 and 8 present the advantages and disadvantages of the hand-dipping and stripping-film techniques, respectively. Subheadings 4.2.1. and 4.2.2. detail the procedures for the two techniques.

4.2.1. HAND-DIPPING PROCEDURE

1. Maintain emulsion at 40–42°C in a water bath throughout dipping procedure.
2. Dip the slides one by one into the liquid emulsion to a depth of 0.67–1.27 cm greater than that required to coat the specimen; keep dipping jar vertical.
3. Withdraw the slide immediately or wait a few seconds to derive a thicker emulsion layer (DO NOT dip each slide repeatedly to achieve a thicker emulsion; instead, reduce withdrawal rate or emulsion temperature). Consult Rogers *(7)* for selecting an appropriate emulsion thickness (1.5–2.0 µm but depends on emulsion and isotope in question), especially if one is contemplating quantitative autoradiography. Move slide away from the emulsion so that excess emulsion does not drip back onto the surface of the emulsion creating bubbles.
4. Place emulsion-coated slide in a horizontal position and wipe the back of the dipped slide with a paper towel or a damp sponge and place the slide flat on a drying rack protected from safelight exposure for 1 h at 45–50% relative humidity (RH). The emulsion will gel in approximately 1 h but can be speeded up by placing the slide on a cold metal plate for 15-

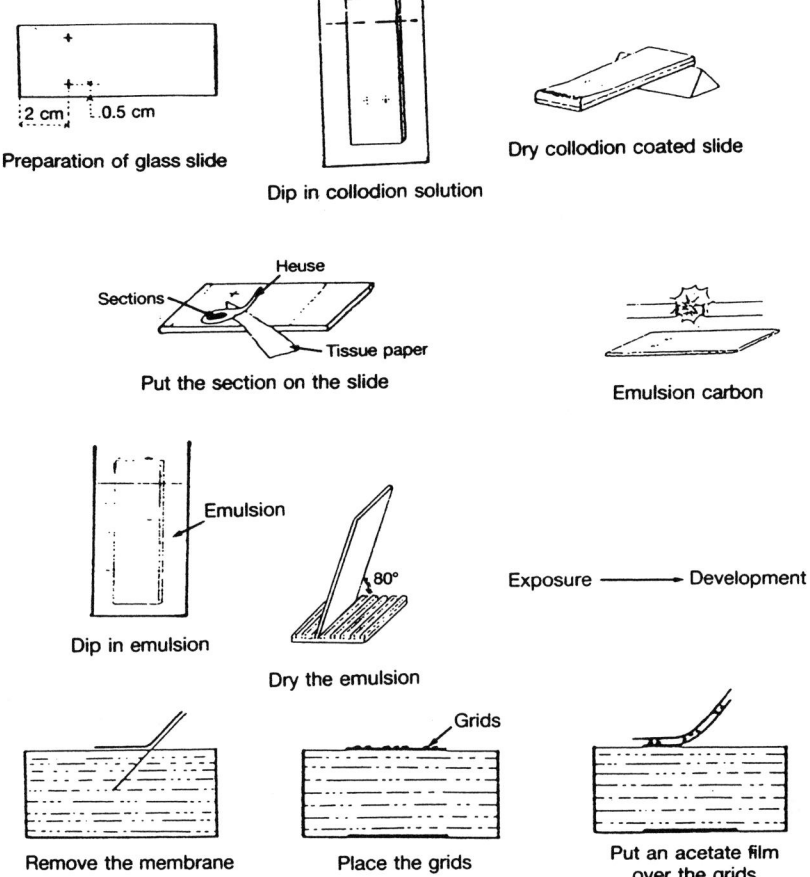

Fig. 1. Emulsion application by dipping in liquid emulsion. From ref. *(21)*, with permission.

20 min; however, this will result in a thicker emulsion layer and decrease the resolution of the specimen.

Note: The dipped slide can be placed vertically for drying but results in an emulsion layer that varies in thickness from top to bottom of the slide and is, therefore, not recommended.

The reader is referred to Rogers *(7)* for differences in dipping techniques for Kodak NTB-2 and Ilford K2 emulsions.

4.2.2. STRIPPING-FILM PROCEDURE[a]

Protocol for the use of Kodak AR-10 stripping film whose shelf-life is about 6 mo at 4°C is contained in its shipping container.

[a] Please consult Rogers *(7)* for a copious supply of details for each step.

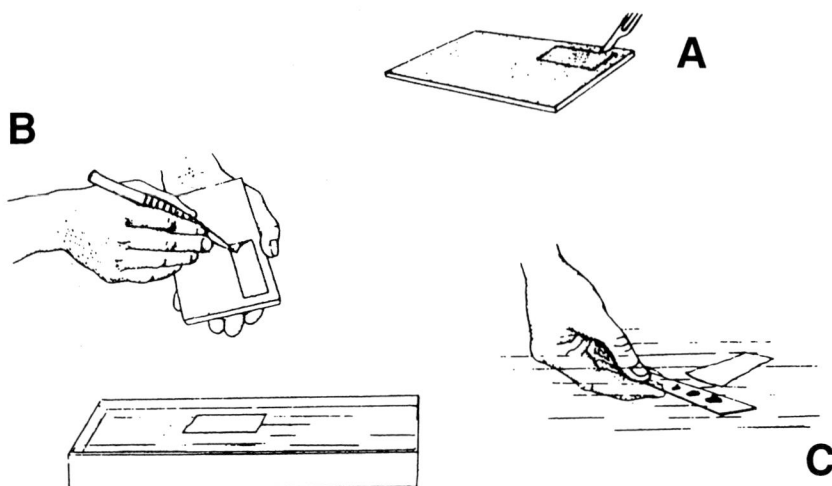

Fig. 2. Preparation of stripping-film autoradiographs. From ref. *(7)*, with permission.

Table 7
Advantages and Disadvantages of Dipping Techniques[a]

Advantages	Disadvantages
Close contact between emulsion and the specimen can be achieved	Difficult to obtain an emulsion layer of constant and reproducible thickness making comparisons of radioactivity with different structures very tenuous [*see* ref. *(7)* for a thorough discussion of such comparisons as a function of isotope employed]
Can obtain autoradiographs with a significantly superior resolution than with stripping film with many isotopes	
Rapid and simple to prepare autoradiographs	
Staining, mounting under a coverslip, viewing and photography are superior to that obtained with stripping film	
Wide range of emulsions commercially available differing in sensitivity and in crystal diameters to achieve optimal resolution.	

[a] Adapted in part from Rogers *(7)* and Rogers *(35)*.

1. The darkroom should be kept at 18–21°C with a relative humidity of 60–65%; emulsion should be handled at least 1–2 *M* from a Kodak No. 1 (red) safelight. Strip the packaged emulsion from its glass support and remove the film from the refrigerator 1–2 h before it is needed to allow it to come to room temperature.

Table 8
Advantages and Disadvantages of Stripping Film

Advantages	Disadvantages
Adequate resolution of 0.5–5 µm can be obtained	Considerable handling required, which increases the chance of contamination
Uniformity of layer is better than what can be obtained by liquid emulsion	Removing the stripping film from its gelatin base can cause visible flakes of static charge resulting in high background fog
Stripping film remains the only basic technique that is applicable to almost the entire range of observations of the light microscope level	Time required for the emulsion to swell results in a limit on how rapidly autoradiographs can be prepared
	Stripping film adheres less firmly than a layer of liquid emulsion, thus film can be displaced from specimen during development
	Presence of the gelatin layer above the emulsion may yield staining problems, as gelatin can become stained, obscuring view of specimen

Adapted in part from Rogers *(7)* and Appelton *(36)*.

2. Place film on the surface of water with the emulsion side down.
3. Attach film to the specimen by holding the slide at the end farthest from the specimen. Next, place slide in the water and move it under the piece of film at an angle about 30° to the horizontal lift film out of the water on the slide.
4. Stand slide with film vertically and dry in a gentle stream of cool air for 20–30 min.
5. Allow slides to remain in the darkroom for an additional 1 h and then transfer to a desiccator at room temperature over dried silica for 18–24 h.
6. Store in light-tight box at 4°C for appropriate exposure time.
7. Develop film in Kodak D19 developer at 18°C for about 5 min.
8. Rinse for 30–60 s in d H_2O.
9. Fix in 30% w/v sodium thiosulfate, 10 min or Kodak Acid Fixer

4.3. Exposure Time

Exposure time is relative to the experimental system, isotope, and dosage. There are no steadfast rules for the appropriate combination of the foregoing variables and, therefore, pilot studies are required to optimize these variables for each experimental situation. In general, for tissues incubated for 24 h in tritiated solution at a concentration of 1 µCi/1 µmol/1 mL should permit a 1–5-µm-thick section of resultant tissue to be adequately exposed to emulsion if it is incubated for 4 wk at 4°C. This should allow for maximal oxidation/reduction of the emulsion to occur with minimal background or latent image formation occurring.

Subheading 5., regarding analyses and interpretation of radioautographs, offers additional information regarding exposure time.

4.4. Specimen Staining

With regard to staining of specimens, one can prestain before applying the emulsion or poststain after photographically processing the emulsion (37). Prestaining has several advantages including elimination of staining through the emulsion layer. Kodak (12) points out that although the emulsion is reactive with some stains, such as celestine blue (38) and gallocyanin–chrome alum (39) in that the former increases background fog grains and the latter may decrease the intensity of the silver grains. A general solution, if prestaining is required, is to use routine hematoxyln and eosin staining as they are not reactive towards the emulsion. In contrast to prestaining, poststaining is usually preformed directly after photographic processing. Poststaining is used most often in experiments involving dry-mounting or apposition procedures. If 1–2-µm-thick sections of resin-embedded material were employed for radioautography, the emulsion-covered sections can be stained with 1% toluidine blue in 1% sodium borax (21). Superior staining results are obtained when the sections are stained after autographic emulsion exposure and development.

5. ANALYSES AND INTERPRETATION OF RADIOAUTOGRAPHS

The number of developed silver grains is affected by exposure time and an algorhythm has been developed for calculating autoradiographic exposures (40). The analysis of the image is perhaps the most time-consuming aspect of light microautoradiography (41). Historically, autoradiographic slides have been quantitated by using a microscope, hemocytometer, counter, and a very patient technician. Improvements using dark-field microscopy and semiautomation were helpful but still did not make the process very efficient. Recent advances in imaging software have improved the process, but there is still work to be done.

An example of a routine slide may involve needing to know how much of a metabolite is accumulated in cells in a representative area of the section. Therefore, between 50–75% of the cells must be counted and the amount of developed silver grains above them quantitated. Most of the software available for autoradiography is designed to analyze eletrophoretograms or whole-body dosing, not cellular silver halide accumulation. Therefore, some modifications are required.

Using various autoradiography software, cells must be distinguished by using a lasso or other tool to outline all of the cells. This may require simultaneously holding down the shift key on the computer to circle multiple cells. Once the cells are enumerated, the software must be asked to count all of them and the number recorded. The second task is to count the silver grains; however, it may be more appropriate to determine how much of the cell is covered by silver grains. To ascertain this, use the shading functions, thresholding, available to change all the colors on the slide to white except the silver grains. At this point, the software can be asked to provide what percentage of each cell is black. Relative to the software package, each cell may need to be circled, shaded, and quantitated. The advantage of the image analysis software is the preciseness of the results provided; the disadvantage is that it is still rather labor intensive. The following suppliers of image-analysis software have autoradiographic applications built into their software, some protocols are as described. Others are more user friendly, such as Molecular Dynamics, Inc. at *www.mdyn.com.* and Loats Associates at *www.loats.com.* The reader is referred to refs. *(42–45)* for methods for intensifying radioautographs.

6. LIGHT MICROSCOPE AUTORADIOGRAPHY

A protocol for the light microscope radioautography of *Lilium longiflorum* pollen tubes labeled with [^{14}C]-proline follows. This protocol, which does not require tissue embedding in paraffin or Paraplast, can be modified for paraffin-embedded tissues (*see* Chapter 2). Thus, by employment of the protocol, together with the preceding introductory information in this chapter, one should be able to derive a protocol applicable to the cells or tissue in question. The performance of the protocol requires approval of an institution's Radiation Safety Officer. An inventory of incoming radionuclides, their presence in secondary containers, and their waste must be carefully recorded. The waste must be further broken down into solid waste, liquid waste, and animal carcasses to aid in its proper disposal.

Chemicals	*Plant material*
Acetic acid	*Lilium longiflorum*, cv. "Ace"
Carnoy's (3 parts ethanol: 1 part	pollen (common Easter lily)
acetic acid)	Sources of *Lilium longiflorum* (lily)
Collodion	pollen—
Ethanol	Greenhouses
Growth medium (Dickinson's)	Local botanical gardens
[^{14}C]-proline (specific activity,	Supermarket at Easter time
214 mCi mmol^{-1})	Researchers investigating lily
Film-processing solutions	pollen—see research literature
Kodak D-19	*Note*: Lily pollen attached to anthers
Kodak Rapid Fixer	can be stored in Petri dishes at 4°C
Permount or Polymount	in a refrigerator. Viability checks
Scintillation cocktail for aqueous	should be performed regularly
samples, e.g., Aquasol	(*see* Stanley and Linskens, ref. 52)

Supplies	*Equipment*
Aluminum foil	Autoclave
Autoradiographic film	Darkroom
Beakers	Tabletop centrifuge
Conical centrifuge tubes	Liquid scintillation counter (optional)
Coverslips 25 × 50 nm	
Frosted-end microscope slides	
Emulsions—one of the following	
Kodak's NTB-2 Nuclear Track	
Amersham's LM-1	
Kodak's AR-10 stripping film	
Erlenmeyer flasks	
Graduated cylinders	
Pasteur pipets	
Petri dishes	
Pipets 1,5, and 10 mL	
Probes	
Propipets	
Low K$^+$ scintillation vials (optional)	
Single-edge razor blades	
Slide box with drierite	
Vials for fixation	

1. Germinate 20 mg fresh weight lots of *Lilium longiflorum*, cv. "Ace" pollen[a] in Petri dishes containing 10 mL medium supplemented with 0.1 μCi[^{14}C]-proline (pro) for 4 h at 27 ± 2°C.

[a]A comprehensive review of pollen biology can be found in Stanley and Linskens *(52)*.

2. To germinate lily pollen, spread 20 mg g fresh weight lots onto the surface of 10-mL sterile SYB *(46,47)* or Dickinson's *(48)* media in sterile Petri dishes.
3. Wash germinated pollen with medium lacking [^{14}C]-pro and count washes via liquid scintillation.
4. Fix germinated pollen in Carnoy's and wash twice (10–15 min each time) with 30%, 50%, 70%, and three times with 100% EtOH (use 200 proof alcohol).
5. Suspend in collodion.
6. Construct a radioautographic sandwich consisting of a slide with a thin layer of collodion containing pollen and another slide to which was affixed a thin layer of Kodak NTB-2 Nuclear Track Emulsion (use appropriate safelight and bulb wattage—*see* Table 1).
7. Incubate for 10 d in the dark in a light-tight slide box containing a desiccant *(12)*.
8. Other emulsions that maybe useful include Amersham's LM-1 and Ilford's K5 and Kodak's AR10 stripping film (Table 1).
9. Separate slides and develop emulsion with Kodak D-19.

Protocol updated from Krolak et al. *(49)* and Dashek and Mills *(50)*.

Controls: (a) Germinate pollen without [^{14}C]-pro and process as before. (b) Construct a radioautographic sandwich with collodion lacking pollen. (c) Expose a slide coated with Kodak NTB-2 Nuclear Track Emulsion to light and develop emulsion with D-19. (Sample autoradiographs are presented in Fig. 3.)

Note: The purpose and use of radionuclides are monitored by the Institution's Radiation Safety Officer and requires his/her approval. Most institutions follow Nuclear Regulatory Agency guidelines. Consult your safety officer for record keeping and disposal. A useful adjunct to this simple experiment involves determination of uptake of [^{14}C] proline via liquid scintillation counting *(49,50)*. Germinated pollen can be collected on filter papers and after washing the pollen with growth medium lacking [^{14}C] proline, the filters can be placed into scintillation vials (low K^{+}-glass or plastic) prefilled with Cab-o-sil [a thixotropic gel (New England Nuclear)] and liquid scintillation cocktail *(51)*. Aliquots (0.1 mL) of the medium can be added to vials containing a cocktail that will accept aqueous solutions, e.g., Aquasol. There are many contemporary, commercially available cocktails that will accept limited aqueous solutions. Samples should be counted three times with a 2% preset error in a liquid scintillation counter equipped with automatic quench calibration for the conversion of cpm to dpms. Many modern liquid scintillation counters possess programs with quench curves in memory.

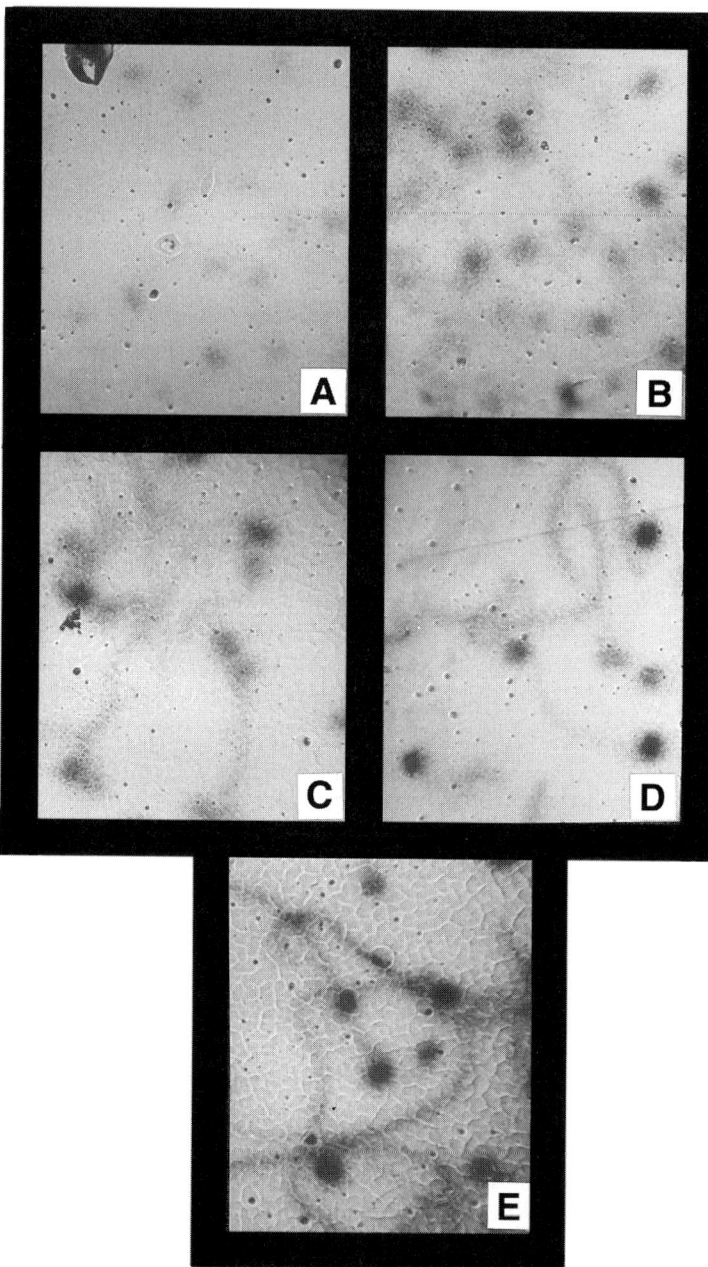

Fig. 3. Light microscope radioautographs of intact tubes from pollen germinated 4 h in a medium containing [^{14}C]-proline with or without the proline analogue, azetidine-2-carboxylic acid (AZC). A = 1.0 mM AZC, B = 0.1 mM AZC, C = 0.01 mM AZC, D = 0.001 mM AZC, E = Control. From Dashek and Mills (50), with permission.

8. CONCLUSIONS

Although radioautography is still widely used in plant cell biology, many investigators prefer to use more recent methods, e.g., immunocytochemistry, as the laboratory can be easily contaminated with radioactivity if meticulous safety procedures are not adhered to. Perhaps, the future of light microscope autoradiography for the plant biologist can be found in the recent applications of animal-oriented cell and medical biologists to the visualization of receptors *(53–55)*.

Improvements in light microscope radioautography are most likely to center around the development of fine-grain film emulsions and their application to sections, double-emulsion techniques *(56,57)*, tests for chemography *(58)*, procedures to minimize latent image fading *(27)*, conditions during exposure *(59)* and photographic processing (*see* literature in *[12]*), and color microradioautography *(60)*.

REFERENCES

1. Gude D. *Autoradiographic Techniques*, Prentice Hall, Engelwood Cliffs, NJ, 1968.
2. Baserga R, Malamud D. *Autoradiography Techniques and Applications*, Harper and Row, New York, 1969.
3. Budd GC. Recent developments in light and electron microscope radioautography. *Int Rev Cytol* 1971; 31: 21–56.
4. Fisher HA, Warner G. *Autoradiography*, Walter de Gruyter, New York, 1971.
5. Gahan PB. *Autoradiography for Biologists*, Academic Press, London, UK, 1972.
6. Luttge U. *Microautoradiography and Electron Probe Analysis: Their Application to Plant Physiology*, Springer-Verlag, New York, 1972.
7. Rogers AW. *Techniques of Autoradiography*, Elsevier, Amsterdam, The Netherlands, 1973.
8. Evans LV, Callow ME. Autoradiography, in *Electron Microscopy and Cytochemistry of Plant Cells* (Hall JL, ed.), Elsevier/North-Holland, Amsterdam, The Netherlands, 1978, pp. 235–277.
9. Williams MA. *Autoradiography and Immunocytochemistry*, Elsevier/North-Holland, Amsterdam, The Netherlands, 1980.
10. Williams MA. Autoradiography: its methodology at the present time. *J Microsc* 1982; 128: 79–94.
11. Baker RJ. *Autoradiography: A Comprehensive Overview*, Oxford University Press, Oxford, UK, 1989.
12. Kodak Scientific Imaging Systems. *Microautoradiography: Autoradiography at the Light Microscope Level*, Eastman Kodak, Rochester, New York, 1994.
13. Hermann RG, Abel WD. Microautography of organic compounds insoluble in a wide range of polar and non-polar solvents, in *Microautography and Electron Probe Analysis. Their Application to Plant Physiology* (Luttge U, ed.), Springer-Verlag, New York, 1972, pp. 123–165.
14. Roth LJ, Stumpf WE. *Autoradiography of Diffusible Substances*, Academic Press, New York, 1969.
15. Williams MA. *Autoradiography and Immunocytochemistry*, North Holland Publishing Co, Amsterdam, Netherlands, 1977.

16. Eschrich W, Fritz E. Microautoradiography of water-soluble inorganic compounds, in *Microautography and Electron Probe Analysis: Their Applications to Plant Physiology* (Luttge, U, ed.), Springer Verlag, New York, 1972, pp. 99–122.

17. Stumpf WE. Techniques for the autoradiography of diffusible compounds. in *Methods in Cell Biology*, Vol. 13, Academic Press, New York, 1976.

18. Inson EJ, Sheridan PJ. Autoradiography of diffusible substances: a practical approach. *Am J Med Technol* 1981; 47: 38–42.

19. Vuillet J. Correlative light and electron microscopic method for the visualization of the same *in vitro* cell using radioautography and serial sectioning, in *Correlative Microscopy in Biology: Instrumentation and Methods* (Hayat MA, ed.), Academic Press, Orlando, FL, 1987, pp. 166–172.

20. Slater R. *Radioisotopes in Biology. A Practical Approach*, IRL Press, Oxford, UK, 1990.

21. Morel G. Electron microscopic autoradiographic techniques, in *Electron Microscopy in Biology. A Practical Approach* (Harris JR, ed.), IRL Press, Oxford, UK, 1994, pp. 83–123.

22. Vanha-Pertulla T, Grimley PM. Loss of proteins and other macromolecules during preparation of cell cultures for high resolution autoradiography quantitation. *Histochem Cytochem* 1970; 18: 565–573.

23. Sirlin JL, Leoning UE. Nucleolar 4s ribonucleic acid in dipteran salivary glands in the presence of inhibition. *Biochem J* 1968; 109: 375–387.

24. Monneran A, Moule Y. Critical evaluation of specificity in electron microscopical radioautography in animal tissues. *Exptl Cell Res* 1969; 56: 179–193.

25. Williams MA. *Adv Opt Electron Microscop* 1969; 3: 219.

26. Appelton TC. Resolving power, sensitivity and latent image fading of soluble-compound autoradiographs. *J Histochem Cytochem* 1966; 14: 414.

27. Boren, HG., Wright, EC, Harris, CC. Quantitative light microscopic autoradiography. Emulsion sensitivity and latent image fading. *J Histochem Cytochem* 1975; 23: 901–909.

28. Kopriwa BM, LeBland CP. Improvements in the coating technique of radioautography. *J Histochem Cytochem* 1962; 101: 269–284.

29. Brock ML, Brock TD. The application of microautoradiographic techniques to ecological studies. *Int Assoc Theor Appl Limnol* 1968; 15: 1–29.

30. Kopriwa BM. A semiautomatic instrument for the radioautographic coating technique. *J Histochem Cytochem* 1966; 14: 923–928.

31. Jerry NL. Quantity dipping and processing of radioautographic slides. *J Biol Photogr Assoc* 1967; 35: 73–82.

32. McGuffee LJ, Hurwitz L, Little SA. A method of coating multiple slides for light microscopic autoradiography. *J Histochem Cytochem* 1977; 25: 1107–1108.

33. Pickering ER. Autoradiography of mobile C^{14}-labeled herbicides in sections of leaf tissue. *Stain Technol* 1966; 41: 131–137.

34. Jenkins EC. Wire-loop application of liquid emulsion to slides for autoradiography in light microscopy. *Stain Technol* 1972; 47: 23–26.

35. Rogers AW. *Techniques of Autoradiography* (3rd ed.), Elsevier, Amsterdam, The Netherlands, 1979.

36. Pelc SR, Appelton TC, Welton ME. State of light autoradiography, in *The Use of Radioautography Protein Synthesis* (LeBlond CP, Wauen KB, ed.), Academic Press, New York, 1965, pp. 9–22.

37. Belanger LF. Staining processed radioautographs. *Stain Technol* 1961; 36: 313–317.

38. Deuchar EM. Staining sections before autoradiographic exposure: excessive background graining caused by celestin blue. *Stain Technol* 1962; 37: 324.

39. Sternam U. Loss of silver grains from radioautographs stained by gallocyanin-chrome alum. *Stain Technol* 1962; 37: 231–234.
40. Duncombe WF. A nomogram for calculating autoradiographic exposures. *Int J Appl Radiat Isot* 1961; 10: 212–213.
41. Domer P. Photometric methods in quantitative autoradiography, in *Microautoradiography and Electron Probe Analysis. Their Application to Plant Physiology* (Luttge U, ed.), Springer Verlag, New York, 1972, pp. 7–48.
42. Kopriwa BM. Quantitative investigation of cellular intensification for light and electron microscope radiautography. *J Histochem Cytochem* 1980; 68: 265–279.
43. Kopriwa BM. Examination of various methods for the intensification of radiautographs by scintillators. *J Histochem Cytochem* 1979; 27: 1524–1526.
44. Panayi GC, Neill WA. Scintillation autoradiography. A rapid technique. *J Immunol Methods* 1972; 2: 115–117.
45. Rogers AW. Scintillation autoradiography at the light microscope level: a review. *Histochem J* 1981; 13: 173–186.
46. Rosen WG, Gawlik SR, Dashek WV, Siegesmund KA. Fine structure and cytochemistry of *Lilium* pollen tubes. *Am J Bot* 1964; 51: 61–67.
47. Dashek WV, Rosen WG. Electron microscopical localization of chemical components in the growth zone of *Lilium* pollen tubes. *Protoplasma* 1966; 61: 191–204.
48. Dickinson DB. Germination of lily pollen. Regulation and tube growth. *Science* 1965; 150: 1818–1819.
49. Krolak JM, Taylor N, Dashek WV, Mills R. Azetidine-2-carboxylic acid-induced suppression of ^{14}C-proline incorporation into cytoplasmic macromolecules and cell wall of *Lilium longiflorum* pollen, in *Advances in Plant Reproductive Physiology* (Malik CP, ed.), Kalyani Publishers, New Delhi, India, 1978, pp. 62–71.
50. Dashek WV, Mills RW. Azetidine-2-carboxylic acid and lily pollen tube elongation. *Ann Bot* 1980; 45: 1–12.
51. Ross HJE, Noakes E, Spalding JD. *Liquid Scintillation Counting and Organic Scintillators*, Lewis, Chelsea, MI, 1991.
52. Stanley RG, Linskens HF. *Pollen Biology Biochemistry Management*, Springer-Verlag, Berlin, Germany, 1974.
53. Wharton JP. *Receptor Autoradiography and Practice. Modern Methods in Pathology Services*, Oxford University Press, New York, 1993.
54. Moise E, Coronas V. *In vitro* localization and pharmacological characterization of receptors by radioligand binding on tissue sections and quantitative radioautography, in *Visualization of Receptors. Methods in Light and Electron Microscopy* (Morel G, ed.), CRC Press, Boca Raton, FL, 1997.
55. Shimada M, Wtanabe M. Whole body and microradioautography of receptors, in *Visualization of Receptors. Methods in Light and Electron Microscopy* (Morel G, ed.), CRC Press, Boca Raton, FL, 1997.
56. Baserga R, Neneroff K. Two emulsion radioautography. *J Histochem Cytochem* 1962; 10: 628–635.
57. Han SS, Kim MK. An improved method for double-isotope and double-emulsion radioautography using epoxy resin sections. *Stain Technol* 1972; 47: 291–296.
58. Robertson GB, Rogers AW. Loss of silver grains from completed autoradiographs. *J Microsc* 1979; 117: 301–303.
59. Siwicki W, Ostrowski K, Rowinski J. Effect of temperature on overall efficiency of autoradiographies and on fogging of emulsion. *Stain Technol* 1968; 43: 35–39.
60. Haase AT, Walker D, Stowing L, et al. Detection of two viral genomes in single cells by double label hybridization *in situ* and color microautoradiography. *Science* 1985; 227: 189–112.

4

Some Fluorescence Microscopical Methods for Use with Algal, Fungal, and Plant Cells

Virginia A. Shepherd

I. INTRODUCTION

The number of research papers employing fluorescence methods has increased exponentially since the 1930s, when such methods were first applied to plant anatomy and physiology. The marriage between the traditional fluorescence microscope and new technologies, primarily confocal laser scanning microscopy (CLSM) and computers with vast storage and image-processing capacity, has shifted our concept of the cell from a highly resolved, but static, view engendered by electron microscopy, to one that is dynamic, vital, and resolved in time. Many new advances concern the dynamic behavior of the cytoskeleton, motile organelles, intra- and intercellular transport processes, and oscillatory behavior of "second messenger" Ca^{2+}. This chapter is not a comprehensive review of such advances, nor of fluorescence microscopy in general. The three volumes by Rost *(1–3)* give a lucid "user-friendly"

From: *Methods in Plant Electron Microscopy and Cytochemistry*
Edited by: W. V. Dashek © Humana Press Inc., Totowa, NJ

account of the history, principles, and detailed methods of fluorescence microscopy. Other volumes give practical details of fluorescence methods with reference to plant cells *(4,5)* or list fluorochromes, their structures and their uses *(6)*. CLSM is an evolving technique *(7)* and the major recent advances based on CLSM in plant cell biology have been comprehensively reviewed *(8)*. CLSM is valuable in studies of three-dimensional structures and calculation of volumes *(9,10)* and has a novel use in scanning autofluorescent resin-embedded specimens *(11)*. Resolution is improved 1.4 times over epifluorescence microscopy *(8)*. Within the cell wall, the cytoplasm is interpenetrated by cytoskeletal elements and encloses endomembrane systems such as vacuoles and endoplasmic reticulum, as well as mitochondria, chloroplasts and other organelles. This chapter focuses on fluorescence microscopical methods for examining properties of the cell wall, the cytoplasm, endomembranes and other organelles of fungal, algal and plant cells.

2. FLUORESCENCE MICROSCOPY OF ALGAL, FUNGAL, AND PLANT CELLS

2.1. Symplastic and Apoplastic Tracing

Historically, the most widespread use of fluorescence microscopy has been in studies of symplastic and apoplastic transport. Symplastic tracing involves introducing charged (impermeant) fluorochromes into living cells and monitoring their movements. Fluorochromes have been used to determine the size exclusion limit (SEL) of plasmodesmata *(12, 13)*, to map symplast domains *(12)*, to observe the process of vacuolar compartmentation *(14,15)*, to study phloem transport *(16,17)*, to determine viability, and in studies of endocytosis *(18,19)*. Fluorescence-recovery after photobleaching (FRAP) enables quantification of rates of fluorochrome diffusion in the cytoplasm. Fluorochromes, especially sulphorhodamine *(20)* and berberine sulphate *(21)*, have been used to determine the places where water from the transpiration stream enters the symplast, and to detect apoplastic barriers such as suberised lamellae of casparian strips.

2.2. Organelle-Specific Fluorescence Staining

Fluorescent stains such as DAPI and ethidium bromide bind to DNA (sometimes even mitochondrial DNA) and have many uses, including location of nuclei and rapid detection of microorganisms or virus formation *(22)*. The lipophilic stain $DiOC_{(6)}$ can be used to stain mitochondria and endoplasmic reticulum in living cells *(23,24)*, and is also a potential-sensitive stain.

2.3. Immunocytochemistry and Fluorescence Tagging

Fluorescent tagging of specific antibodies can pinpoint the location of specific molecules, for example, cytoskeletal elements. Fluorescence-labeled lectin probes are used to detect sugar residues in mucilages, especially in studies of symbioses such as mycorrhizae, nodules, and associations between fungi and cyanobacteria *(25)*. The intrinsic fluorescence of *Aequorea victoria* green fluorescent protein makes it useful as a transcriptional reporter or as a tag for proteins (such as cytoskeletal proteins) and it can be used to monitor gene expression or protein dynamics *(26,27)*. The endoplasmic reticulum appears to be intimately associated with actin and myosin of the cytoskeleton, as well as the putative membrane-wall linkers spectrin and integrin *(24)*. Actin and tubulin have been localized using immunofluorescence methods following freeze-substitution and methacrylate embedding *(28)*. Actin has commonly been visualized by its specific binding to fluorescence-labeled phallotoxins as well as by immunofluorescence*(29–31)*.

2.4. Ion Concentrations and Ratio Imaging

Aspects of the inner physical and chemical environment of cells, such as pH and cytoplasmic Ca^{2+} ion concentrations, $(Ca^{2+}_{(cyt)})$, can be measured using fluorescence ratio-imaging methods. Ratio imaging utilizes the concentration-dependent change in fluorescence emission of a dye that has two excitation peaks. For example, fura-2, the Ca^{2+} indicator, has two absorbance peaks, at 340 nm and 380 nm, and the ratio of emission intensity depends on the Ca^{2+} ion concentration. Fura-2 was used to visualize the Ca^{2+} transient occurring on hypotonic shock in *Fucus* rhizoids *(32)*, whereas indo-1 was used to show transient increases in $Ca^{2+}_{(cyt)}$ in *Chara fragilis* rhizoids following changes in external K^+ or Ca^{2+} concentrations *(33)*. Light-mediated oscillations in $Ca^{2+}_{(cyt)}$ were visualized using quin-2 *(34)* and ratio imaging of calcium green was used to demonstrate stimulus-induced oscillations in $Ca^{2+}_{(cyt)}$ in guard cells *(35)*. Increases in $Ca^{2+}_{(cyt)}$ temporally associated with the action potential were visualized nonratiometrically using the Ca^{2+}-sensitive photoprotein aequorin *(36)*. Most of these indicators are compartmented by cell vacuoles, although conjugation to large-molecular-weight dextrans is likely to resolve this problem. Methods for fluorescence measurement of Ca^{2+}, especially in fungal cells, have been comprehensively reviewed*(37)*. The pH dependence of fluorescein isothiocyanate (FITC) derivatives (such as BCECF or carboxyfluorescein) has been exploited to determine pH of cell compartments such as the small (approx 5×3 μm) spherical vacuoles that are part of the motile tubule–vacuole system in fungal hyphae, using the pH dependent difference in excitation at blue and violet wavelengths *(38)*.

2.5. Intrinsic Autofluorescence

Intrinsic autofluorescence can be used to show the arrangement or presence of chloroplasts and cuticles, or the location of phenolics, lignin, and suberin deposition in thick hand sections or whole cells. Autofluorescence of sporopollenin can pinpoint pollen grains in coal *(4)*. Phenolic compounds play many roles in plants and algae, including signaling (between plants or plants and their symbionts or pathogens) *(39)* or in the response to wounding in giant-celled algae *(40)*. A recent paper utilizing CLSM shows many beautiful images of phenolic autofluorescence in plant tissues *(39)*.

3. SYMPLASTIC TRACING STUDIES

3.1. Methods for Introducing Membrane-Impermeant Fluorochromes into Living Cells

Membrane-impermeant fluorochromes are insoluble in lipids and highly charged at cellular pH (approx 7.6 for plant cytoplasm, approx 5 for the vacuole and cell wall). Once inside the cell, these fluorochromes tend to be retained by lipid membranes against high concentration gradients by electrostatic forces. Exceptions to this have been reported in protoplasts *(19)*. Such fluorochromes can be introduced into living cells by microinjection, permeabilization, or ester loading. Microinjection methods are used for studies of intra- and intercellular transport and for mapping the symplast. Superb images of the cytoskeleton in charophyte cells *(29)* and spectacular images of the dynamics of F-actin during mitosis in plant cells *(30,31)* have been obtained by microinjection of low concentrations (0.66–3.3 μM) of fluorescence-labeled phalloidin. Permeabilization is also used to introduce large fluorochromes (e.g., rhodamine-labeled phalloidin) into living cells. Ester loading is useful for introducing Ca^{2+}- and pH-sensitive fluorochromes into cells and for studies of vacuole dynamics in fungal hyphae *(38,41–43)*. FITC conjugates are ideal for symplastic tracing studies, as they have low toxicity, a negative charge, and a pK_a close to 6.3. FITC has its main absorption peak at 490 nm and maximal emission at 520 nm. Blue excitation produces intense yellow–green fluorescence that is easily distinguished from the red autofluorescence of chloroplasts. FITC is readily conjugated to other molecules such as amino acids, dextrans, proteins, and even viruses *(44)* and the excitation and emission spectra are hardly changed by conjugation. 6-Carboxyfluorescein (6 CF, pK 6.3) is widely used as a symplastic tracer. It is poorly soluble in water and has to be dissolved in a small volume of 0.3 M KOH. FITC-conjugated peptides and immunochemicals can be purchased commercially (e.g., from Molecular Probes, Eugene, OR) (6). However, it is useful to be able to FITC-label molecules for individual experiments.

3.1.1. MICROINJECTION METHODS

Microelectrode methods have been described in detail (45). Fluorochromes can be introduced into cells by iontophoresis (current injection) or by pressure injection. Both current and pressure injection can cause loss of viability and cells which undergo cyclosis give useful feedback. Microelectrodes are pulled on commercial pullers so that tip size and resistance can be customized for each cell type. Alumino-silicate capillaries (1–1.5 mm) with inner filaments (for backfilling) are tougher than boro-silicate capillaries. The size of microelectrode tips, the magnitude of injecting current or pressure, and the dye concentration in the tip depend on the size of the cells. A 1-µm tip microelectrode, 300–1000 kPa pressure, or a negative current of 400–800 nA for up to 1 min, and fluorochrome tip concentration of approx 50 mM are suitable for giant cells (e.g., a 5-mm *Chara* internode). A maximum microelectrode tip size of 0.8 µm, 100 kPa pressure, or a negative current of 40–100 nA for 30 s, and a 10-mM tip concentration of fluorochrome are suitable for smaller cells (e.g., an *Egeria* leaf epidermal cell). A drawn-out Pasteur pipet is used to backfill about 5–20 µL of fluorochrome into the microelectrode tip, followed by 0.1 M KCl, which gives reasonable electrical stability. The capillary is then inserted into standard Ag/AgCl microelectrode holders (World Precision Instrument, New Haven, CT) filled with 3 M KCl, and connected to microprobes mounted on micromanipulators with x, y, z maneuverability. A reference microelectrode is placed in the bathing medium, another is used for microinjection/measuring cell membrane potential. An audio output is useful so that a drop in noise gives feedback when the cell is impaled. Current is injected via a "breakaway box" using a constant-current generator. Negative pulse trains are used to inject negatively charged fluorochromes and positive pulse trains to inject positively charged fluorochromes. Positive current injection usually initiates action potentials in excitable cells, with depolarization and transient increase in $Ca^{2+}_{(cyt)}$. Fortunately, most fluorochromes are anions and negative current injection causes little discernable effect (as judged by cytoplasmic streaming). The output from the microelectrodes is monitored with an oscilloscope, chart recorder, and/or computer-linked data logger. There are now available fluorescence microscopes with large stages and specialized lenses that are suitable for simultaneous microinjection, electrophysiology, and fluorescence tracing (e.g., The Nikon Physiostation, cat. No. E600FN).

3.1.2. PROCEDURE FOR FITC-LABELING AMINO ACIDS, PEPTIDES, AND PROTEINS (13)

3.1.2.1. Conjugation. To a small (25 mL) conical flask add 20 mg KHCO$_3$ and 9 mL purified water, and dissolve. Add 3.125×10^{-5} mol or 8.1 mg of large peptide, (e.g., leucyl-diglutamyl-leucine) or 1.25×10^{-5} mol (5 mg) of a smaller peptide (e.g., glutamyl-glutamine). In a darkened room, add 10 mg FITC (isomer 1) to 1 mL dry acetone (keep over molecular sieve). Dissolve. The solubility depends on the acetone being completely dry. Using a finely drawn out Pasteur pipet, slowly (over 10–20 min) add the FITC to the peptide solution mixing by swirling gently. Leave 7 d in darkness at room temperature. Freeze dry to a volume of 0.3 mL.

3.1.2.2. Purification. Cut Whatman 3MM chromatography paper to fit chromatography tank leaving a foldover at the top of about 6 cm (e.g., approx 60 cm × 30 cm). Cut serrations into the bottom. Mark a line where the FITC-conjugate will be loaded. Load conjugate using a finely drawn-out Pasteur pipet, drying with a hair dryer on cool setting after each streak is added. Leave overnight. Elute in a cold room using descending chromatography with 50 mM KHCO$_3$, which takes about 5 h in the tank described. Examine dry paper under a UV light box. The conjugate appears yellow, and the unconjugated product orange. Cut out the yellow part of the paper and elute again using purified water. Freeze-dry and store in a dessicator in the dark. When needed, redissolve in distilled water. The pH is close to cytoplasmic pH (7.4–7.6) depending on peptide used.

3.1.2.3. Purity and Testing for Breakdown Using Thin-Layer Chromatography. Load FITC (dissolve in 0.3 M KHCO$_3$) and conjugates as spots onto Kieselgel 60 thin-layer chromatography plates using a drawn-out Pasteur pipet. Run in a fume hood using the solvent mixture n-butanol:acetic acid:pyridine:water in a 15:10:3:12 ratio. Examine under a UV light box. FITC has an R_f of 0.9, F-leucyl-diglutamyl-leucine (874 Dalton) has an R_f of 0.85, and F-glutamyl-glutamine (666 Dalton) has an R_f of 0.42.

3.1.2.4. Dye Concentration and Degree of Conjugation. The molar ratio (moles of peptide:moles of FITC) can be calculated from the relation: $(2.87 \times$ OD at 495 nm$)/C$, where C is the concentration of the peptide (mg/mL), obtained from the relation: (OD at 280 nm) $- (0.35 \times$ OD at 495 nm)/1.4, and OD is the optical density at 495 or 280 nm.

3.1.2.5. Controls for Membrane Impermeance. Some fluoro-chromes can enter protoplasts via probenecid-sensitive anion channels *(19)* and control experiments must establish membrane impermeance. Soak the tissues in the fluorochrome at the concentration used for up to 12 h, rinsing and examining the tissues for fluorescence at hourly intervals. Cell viability is assessed by rate of cytoplasmic streaming, or by chloroplast autofluorescence, which declines in dead cells.

3.1.3. CELL PERMEABILIZATION

It may be useful to introduce impermeant fluorocromes without micro-injection. Permeabilization is suitable for small to medium cells. Large cells (e.g., *Chara*) can be permeabilized osmotically *(46)* and become permeable to the fluorescent analogue of ATP, e-ATP *(46)*. Vacuolar compartmentation still occurs, but the cell wall is porous to molecules of only 4158 Dalton (approx 4.8 nm). Internal fluorescence concentration increases only slowly because of the large volume of cytoplasm.

3.1.3.1. Permeabilization for Introducing Large Molecules (e.g., Rhodamine-Labeled Phalloidin Toxin) into Cells. F-actin microfilaments are involved with many processes, such as organelle motility, cytoplasmic streaming, cell division *(30,31)*, stomatal opening and K^+ channel activities *(47)*, and cell-to-cell communication through plasmodesmata *(48)*. Actin filaments can be specifically labeled with phalloidin toxin derived from the death-cap fungus *Amanita phalloides*. The labeled toxin is too large to enter cells passively. Permeabilization methods cause artifacts *(49)*, but an improved method has been recently devised *(50)*.

3.1.3.2. Permeabilization for F-Actin Staining and Colocalization of Mitochondria. Make a large epidermal peel (at least 1 cm in largest direction) by cutting lightly into the undersurface of a leaf, grasping the thin epidermis with fine forceps, and pulling. The thin sheet of the epidermis consists of epidermal cells (lacking chloroplasts) and guard cells (with some chloroplasts). Immerse in actin buffer (10 mM EGTA, 5 mM MgSO$_4$, 0.3 M mannitol, and 100 mM PIPES at pH 6.9), containing 1–2% glycerol (pure), as well as rhodamine-phalloidin (make stock in methanol) at 0.1–0.2 µM concentration. Add DiOC$_{(6)}$ to a concentration of 0.35 µM to label mitochondria. Traditional methods use detergent (0.05–0.5% v/v Nonidet P-40, Sigma, St. Louis, MO) and 1–5% v/v DMSO, instead of glycerol. Examine after 35 min with blue/green (554 nm) excitation (for rhodamine-phalloidin) and blue (450–490 nm) excitation for DiOC$_{(6)}$. Actin stains orange-red (573 nm) and ER yellow (520 nm). This method was modified from Olyslaegers and Verbelin *(50)*.

3.1.4. Ester-Loading Fluorochromes

Both vacuoles and cytoplasm can be visualized by ester-loading impermeant fluorochromes. The fluorochrome 6CF can be introduced by "ester-loading" with 6CF-diacetate, which is not fluorescent or polar and readily permeates cells. Once inside the cell, it is cleaved into the highly fluorescent and charged anion 6CF, which is "ion-trapped." Lipophilic FITC derivatives are compartmented in patterns that depend on the subcellular location of esterases, and in different cells may be compartmented by cytoplasm or vacuole *(51)*.

3.2. Example of Symplastic Tracing Using Microinjection Methods: Mapping the Symplast of the Giant-Celled Alga Chara

Charophyte algae have been favored in electrophysiological research because of their large size and the comparative ease in introducing microelectrodes. The charophytes are tip-growing organisms, technically filamentous, although the cells of the nodal complexes are totipotent and behave analogously to meristems. Although usually considered as a "simple" model of higher plants, the symplast of *Chara corallina* is highly structured *(52)*. Internodal cells of vegetative plants have low (approx −120 mV) resting PD (membrane potential difference) and vegetative branches have restricted symplast domains consisting of an internode and some nodal cells, as shown by microinjection of fluorochromes (Fig. 1A). When plants become fertile, these restricted symplast domains are expanded to include the entire apex of the plant (Fig. 1B), and internodal cell membrane PDs hyperpolarize (to approx −210 mV) *(52)*.

The symplast of single internodal cells is highly specialized into separate acropetal and basipetal transport streams. Each internodal cell has a helical pattern of cyclosis. Fluorochromes injected into a single stream (ascending or descending) remained in that stream and entered nodal cells on that side of the plant several minutes before entering the opposite stream (Fig. 1A) *(52)*. This is related to development, as reproductive structures (antheridia) develop only from nodal cells on the side of the descending cytoplasmic stream *(52)*. Solutes can be preferentially directed towards apex, rhizoid, or reproductive structures. The period of expanded symplast domains remains until antheridial maturation, when internal spermatogenous filaments contain 32 cells. Symplast domains then become evident in antheridia *(53)*. Following maturation, the pattern of restricted symplast domains returns *(53)*. $Ca^{2+}_{(cyt)}$ is involved in the extensive or restricted symplast domains *(52)*. Microinjected inhibitors

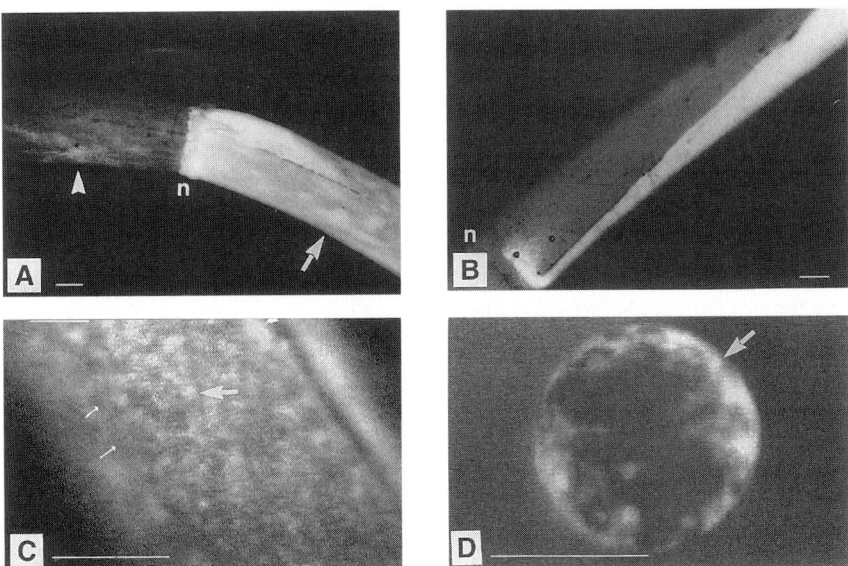

Fig. 1 (A,B). Microinjection of six carboxyfluorescein into internodal cells of *Chara corallina* showing symplast domains in fertile (**A**) and vegetative (**B**) branches. In (**A**) the microinjected cell shows "domains" (arrow) in the streaming cytoplasm. The 6CF has entered the nodal complex (n) and has been transported into the adjoining internodal cell (arrowhead). In (**B**), a vegetative cell, 6CF does not enter the node complex (n) or the adjoining internodal cell. The specialization into separate acropetal and basipetal transport streams is seen, as the 6CF remains in the injected basipetal stream, without diffusing into the opposing stream, which it enters only when the stream changes direction at the base of the cell. Bars = 100 μm. (**C,D**) Video images made with a cool CCD camera of 6CF ester loaded into giant algal cells at pH 5. With this method, the 6CF is retained by the cytoplasm and the irregular surface of the vacuole is contrasted as nonfluorescent corrugations against fluorescing cytoplasm. (**C**) The cytoplasmic structure in *Lamprothamnium*, a relative of *Chara*, and (**D**) in *Valonia* (*Ventricaria*), a giant marine algal cell. The coherent regions of streaming cytoplasm (arrow) in (**A**) are highlighted against corrugations or ripples of the underlying vacuole (small arrows). Similarly, in *Valonia*, which does not stream, the fluorescent cytoplasm forms a sponge-like structure (arrow) into which corrugations of the vacuole are projected. Bars = 100 μm.

of Ca^{2+} channels (e.g., La^{3+}) converted the restricted symplast domains into the extensive form characteristic of fertile plants, whereas microinjection of Ca^{2+} converted the extensive symplast domains into the restricted form characteristic of vegetative plants *(52)*. Intercellular fluorochrome transport was transiently interrupted during the peak of the action potential, when Ca^{2+} channels open and streaming ceases *(52)*.

3.3. Vacuolar Compartmentation of Fluorochromes

Most anionic FITC-labeled fluorochromes microinjected into the cytoplasm are compartmented by the plant cell vacuoles at rates that depend on their molecular size *(14,15)*.

3.3.1. STRUCTURE WITHIN THE CYTOPLASM AND THE WAVEFORM NATURE OF THE INTERFACE BETWEEN VACUOLE AND CYTOPLASM

Flurochromes injected into the *Chara* cytoplasm were compartmented by the vacuole as occurred in the higher plant *Egeria densa (14,52)*. The streaming cytoplasm consisted of "domains," irregular but coherent fluorescing regions *(52)*. The gel cytoplasm situated closest to the plasmalemma appeared to sort molecules for size, with an SEL similar to that of plasmodesmata *(46)*. The vacuole was in constant motion, "massaged" by the streaming cytoplasm *(52)*. Ester loading with 6CF-diacetate at low pH (approx pH 5), results in long-term labeling of similar fluorescing cytoplasmic domains in the related charophyte *Lamprothamnium (54)*. Viewed with a cooled charge-coupled device (CCD) camera, these domains are clearly regions of cytoplasm into which the wavy profile of the vacuole projects (Fig. 1C). Thus, the interface between vacuole and cytoplasm resembles the surface of a wind-ruffled sea rather than a smooth surface. In the giant-celled marine alga, *Valonia (Ventricaria)*, ester loading of 6CF also reveals highly structured cytoplasm, which resembles the texture of a sponge, where the holes in the sponge are occupied by projections of the vacuole (Fig. 1D). A recent article concludes that such vacuole "ripples" are common in plant cells *(55)*.

3.4. Example of Ester-Loading Methods for Symplastic Tracing: Intra- and Intercellular Transport in the Motile Tubule–Vacuole System

Growing tip cells of fungal hyphae accumulate 6CF (supplied as 6CF-diacetate) into a complex and dynamic vacuole system consisting of motile tubules and spherical vacuoles *(4)*. The behavior of the system can be imaged with epifluorescence microscopy (Fig. 2) or at greater resolution with CLSM (Fig. 3). The motile tubule–vacuole system comprises an active intracellular transport system, which can move bidirectionally, and at different rates and direction to cytoplasmic streaming *(41)*. Clusters of spherical vacuoles are located in zones along the hyphal tip cell and are transiently linked by motile tubules that extend from one cluster of vacuoles to the next, traversing distances as great as 60 μm in a cell 120 μm long and 5 μm wide *(41)*. Pulsations of the tubules

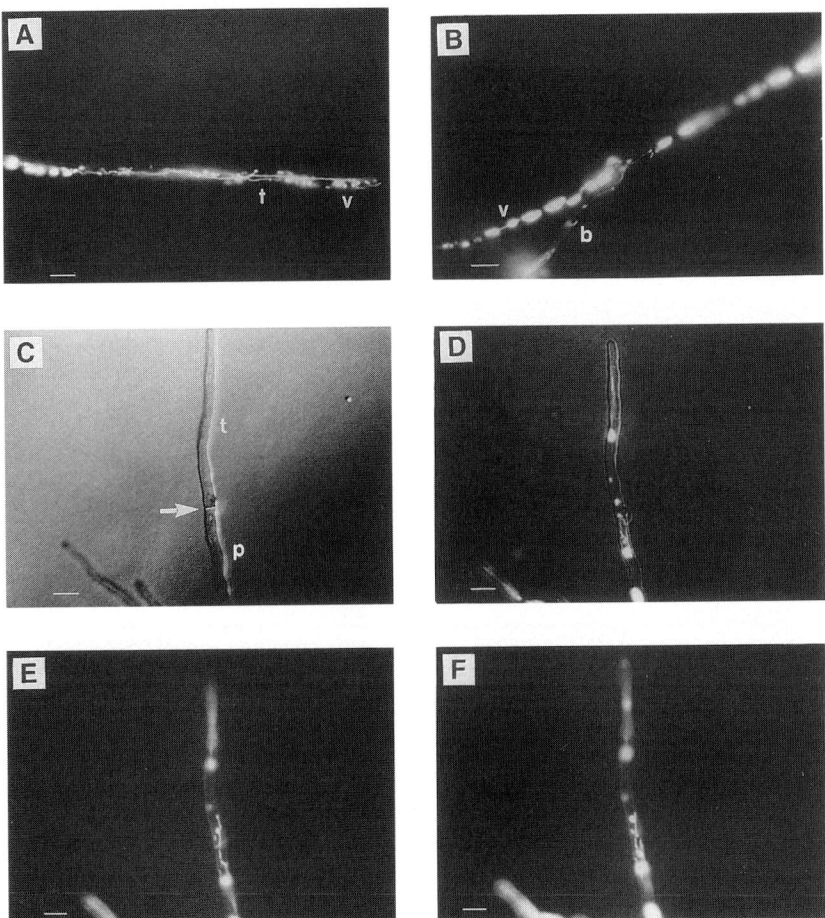

Fig. 2 (A–F). Epifluorescence images of hyphae of the fungus *Pisolithus tinctorius*, ester loaded with the fluorochrome 6CF. The fluorochrome is compartmented by the motile tubule–vacuole system. (**A**) Epifluorescence micrograph of a tip cell with tubules (t) and clusters of spherical vacuoles (v). (**B**) Older cells accumulating the 6CF into large vacuoles (v) interconnected by tubular projections. Branches (b) are tip cells that contain the motile system, and feed into the older cells. (**C–F**) Images selected from a 30-min sequence shot at 4-s intervals, showing intercellular transport via motile tubules. (**C**) Differential interference contrast image, showing position of the hyphal septum (arrow) between tip (t) and penultimate (p) cell. (**D**) Combined DIC and epifluorescence image of the commencement of intercellular transport-tubules have not yet crossed the septum. In (**E**), a tubule from the tip cell crosses into the penultimate cell and contacts tubules and vacuoles in that cell. (**F**) The process 12 s later, and transfer of fluorescent material continued for 30 min. Bars = 10 μm.

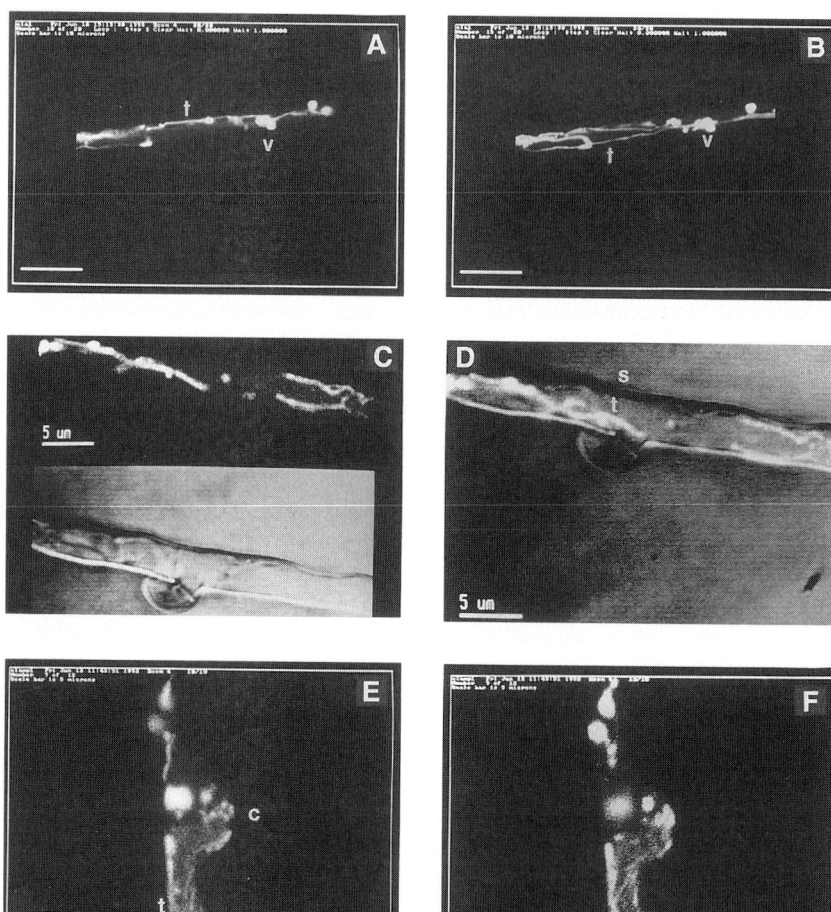

Fig. 3 (A–F). Greater resolution and detail are obtained with CLSM of the motile tubule-vacuole system in *Pisolithus tinctorius* hyphae. (**A,B**) Optical sections from a series made at 0.2 μm intervals of the region containing the two nuclei of the dikaryotic cells. This region of the cell contains mainly tubular elements. (**A**) Section close to the surface of the cell, showing tubules (t) interacting with vacuoles (v). (**B**) is 0.6 μm deeper into the cell, and shows further detail of the tubules (t) interacting with clusters of spherical vacuoles (v). Bar = 10 μm. Images were made with a Bio-Rad MRC-600 CLSM (Hercules, CA). (**C,D**) The process of cell-to-cell transport across a septum via motile tubules. (**C**) Confocal image of fluorescent tubules; the image below shows a nonconfocal bright-field image made with the pinhole aperture wide open. (**D**) Superimposition of fluorescent confocal and bright-field images using computer software, so that the septum (s) and tubule (t) crossing it are both visible. The resolution was approx 170 nm, and what appears to be a single tubule is actually a bundle of finer strands passing through pores in the dolipore septum. Images made with a Leica CLSM. (**E,F**) Intercellular transport across the septum between older cells, where cell

apparently transfer material between vacuoles *(41)*. Older fungal cells develop larger vacuoles with tubular projections that are fed by the development of branches. The motile tubule–vacuole system is widespread across all members of the Eumycota, Basidiomycotina, Ascomycotina, Zygomycotina, and some Deuteromycotina *(42)*. Microspectrofluorimetry was used to determine the pH of the spherical vacuoles in tip cells, based on the pH dependence of carboxyfluorescein fluorescence emission at blue and violet wavelengths *(38)*. The larger vacuoles of older cells were more acid (modal pH 5.5–6.0) than the smaller vacuoles in tip cells (modal pH 7.0–7.5) *(38)*.

The tubules form a unique, intermittent means of cell-to-cell transport (Figs. 2 and 3). Motile tubules are able to cross the septal pore between the tip and penultimate cells following nuclear division in a basidiomycete *(43)*. The tubules transfer fluorescent material between vacuoles in separate cells and thus form a previously unrecognized form of cell-to-cell communication *(43)*. Similar phenomena occur in higher plant cells. Recently, a very similar complex vacuole system was demonstrated in trichomes by labeling with lucifer yellow *(56)*. This also consisted of motile tubules and vacuoles, and tubules were observed crossing the cell wall, presumably through clusters of plasmodesmata. This system also forms a rapid solute delivery system between the base and tip of the trichome. It may be that some of the membranous material traversing plasmodesmata and interpreted as endoplasmic reticulum is in fact elements of the tubule–vacuole system *(43,56)*.

3.4.1. Procedure for Fluorescence Labeling the Motile Tubular– Vacuole System in Fungal Hyphae with 6CF *(41–43)*

Make up a small amount (e.g., 10 mL) of 10 mg/mL stock solution of 6CF-diacetate in dry acetone (keep over molecular sieve). The solution should be clear. If it is cloudy, the acetone was not dry and the method will not work. The solution keeps in the dark for about a week. Dilute to a 20-μg/mL solution in distilled water (for fungi growing on agar medium). Cut a shallow triangular slice (approx 4 mm × 4 mm) including the growing edge (broad face of triangle) of the fungal colony and float it on

---

division and formation of the clamp connection (c) is complete. Optical sections were made at 0.2 μm intervals. (**E**) Fluorescent cytoplasm and fine tubules (t) in the tip cell separated from the penultimate cell by the septum (s), and details of the vacuoles inside the clamp cell (c). In (**F**), an optical section 0.4 μm deeper into the cell, the fluorescent projection adjacent to the septum in the tip cell is shown to be aligned with a fluorescent projection in the penultimate cell. Bar = 5 μm. Images were made with a Bio-Rad MRC-600 CLSM.

the 6CF-diacetate, face down for hydrophobic cultures. Allow the cells to take up the diacetate for 10–15 min. Transfer to distilled water for 30 min in the dark. Mount face-side up by pressing a size 0 coverslip firmly onto the slice so that the agar squeezes out. This leaves the growing tip hyphae splayed out and viable. View using FITC filter blocks. After an initial greenish fluorescence in the cytoplasm, the 6CF is compartmented by the motile tubule–vacuole system. Cell morphology can be imaged by using simultaneous fluorescence and bright-field/differential interference contrast optics, or by CLSM using computer software to superimpose a bright-field image (obtained with wide-open pinhole) and fluorescence image. A similar procedure can be used to visualize a motile–tubule vacuole system in guard cell protoplasts.

3.5. Methods for Fluorescence-Labeling Endoplasmic Reticulum and Nuclei

3.5.1. Staining Endoplasmic Reticulum and Mitochondria with DiOC$_{(6)}$

DiOC$_{(6)}$ is commonly used with plant cells, and spectacular images of the ER can be obtained (23) as well as staining of the mitochondria. In fungal cells, the reticulate ER and mitochondria and hydrophobic extracellular material are readily stained with DiOC$_{(6)}$. In the giant alga, *Acetabularia,* a DiOC$_{(6)}$ concentration suboptimal for staining ER highlighted a previously unreported tubular membrane system that interconnects chloroplasts (the plastidom) (57). DiOC$_{(6)}$ has been used as a potential-sensitive stain in *Acetabularia* (58) and to identify the origin of vesicles appearing in cytoplasm-enriched fragments of *Chara* (59).

Dissolve DiOC$_{(6)}$ powder in a small amount of DMSO or 100% ethanol, and dilute to make a stock solution in distilled water. The solutions should be freshly made. For staining ER in giant algal cells such as *Chara,* the final staining solution varies from 10 µg/mL–50 µg/mL concentration (68). For smaller cells the concentration varies between 1 and 10 µg/mL. Allow to stain for at least 10 min (small cells) or several hours (giant algal cells). The mitochondria stain first. In fungal cells, the mitochondria, ER, and motile tubular–vacuole system can be difficult to distinguish, as all are reticulate. Very low concentrations (<0.1 µg/mL) stain chloroplasts, mitochondria, and the plastidom of giant algae (57). View using blue excitation. DiOC$_{(6)}$-stained organelles appear bright yellow/green. The ER in *Chara* and in onion epidermal cells appears to have two forms, the first a tubular semistationary reticulum and the second a motile system of tubular elements that intertwine and pull apart.

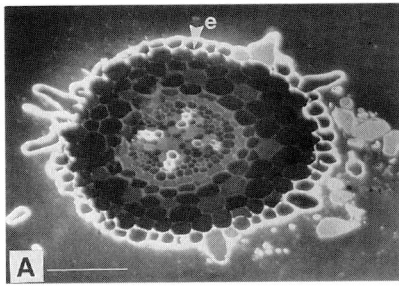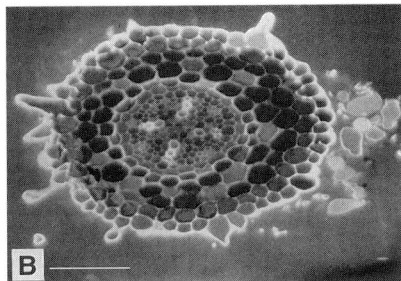

Fig. 4 (A,B). Sections of a young air-grown *Eucalyptus* root fluorescence labeled by flooding with 0.02% solution of SR, freeze-substituted and resin embedded, and serially sectioned at 2 μm intervals. (**A**) Section photographed under green excitation to reduce autofluorescence. The exodermis (e) is clearly seen to be an impermeable apoplastic barrier. Some tissue autofluorescence of the endodermis xylem vessels, and phenolic-containing stelar parenchyma cells is still visible. (**B**) Section of the same root, photographed with blue excitation, showing increased autofluorescence of cortical and parenchyma cell contents, as well as SR labeling of the walls of the exodermis. Bars = 100 μm.

3.5.2. Method for Staining Nuclei with DAPI

4',6-diamidino-2-phenylindole (DAPI) is soluble in distilled water. Make up a small volume (approx 5–10 mL) of stock solution at a concentration of 1 mg/mL (w:v). Keep in foil-covered bottle in the refrigerator. Dilute to 0.5–10 μg/mL depending on size of cells and expose cells for 10–20 min. Rinse in medium and examine with UV (358 nm) excitation. DNA appears blue/white (emission at 461 nm). Even DNA in chloroplasts, mitochondria, or mycoplasmas/bacteria will stain.

4. APOPLASTIC TRACING EXPERIMENTS

Fluorescent tracers cannot trace pathways of water movement *(20)*, but they can locate impermeable barriers (e.g., suberized lamellae of casparian strips in the endodermis and exodermis). Suberized lamellae of the endodermal or exodermal casparian strip in roots create an impermeable barrier that forces solutes and water to enter the symplast *(60)* (Fig. 4). In angiosperm leaves, the bundle sheath acts as an endodermis, which can develop suberized lamellae with an analogous function *(20)*. The ideal apoplastic fluorescent tracer is highly ionized at pH 7, does not readily enter living cells or bind to cell walls, and is nontoxic. Sulphorhodamine G (SR) has been used extensively by Canny and co-workers to delineate the positions in leaves where water flow to the symplast begins. SR is only sparingly soluble (0.4%) in water and forms crystals where it is concentrated in walls by water flow to the symplast *(20)*. It can locate proton extrusion pumps, as it does enter cells where the wall pH is acidified.

4.1. Apoplastic Tracing Using Sulphorhodamine G

Cut a shoot under a 0.02% solution of SR (method from Canny [20]). Encourage transpiration for 1–2 h or add SR to air- or liquid-grown roots. Air-grown roots form an exodermis. Freeze-substitute and embed (see Subheading 4.2.), or cut hand sections under paraffin oil. Examine under green excitation (530 nm). This minimizes tissue autofluorescence (faint red). SR appears bright orange/yellow (emission 555 nm). Blue excitation shows autofluorescent cellular detail and SR appears yellow. SR crystals show the place where water enters the symplast and do not fluoresce. They are visible with bright-field optics or using polarized light (through birefringence).

4.2. Immobilizing SR through Freeze-Substitution and Anhydrous Embedding

Ultrastructural ice-crystal damage occurs deeper than 10 mm into the tissue but freeze-substitution still works well for fluorescence micros-copy (see [20]). Place a stand supporting a small aluminium cup inside a wide-mouthed thermos flask. To the cup add cyclohexane/isopentane (1:1 by volume). Add liquid nitrogen to the thermos flask. The cyclohex-ane/isopentane remains viscous at −160°C. Half-fill polythene vials with dry ether (keep over 5 Å molecular sieve) and place in a rack immersed in liquid nitrogen. Excise the roots or leaves, rapidly hand section (pieces approx 5 mm maximum size) and plunge them into the viscous cyclo-hexane/isopentane using Teflon-coated forceps. Select a vial and allow the surface ether to melt forming a small liquid layer. Using a wire loop with a Teflon- coated handle, retrieve sections, and place in the liquid ether, immediately refreezing in the liquid nitrogen. Place vials in deep freeze at −80°C and allow to freeze-substitute for 3 wk. Bring to room temperature slowly using −4°C and 20°C steps. Infiltrate and embed in Spurr's resin or L.R. White, using an anhydrous embedding box filled with nitrogen gas. Cut dry sections (2–4 µm) and flatten onto gelatin-coated slides, mount in fluorescence-free immersion oil. Examine under green and blue excitation (Fig. 4).

4.3. Fluorescence-Labeling Suberized Lamellae with Berberine–Aniline Blue

Suberized lamellae in casparian strips can be located using the berbe-rine-aniline blue fluorescence staining method, which works equally well on fresh and resin-embedded sections (method adapted from ref. *[21]*). Berberine is a basic fluorochrome derived from the barberry and is the only known natural cationic stain *(3)*. This method has been used to detect extrarapid suberization of the endodermis due to increased salinity *(61,62)*, to demonstrate a cork apoplastic barrier *(63)*, an endodermis with casparian bands in roots of seedless vascular plants *(64)*, and to demonstrate that *Fusarium culmorum* enters seedling roots where the casparian strip lacked suberin lamellae *(66)*. Aniline blue quenches some of the vivid berberine staining, enabling visualization of fainter staining of suberized lamellae *(21)*. Aniline blue staining of callose was due to a fluorescent impurity (available as "Sirofluor"), which is removed from purified batches *(3)*.

4.3.1. THE BERBERINE–ANILINE BLUE FLUORESCENCE METHOD
FOR SUBERIZED LAMELLAE

Either embed in L.R. White and section or use hand sections as described above. (Anhydrous processing is not necessary). Dry sections onto clean slides. Staining solutions can be added to the sections as droplets and sucked off using twists of tissue paper. Make up 0.1% (w/v) berberine hemisulphate in water. Berberine is only sparingly soluble in cold water but dissolves in ethanol or hot water without loss of fluorescence. Stain hand sections for 1 h, and rinse twice in tap water. Make up 0.5% aniline blue, water soluble (WS), in distilled water. This solution quenches background fluorescence and nonspecific staining. Stain sections for 30 min and rinse through two changes of water. Make up 0.1% (w/v) $FeCl_3$ in 50% (w/v) glycerin, adding the glycerin to aqueous $FeCl_3$. This solution is to prevent section destaining. Mount sections in the $FeCl_3$/glycerin solution. Examine under UV excitation (berberine excites at 343 and 420 nm). Casparian strips of the endodermis, exodermis, and bundle sheath appear yellow and the suberized lamellae, if present, appear pale blue/white.

5. CONCLUSIONS

It is always difficult to anticipate the future. The coupling of laser and computer technology, which has had such an impact on plant cell biology, was unimaginable when, for example, the laser was invented. At the turn of the 21st century we stand at the beginning of a technological revolution in fluorescence microscopy where 35 mm film will soon be replaced by digital storage, CLSM will evolve into multiple-photon imaging, using more continuous wavelengths, and where the images produced will make those of today seem as crude as early electron micrographs.

ACKNOWLEDGMENTS

The author is grateful to the following laboratories and people for their resources. Figure 1(A,B) was made in Dr. Peter Goodwin's laboratory in the School of Crop Sciences, University of Sydney. Figure 1(B,C) was made in Dr. Mary Beilby's laboratory at The University of New South Wales using microscope facilities of Fucell Pty. Ltd. Figure 2(A–F) and Figure 4(A,B) were made in Professor Anne Ashford's laboratory, School of Biological Sciences, The University of New South Wales. Figure 3(A,B and E,F) were made in the Electron Microscope Unit of the University of Sydney. Figure 3(C,D) was made in the School of Anatomy, The University of New South Wales.

REFERENCES

1. Rost FWD. *Quantitative Fluorescence Microscopy*, Cambridge University Press, Cambridge, UK, 1991.
2. Rost FWD. *Fluorescence Microscopy*, Vol. 1, Cambridge University Press, Cambridge, UK, 1992.
3. Rost FWD. *Fluorescence Microscopy*, Vol. 2, Cambridge University Press, Cambridge, UK, 1995.
4. O'Brien TP, McCully ME. *The Study of Plant Structure: Principles and Selected Methods*, Termarcarphi, Melbourne, Australia, 1981.
5. Harris N, Oparka K. *Plant Cell Biology. A Practical Approach*, Oxford University Press, New York, 1994.
6. Haughland RP. *Handbook of Fluorescent Probes and Research Chemicals*, 6th ed., Molecular Probes, Eugene, OR, 1996.
7. Cox G. Trends in confocal microscopy. *Micron* 1993; 24: 237–247.
8. Hepler PK, Gunning BES. Confocal fluorescence microscopy of plant cells. *Protoplasma 1998;* 201: 121–157.
9. Lucas L, Gilbert N, Ploton D, Bonnet N. Visualisation of volume data in confocal microscopy: comparison and improvements of volume-rendering methods. *J Microsc* 1996; 181: 238–252.
10. Delorme R, Benchaib M, Bryon PA, Souchier C. Measurement accuracy in confocal microscopy. *J Microsc* 1998; 192: 151–162.
11. Prior DAM, Oparka KJ, Roberts IM. *En bloc* optical sectioning of resin-embedded specimens using a confocal laser scanning microscope. *J Microsc* 1998; 193: 20–27.
12. Lucas WJ, Ding B, van der Schoot C. Plasmodesmata and the supracellular nature of plants. *New Phytol* 1993; 125: 435–476,
13. Goodwin PB. Molecular size exclusion limit for movement in the symplast of the *Elodea* leaf. *Planta* 1983; 157: 124–130.
14. Goodwin PB, Shepherd V, Erwee MG. Compartmentation of fluorescent tracers injected into the epidermal cells of *Egeria densa* leaves. *Planta* 1990; 181: 129–136.
15. Oparka KJ. Uptake and compartmentation of fluorescent probes in plant cells. *J Exp Bot* 1991; 238: 565–579.
16. Van Bel AJE, Gamelai YV, Ammerlan A, Bik LPM. Dissimilar phloem loading in leaves with symplasmic and apoplasmic minor-vein configurations. *Planta* 1992; 186: 518–525.

17. Van der Schoot C, van Bel AJC. Mapping membrane potential differences and dye coupling in internodal tissues of tomato. *Planta* 1990; 182: 9–21.
18. Roscak R, Rambour S. Uptake of lucifer yellow by plant cells in the presence of endocytotic inhibitors. *Protoplasma* 1997; 199: 198–207.
19. Wright KM, Oparka KJ. Uptake of Lucifer Yellow CH into plant-cell protoplasts: a quantitative assessment of fluid-phase endocytosis. *Planta* 1989; 179: 257–264.
20. Canny MJ. What becomes of the transpiration stream? *New Phytol* 1990; 114: 341–368.
21. Brundrett MC, Enstone DE, Peterson CA. A berberine-aniline blue fluorescent staining procedure for suberin, lignin and callose in plant tissues. *Protoplasma* 1988; 146: 133–142.
22. Wolf S, Maier I, Katsaros C, Muller DG. Virus assembly in *Hincksia hincksiae* (Ectocarpales, Phaeophyceae). An electron and fluorescence microscopic study. *Protoplasma* 1998; 203: 153–167.
23. Quader H. Formation and disintegration of cisternae of the endoplasmic reticulum visualised in live cells by conventional fluorescence and confocal laser scanning microscopy: evidence for the involvement of Ca^{2+} and the cytoskeleton. *Protoplasma* 1990; 155: 166–175.
24. Reuzeau C, Doolittle KW, McNally JG, Pickard BG. Covisualisation in living onion cells of putative integrin, putative spectrin, actin, putative intermediate filaments, and other proteins at the cell membrane and in an endomembrane sheath. *Protoplasma* 1997; 199: 173–197.
25. Schußler A, Bonfante P, Schnepf E, Mollenhauer D, Kluge M. Characterisation of the *Geosiphon pyriforme* symbiosome by affinity techniques: confocal laser scanning (CLSM) and electron microscopy. *Protoplasma* 1996; 190: 53–67.
26. Ludin B, Matus A. GFP illuminates the cytoskeleton. *Trends Cell Biol* 1998; 8: 72–77.
27. Hedtke B, Meixner M, Gillandnt S, Richter E, Borner T, Weihe A. Green fluorescent protein as a marker to investigate targeting of organellar RNA polymerases of higher plants in vivo. *Plant J* 1999; 17: 557–561.
28. Czymmek KJ, Bourett TM, Howard RJ. Immunolocalisation of tubulin and actin in thick-sectioned fungal hyphae after freeze-substitution fixation and methacrylate de-embedment. *J Microsc* 1996; 181: 153–161.
29. Wasteneys GO, Collings DA, Gunning BES, Hepler PK, Menzel D. Actin in living and fixed characean internodal cells: identification of a cortical array of fine strands and chloroplast actin rings. *Protoplasma* 1996; 190: 25–38.
30. Schmit A-C, Lambert A-M. Microinjected fluorescent phalloidin in vivo reveals the F-actin dynamics and assembly in higher plant mitotic cells. *Plant Cell* 1990; 2: 129–138.
31. Cleary AL. F-actin redistributions at the division site in living *Tradescantia* stomatal complexes as revealed by microinjection of rhodamine-phalloidin. *Protoplasma* 1995; 185: 152–165.
32. Taylor AR, Manison NFH, Fernandez C, Wood J, Brownlee C. Spatial organisation of calcium signalling involved in cell volume control in the *Fucus* rhizoid. *Plant Cell* 1996; 8: 2015–2031.
33. Hodick D, Gilroy S, Fricker MD, Trewavas AJ. Cytosolic concentrations and distributions Ca^{2+} in rhizoids of *Chara fragilis* determined by ratio analysis of the fluorescent probe Indo-1. *Bot Acta* 1991; 104: 222–228.
34. Volotovski ID, Sokolovsky SG, Nikiforov EL, Zinchenko VP. Calcium oscillations in plant cell cytoplasm induced by red and far-red light irradiation. *J Photochem Photobiol* 1993; 20: 95–100.

35. McAinsh MR, Webb AAR, Taylor JE, Hetherington AM. Stimulus-induced oscillations in guard cell cytosolic free calcium. *Plant Cell* 1995; 7: 1207–1219.
36. Kikuyama M, Tazawa M. Temporal relationship between action potential and Ca^{2+} transient in Characean cells. *Plant Cell Physiol* 1998; 39: 1359–1366.
37. Read ND, Allan WTG, Knight H, Knight MR, Malho R, Russell A, Shacklock PS, Trewavas AJ. Imaging and measurement of cytosolic free calcium in plant and fungal cells. *J Microsc* 1992; 166: 57–86.
38. Rost FWD, Shepherd VA, Ashford AE. Estimation of the vacuolar pH in actively growing hyphae of the fungus *Pisolithus tinctorius*. *Mycol Res* 1995; 99: 549–553.
39. Hutler P, Fischbach R, Heller W, Jungblut TP, Reuber S, Schmitz R, Veit M, Weissenbock G, Schnitzler J-P. Tissue localisation of phenolic compounds in plants by confocal laser scanning microscopy. *J Exp Bot* 1998; 49: 953–965.
40. Menzel D. Wounding in giant plant cells. *Protoplasma* 1988; 144: 73–91.
41. Shepherd VA, Orlovich DA, Ashford AE. A dynamic continuum of pleiomorphic tubules and vacuoles in growing hyphae of a fungus. *J Cell Sci* 1993; 104: 495–507.
42. Rees B, Shepherd VA, Ashford AE. Presence of a motile tubular vacuole system in different phyla of fungi. *Mycol Res* 1994; 98: 985–992.
43. Shepherd VA, Orlovich DA, Ashford AE. Cell-to-cell transport via motile tubules in growing hyphae of a fungus. *J Cell Sci* 1993; 105: 1173–1178.
44. Woo Y-M, Itaya A, Owens RA, Tang L, Hammond RW, Chou HC, Lai MMC, Ding B. Characterisation of nuclear import of potato spindle tuber viroid RNA in permeabilised protoplasts. *Plant J* 1999; 17: 627–635.
45. Purves RD. *Microelectrode Methods for Intracellular Recording and Iontophoresis*, Academic Press, London, UK, 1981.
46. Shepherd VA, Goodwin PB. The porosity of permeabilised *Chara* cells. *Aust J Plant Physiol* 1989; 16: 231–239.
47. Hwang J-U, Suh SS, Yi H, Kim J, Lee Y. Actin filaments modulate both stomatal opening and inward K^+ channel activities in guard cells of *Vicia faba*. *Plant Physiol* 1997; 115: 335–342.
48. White RG, Badelt K, Overall RL, Vesk M. Actin associated with plasmodesmata, *Protoplasma* 1994; 180: 169–184.
49. Doris FP, Steer MW. Effects of fixatives and permeabilisation buffers on pollen tubes: implications for localisation of actin microfilaments using phalloidin staining, *Protoplasma*, 1996; 195: 25–36.
50. Olyslaegers G, Verbelin J-P. Improved staining of F-actin and co-localisation of mitochondria in plant cells. *J Microsc* 1998; 192: 73–77.
51. Brauer D, Uknalis J, Triana R, Tu S-I. Subcellular compartmentation of different fluorescein derivatives in maize root epidermal cells. *Protoplasma* 1996; 192: 70–79.
52. Shepherd VA, Goodwin PB. Seasonal patterns of cell-to-cell communication in *Chara corallina* Klein ex Willd. 1. Cell-to-cell communication in vegetative lateral branches during winter and spring. *Plant Cell Environ* 1992; 15: 137–150.
53. Shepherd VA, Goodwin PB. Seasonal patterns of cell-to-cell communication in *Chara corallina* Klein ex Willd. 1. Cell-to-cell communication during the development of antheridia. *Plant Cell Environ* 1992; 15: 151–162.
54. Beilby MJ, Cherry C, Shepherd VA. Dual mechanisms of turgor regulation in *Lamprothamnium papulosum*. *Plant Cell Environ* 1999; 22: 347–361.
55. Verbelen JP. Tao W. Mobile vacuole ripples are common in plant cells. *Plant Cell Reports* 1998; 17: 917–920.
56. Lazzaro MD, Thomson WW. The vacuolar-tubular continuum in living trichomes of chickpea (*Cicer arietinum*) provides a rapid means of solute delivery from base to tip. *Protoplasma* 1996; 193: 181–190.

57. Menzel D. An interconnected plastidom in *Acetabularia*: implications for the mechanism of chloroplast motility. *Protoplasma* 1994; 179: 166–171.

58. Matzke MA, Matzke AJM, Neuhaus G. Age-related differences in the interaction of a potential sensitive fluorescent dye with nuclear envelopes of *Acetabularia mediterranea*. *Plant Cell Environ* 1988; 11: 157–163.

59. Beilby MJ, Shepherd VA. Cytoplasm-enriched fragments of *Chara*: structure and electrophysiology. *Protoplasma* 1989; 148: 150–163.

60. Steudle E, Peterson CA. How does water get through roots? *J Exp Bot* 1988; 49: 775–788.

61. Reinhardt DH, Rost TL. Salinity accelerates endodermal development and induces exodermis in cotton seedling roots. *Env Exp Bot* 1995; 35: 563–574.

62. Valenti SG, Riveros F. The use of berberine-aniline blue for improving visualisation of lignin and suberin in the tissues of *Prosopis juliflora* (SW). Leguminosae growing in saline medium. *Inf Bot Ital* 1996; 27: 253–256.

63. McKenzie BE, Peterson CA. Root browning in *Pinus banksiana* Lamb. and *Eucalyptus pilularis* Sm. 2. Anatomy and permeability of the cork zone. *Bot Acta* 1995; 108: 138–143.

64. Damus M, Peterson DA, Enstone DE, Peterson CA. Modifications of cortical cell walls in roots of seedless vascular plants. *Bot Acta* 1997; 110: 190–195.

65. Kamula SA, Peterson CA, Mayfield CI. Impact of the exodermis on infection of roots by *Fusarium culmorum*. *Plant Soil* 1994; 167: 121–126.

5

Fluorescence Microscopy
of Aniline Blue Stained Pistils

Renata Śnieżko

CONTENTS

1. INTRODUCTION

Fluorescence microscopy after aniline blue staining has been used for years to detect "callose-like" substances in plant tissues. Callose-like substances can be polysaccharides or glycoproteins—polysaccharide chains of glucose residues covalently linked to one another by $\beta 1 \rightarrow 3$ glycosidic bonds are important. Aniline blue is absorbed by such compounds and after exposure to UV light of 430 nm, it emits a yellow–green fluorescence. The method is not chemically specific and many glycoproteins or proteoglycans can be visualized by it (1). Fluorescence is strong and bright only at pH 9.0–10.0. Every change of acidity in the sample has a strong influence on the resulting fluorescence, which can become pale or disappear. The method cannot be taken for quantitative estimation of the callose content, but is sufficient for the morphological or histological investigations applied to anatomy and embryology of plants (2–5).

The pollen tube walls contain callose-like polysaccharides that absorb aniline blue. That is why this method is very useful to detect pollen tube elongation and to understand how the tube penetrates the inside of the pistil tissue. The pollen tubes grow from the surface of the stigma through

From: *Methods in Plant Electron Microscopy and Cytochemistry*
Edited by: W. V. Dashek © Humana Press Inc., Totowa, NJ

the style to the ovary, and there they are directed to the micropyle of fertile ovules. On the basis of observations of pollen tube growth in the ovary of different plants, we can assume that penetration of the pollen tube into the micropyle of the ovule is a clear sign of ovule fertility *(5–7)*. It is important for breeders to obtain information concerning fertility of the mother-plant at the time of flowering (choosing plants for crossing); fluorescence microscopy after aniline blue staining can be a helpful tool for such a purpose *(1–7)*.

2. PROTOCOL FOR STAINING PISTILS WITH ANILINE BLUE

2.1. Materials

Pistils chosen for fluorescence microscopy can be fresh or fixed with a 3:1 mixture of ethyl alcohol and acetic acid (both ingredients highly concentrated, minimum 70%). Fixed pistils can be kept in 70% alcohol for a long time.

2.2. Maceration

Depending on the purpose of the experiment, pistils can be divided into pieces, e.g., stigma, style and ovary, or stigma and style together, and placenta with ovules without ovary walls. The researcher has to decide if the tissue is soft or hard. This is important for choosing the proper time for maceration. Younger tissues are usually softer than older tissues. The placenta is softer than the style and ovary walls. Ovules are relatively compact even when they are young. The best time for maceration is ½ h at room temperature in 1 N HCl or ½ h in 1 N NaOH at 60°C.

The medium for maceration (HCl or NaOH) depends on the material and purpose of staining. If fresh material is at hand, NaOH is best. It keeps material at the proper pH even after washing a few times in water. It is very important to achieve a pH of 9.0–10.0 to obtain the proper fluorescence of callose after staining with aniline blue.

The fixed material is already acidic and can be macerated in HCl. After maceration, material has to be rinsed in distilled water and left in phosphate buffer for a few hours to achieve a pH of about 9.5. If the pH is lower in the specimen, the fluorescence of callose disappears.

2.3. Preparing the Staining Solution

1. Phosphate buffer, pH 9.5: Prepare two solutions: A, in 1.160 g K_2HPO_4 in 100 mL distilled water and B, in 1.142 g K_3PO_4 in 100 mL distilled water. Mix solutions A and B in 3:1 proportions and dilute with distilled

water in 1:1 proportions (e.g., mix 75 mL K_2HPO_4 with 25 mL K_3PO_4 and add 100 mL distilled water) *(8)*.
2. Solution of aniline blue: 0.005% or 0.01% of water-soluble stain in phosphate buffer, pH 9.5.
3. Mix aniline blue solution with glycerin in 1:1 proportions.

The ready mixture is a transparent, pale blue fluid. If kept in a dark bottle, it can be used for a long time, but sometimes some microorganisms develop in it and the stain should be filtered before use.

2.4. Staining Schedule

Put the specimen on a slide, add one drop of staining mixture, and cover the slide with a coverslip pressing gently to spread the material into a thin layer. During preparation, it is important to remove all the unnecessary tissue before staining, which can disturb the spreading of material into a thin layer. A thick layer of the tissue and an abundant amount of staining mixture tend to flow out of the slide and spill onto the coverslip. A dirty slide interferes with observations and photomicrography becomes impossible.

The aniline blue stain penetrates macerated tissue very quickly and in a few minutes the slide is ready to be observed. Slides do not dry out for several hours. They can be stored for a longer time (about 2 wk) in a cool, dark place if sealed around the coverslip with parafilm or Vaseline.

2.5. Observations

Slides should be kept in darkness, and exposed to UV light of 430 nm for not longer than half an hour, because the fluorescence is maximum after 1–10 min of exposure in one place. The color of the callose fluorescence should be bright blue-green, but brightness depends on the pH in the specimen. Cell walls with callose are easy to distinguish from other fluorescing substances that are more yellowish. Pollen tubes are well visible; however, some similarity between pollen tubes and sieve tubes can be detected for an inexperienced observer. The fluorescence of vascular elements is always more yellow, and spiral wall thickenings are present only in sieve tubes *(9–11)*. In the pollen tubes, strongly fluorescing callose plugs are a characteristic feature and are easy to locate in the slide preparation. Location of the pollen tubes in the style and ovary is also different from vascular elements. The pollen tubes are located in the middle part of the style and inside or on the surface of the placenta. Sieve tubes are very close to the xylem elements and located in the ovary wall or in the outer layers of the stylar tissue.

This method gives a bright fluorescence of callose but also a side effect due to the presence of chlorophyll. Chlorophyll dispersed in the specimen yields a red fluorescence all around the tissue, especially a red layer on the surface of the organs due to chlorophyll deposition in any lipid substances, such as a cuticle.

3. CONCLUSIONS

3.1. Purposes of the Application of Fluorescence Microscopy to Pistils Stained by Aniline Blue

The method of aniline blue staining is relatively simple and quick and can be applied in checking pollination and fertilization at the time of plant flowering, thus it has a scientific and practical meaning (1–7,12). Pollen grains are well visible on the stigma and it is possible to follow every stage of pollen tube germination. Pollen grain poruses, which start to open, are seen as blue-greenish spots fluorescing in swollen sporoderma. Short, just-germinating pollen tubes are well visible in contact with stigmatic cells. In case of self-incompatibility, a bright callose patch develops where the pollen tube and stigmatic papilla contact. The method allows for detecting self-incompatibility cases beneath the stigma —the pollen tubes' endings are wider and surrounded by a thick callose wall (3).

In many plant species compatible pollen tubes grow directly down inside the style as a bundle of long smooth tubes with bright callose plugs. Only in female sterile plants of *Oenothera* mut. *brevistylis* was branching of pollen tubes found in the style and ovary (7).

Pollen tubes in the ovary become dispersed among the ovules, but they look so clear that a researcher cannot have any doubts about the subject of observation. The method allows one to observe the events of fertilization (1,2,5,6,12). Pollen tubes grow in the direction of fertile ovules. The picture when a pollen tube grows inside the micropyle is very impressive. In many species, it is possible to see the pollen tube through the integuments and sometimes nucellar cells penetrating up to the embryo sac (6,7,12,13). The callose spot can be seen on the micropylar pole of the embryo sac (6,7). In *Brassicaceae*, this spot represents the filiform apparatus in synergids and some polysaccharides derived from degenerated cells of nucellus. The ovules penetrated by the pollen tube can be taken as fertilized and it makes it possible to judge the fertility of the plant (14,15).

This kind of fluorescence microscopy application would be useful for breeders when they have to estimate fertility of mother plants before crossing. Counting fertilized ovules allows one to make probable expectations for the future crop of seeds.

3.2. Particular Observations
of Pollen Tube Growth and Its Interaction with Ovules

Fluorescence microscopy after aniline blue staining was successfully applied to following particular patterns of growth of the pollen tube inside the ovary. This observation showed that *Oenothera* pollen tubes can split into many branches, especially near the ovules. After spotting the fertile ovule, one branch penetrates the micropyle up to the embryo sac; other—shorter—branches grow into the top part of the integument. Roles of these short branches can be (1) haustoria taking nutrients from the integument or (2) strengthening the main branch when it has to withstand the resistance of many layers of nucellar tissue *(6,13)*.

Another problem is the direction of pollen tube growth, which should be considered as tropism, related to attraction by the fertile ovules. Many observations of pollen tubes in the fluorescence microscope supported this suggestion—pollen tubes pass by the sterile ovules and grow in the direction of the fertile ones. The number of ovules penetrated by a pollen tube is correlated with the number of developing seeds, which supports the hypothesis about interaction (attraction) between the ovule and pollen tube *(1,2,5)*.

REFERENCES

1. Anvari SF, Stösser, R. Fluoreszenmikroskopische Untersuchungen des Pollenschlauchwachstums der Samenanlagen bei Sauerkirschen. *Mitt Klosterneuburg* 1978; 28: 22–30.
2. Anvari SF, Stösser, R. Eine neue fluoreszenmikroskopische Methode zur Beurteilung der Befruntungs-Fähigkeit der Samenanlagen bei *Prunus. Z Pfl Zücht* 1978; 81: 333–336.
3. Dumas C, Knox RB. Callose and determination of pistil viability and incompatibility. *Theor Appl Genet* 1983; 67: 1–10.
4. Vishnyakova MA. Callose as an indicator of sterile ovules. *Phytomorphology* 1991; 41: 245–252.
5. Williams EG, Knox RB. Quantitative analysis of pollen tube growth in *Lycopersicon peruvianum. J Palynol* 1982; 18: 65–74.
6. Şniezko R. Postulated interaction between branching pollen tubes and ovules in *Oenothera hookeri* de Vries (Onagraceae). *J Plant Res* 1997; 110: 411–416.
7. Şniezko R, Winiarczyk K. Pollen tube growth in pistils of female-sterile and fertile plants of *Oenothera* mut. *brevistylis. Protoplasma* 1995; 187: 31–38.
8. Clark G. (ed.) *Staining Procedures,* Williams & Wilkins, Baltimore, MD, 1981.
9. Currier HB, Shih CY. Sieve tubes and callose in Elodea leaves. *Am J Bot* 1968; 55: 145–152.
10. Currier HB, Strugger S. Aniline blue and fluorescence microscopy of callose in bulb scales of *Allium cepa* L. *Protoplasma* 1956; 4: 552–559.

11. Evert RE, Derr WF. Callose substance in sieve elements. *Am J Bot* 1964; 51: 552–559.
12. Wilms HJ. Branching of pollen tubes in spinach, in *Fertilization in Higher Plants* (Linśkens HF, ed.), North-Holland Publishing, Amsterdam, 1974, pp. 155–160.
13. Sniezko R. Pollen tube branching in the ovary of five species of *Oenothera. Acta Soc Bot Pol* 1996;65: 111–116.
14. Sniezko R, Sadaj A. Pattern of pollen tube growth in the *Capsella bursa pastiris* and *Sisymbrium loeselii* (Brassicaceae). VIII Conference of Plant Embryologists, 16–18 September, Gdańsk, Poland, 1997.
15. Chudzik B, Sniezko R. Similarities in some phenomena connected with maturation of crassi- and tenuinucellar ovules (*Oenothera hookeri* and *brevistylis, Sinapis alba, Brassica napus, Galanthus nivalis*). XV International Congress on Sexual Plant Reproduction, August 16–21, Wageningen, Germany, 1998.

6

A Short Introduction to Immunocytochemistry and a Protocol for Immunovisualization of Proteins with Alkaline Phosphatase

John E. Nielsen

CONTENTS

1. INTRODUCTION

Immunocytochemical methods have been widely applied to visualize proteins, carbohydrates, or lipids in sectioned material. The advantage of using immunocytochemistry is to be able to localize the molecules of interest within the tissue. Several procedures have been described. Basically, these procedures can be split into four main steps that are described in subheadings: (1) tissue preparation, (2) the primary antibodies, (3) the visualization of the target, and (4) enhancement of signals with antibody complexes. In addition, a protocol for alkaline phosphatase will be presented in detail in Subheading 5. The terms "primary" and "secondary" antibodies refer to the order in which they are applied to the target. The immunocytochemical procedures are not limited to sectioned

From: *Mehods in Plant Electron Microscopy and Cytochemistry*
Edited by: W. V. Dashek © Humana Press Inc., Totowa, NJ

material. The immunocytochemical procedures can also be used for fixed and enzymatically permeabilized cells of *Saccharomyces cerevisiae (1,2), Schizosaccharomyces pombe (3),* and *Chara (4)* protoplasts and organelles *(5).* The immunocytochemical techniques are used in combination with other techniques: flow cytometry *(6),* tissue prints *(7), in situ* hybridization *(8–10),* and *in situ* PCR (polymerase chain reaction) *(11–13).*

1.1. Tissue Preparation

The standard procedures used for fixation involve the use of various organic solvents before the material of interest is embedded in paraffin or plastic and sectioned. These organic solvents are also used for deparaffination and in various staining methods. These treatments might change the epitopes, e. g., the binding site of the antibody. Cryosectioning, however, can solve these problems *(14).* Detailed advice for fixation of plant tissue used for immunocytochemistry is given by Sing et al. *(5)* and standard plant fixations are provided by O'Brien and McCully *(15).*

2. THE PRIMARY ANTIBODIES

The structure and classes of antibodies are described in many textbooks and as an introduction to immunology in method collections *(16,17).*

2.1. The Polyclonal Antibodies

The primary antibodies can be divided into two main groups: the monoclonal and polyclonal antibodies. Polyclonal antibodies are produced from a wide range of animals; however, the rabbit is most often used. In order to produce polyclonal antibodies, an immunization of an animal is carried out. The molecule of interest (the antigen), together with a nonspecific stimulant of the immune response (an adjuvant), is injected under the skin of animals. The plasma cells of the animal respond to the antigen by producing immunoglobulins (IgG) recognizing the foreign molecules as a step in the animal's immunodeficiency system. The animal produces IgGs from numerous plasma cell lines, which give rise to polyclonal antibodies. These IgGs are different with respect to the site binding to the antigen (the epitope) as well as in other binding properties. As the polyclonal antibodies are very heterogeneous, it is useful to purify the antibodies by column chromatography *(18,19).*

Some of the polyclonal antibodies or the antibodies produced by the animal before the immunization might cause unspecific bindings. To control this, it is important to obtain a preimmune bleed from the same animal. The preimmune serum should also be used as a guide in the initial experiments to determine the dilution of the antibody, and later on as a control of the experiments. A test of the specificity of polyclonal antibodies is necessary. Therefore, they are tested for specificity on sodium dodecyl sulfate-polyacrylamide gel electrophoresis–Western blots of extracts of total cellular proteins. For example, polyclonal antibodies raised against a glycoprotein might bind to other glycoproteins with similar carbohydrates attached and, thus, result in unspecific binding. The primary antibody can be purified in order to remove antibodies responsible for undesired or unspecific bindings on carbohydrate affinity columns (20). Deglycosylation of the glycoproteins before immunization is another approach (21). The ultimate test of antibody specificity is preincubation of the antibodies with an appropriate concentration of the antigen, and then incubation of the tissue of interest with such preabsorbed serum. The result should be a total loss of tissue reaction (17). Details regarding production and purification of polyclonal antibodies are found in De Mey and Moeremans (22), Beltz and Burd (17), and Johnstone and Thorpe (6).

2.2. The Monoclonal Antibodies

Many of the problems with polyclonal antibodies can be circumvented by using monoclonal antibodies. Monoclonal antibodies are produced by culturing cell lines producing only a single IgG. The cell lines are obtained by fusion of mortal spleen cells from an immunized animal with drug-sensitive immortal myeloma cells. The fusion product is called hybridoma cells. After selection and cloning, each cell line produces only one type of IgG, resulting in monoclonal antibodies that can be harvested from the supernatant of the hybridoma culture. The monoclonal antibodies are of high homogeneity. As hybridoma cell lines can be continued forever, no variability of antibodies is observed. This gives another advantage compared to polyclonal antibodies. A detailed description of the technique is described in Ritter (23) and Johnstone and Thorpe (6). The technique was developed by Kohler and Milstein (24). The development of monoclonal antibodies has significantly increased the specificity of antibodies and the amount of antibodies available. A good example of monoclonal antibodies against carbohydrates is the collection of antibodies raised against specific sugar components in the pectin molecule (25). A review of antibodies and the study of plant cell walls is presented by Knox (25).

2.3. Phage Display Library
and Recombinant DNA Technology

The diversity of antibodies can be increased further with phage display libraries and recombinant DNA technology. In recombinant DNA technology, advantage is taken of hybridoma cells producing monoclonal antibodies. The mRNA responsible for the variable and binding areas on the antibodies is isolated, and the corresponding cDNA can be modified in various ways to produce recombinant antibodies and antibody fragments in prokaryotic and eukaryotic expression systems.

In phage display *(26–28)*, a library of hundreds of millions of different random peptide sequences in the hypervariable regions of Fv fragments are expressed as a coat fusion protein on the surface of bachteriophage, so that each different cDNA sequence is in a different phage. Each phage has the potential to express a unique antibody on its surface, and the phage genotype is responsible for the displayed molecule. The library is then screened to detect reaction of any of the individual phages with the desired antigen. Positive phages are then selected, reinfected into bacteria, cloned, screened again, and characterized. The phage antibodies can be visualized with secondary antibodies *(29)*. In phage display, cDNA clones from a preparation of the entire collection of lymphocytes of the animal are used and no initial information of the animal's immune system is needed.

2.4. Lectins

Antibodies against sugars (carbohydrate residues) can be difficult to obtain and lectins are a solution to these problems. Lectins are naturally occurring plant and animal proteins or glycoproteins that selectively bind noncovalently to carbohydrate residues. Lectins can be labeled directly or secondary antibodies against lectins enables the use of other immuno techniques *(30)* including electron microscopy *(31)*.

3. VISUALIZATION OF THE TARGET/STAINING METHODS

Visualization of the antibodies after their binding to the target can be accomplished by labeling the primary, or more often, the secondary antibodies with fluorescent dyes, enzymes, or colloidal gold particles. Radioactive labeling followed by autoradiography is possible but seldom used. The foregoing techniques were developed for quantification of immunocytochemistry *(32)*. The most simple immunoprocedures only use one antibody, which is labeled directly (*see* direct methods, Fig. 1A). The advantage is a fast and simple procedure; the disadvantage is that it is less sensitive than the more complex methods and each new antibody has to be labeled, which involves considerable work and loss of antibody because of purification and labeling procedures. Therefore, secondary antibodies against the primary antibodies are used (*see* indi-

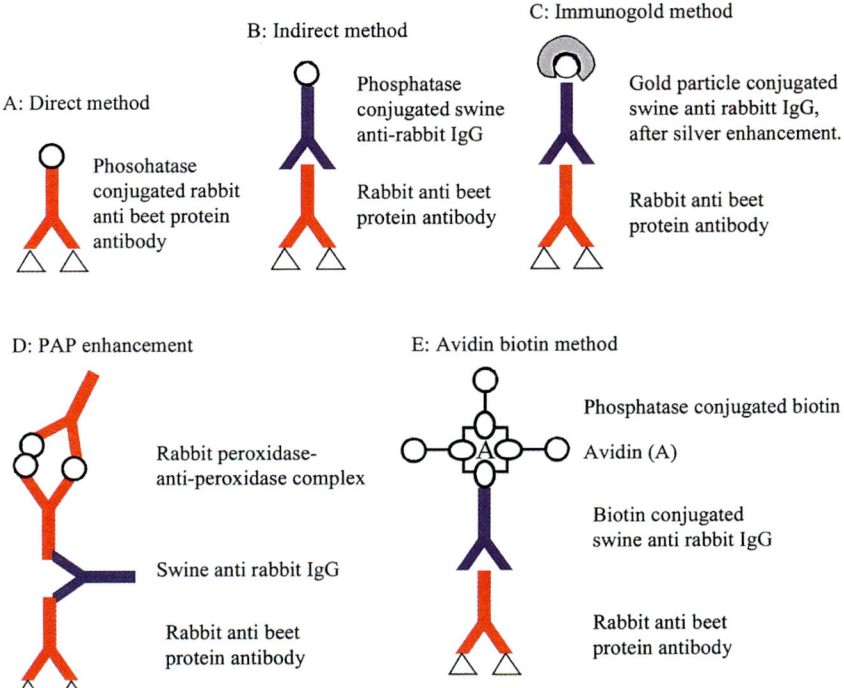

Fig. 1. Schematic drawing of various immune methods for localization of antigens. (**A**) Direct method. (**B**) Indirect method. (**C**) Immunogold with silver enhancement. (**D**) Peroxidase–antiperoxidase method. (**E**) Avidin–biotin system.

rect methods, Fig. 1B). If the primary antibody is made in a rabbit, commercially swine anti-rabbit immunoglobulins coupled with alkaline phosphatase can be used as a secondary antibody. The use of secondary antibodies increases sensitivity and reduces work and the amount of used primary antibody. The increase in sensitivity is caused by the binding of several secondary antibodies to the primary antibody. Sensitivity can be further increased by the use of enzyme–antienzyme methods such as alkaline phosphatase anti-alkaline phosphatase (APAAP) and peroxidase/antiperoxidase (PAP) or the biotin–avidin system (*see* Subheading 4.).

3.1. The Fluorescent Dyes

The advantage of fluorescent dyes coupled to primary antibodies is the fast result obtained with the direct method. Another advantage is the possibility to label more than one antigen at the same time. In plant material, it is important to take into account the possible autofluorescence of the tissue. Disadvantages can be the lack of orientation in the tissue and photobleaching of the dyes. The latter problem can often be

solved by applying antifade. Typical antifades are p-phenylenediamine
(33) and 1,4 diazabiclyclo-(2,2,2)-octane (DABCO) *(16)*. Methods for
coupling the fluorescence dyes to antibodies are described in Johnstone
and Thorpe *(6)*. Fluorescent dyes are also used on secondary antibodies
in indirect methods, giving more intense labeling and saving labor and
primary antibody as mentioned in Subheading 3.

3.2. Enzymatic Labeling

The enzymatic labeling is based on an enzyme reaction that gives
an insoluble colored product in water. These labels often give distinct
colors and the possibility of localization and orientation in relation to
important structures in the tissue, with or without staining. The enzyme
labeling is most frequently used on secondary antibodies and, thereby,
the two-step indirect methods. The enzymes used should have a low
molecular weight as they must be coupled to the antibodies, and, in addition, they should have a high turnover to give a high yield of product.
Ideally no similar endogenous enzyme activity exists or is retained in
the tissue. Specific inhibition and the use of pH values far from the optimal conditions of the endogenous enzymes are used. If the product is
insoluble in ethanol and xylene, coverslips can be mounted conventionally. The most frequently used enzymatic labelings used for plant immunolocalization are alkaline phosphatase and horseradish peroxidase;
glucose oxidase and galactosidase are alternatives *(14,32)*. The substrate for alkaline phosphatase is usually naphthol phosphate and a diazonium salt. The phosphatase produces naphthol, which reacts with the
diazonium salt to produce a colored precipitate. Red or blue precipitates
can be obtained by using the diazonium salts, Fast Red TR, or Fast Blue
BB. The red color produced by Fast Red is very easily detected in the
tissue; the blue color appears more faint. Endogenous phosphatase activity is blocked by addition of the phosphatase inhibitor, levamisol. Levamisol inhibits alkaline phosphatase except intestine alkaline phosphatases.
The alkaline phosphatase coupled to the secondary antibody comes from
calf intestine. Horseradish peroxidase can also be used with a number of
different substrates. An overview is found in Robinson et al. *(32)*. The substrates are 3,3-diaminobenzidine tetrahydrochloride(DAB), 3-amino-9-
ethylcarbazole (AEC), 4-chloro-1-naphthol (CN), and p-phenyldiamione
dehydrochloride. Methods for coupling these enzymes to antibodies are
found in Johnstone and Thorpe *(6)*.

3.3. Labeling with Colloidal Gold

Labeling antibodies with colloidal gold was developed for electron
microscopy, but the procedure can be used for light microscopy as well.

Particles must be greater than 20 nm to be visualized by light microscopy. The high weight of the particles reduces the sensitivity of these
conjugates and they are not stable *(32)*. This problem is solved by precipitation of silver (silver enhancement) on smaller gold particles (1-5 nm)
after the binding the gold-labeled antibodies to the epitopes in the tissue
(Fig. 1C). Details and protocols occur in Verkleij and Leunissen *(34)*,
Vandenbosch *(35)*, and Monaghan et al. *(36)*. A selection of protocols
for plant cells is found in Singh et al. *(5)*. Using different sizes of gold
particles, two or three different antibodies can be visualized at the same
time by electron microscopy.

3.4. Visualizing Two or More Antigens on the Same Section

Visualizing more than one epitope on one section can be accomplished by different fluorescence labeling or different sizes of colloidal
gold coupled to primary or secondary antibodies. Primary antibodies
from different species and adequate secondary antibodies labeled differ-
ently can be used. In case of primary antibodies from the same species,
the hapten technique can be applied. A hapten is a small molecule that
can be bound to antibodies; dinitrophenol and arsinilate are typically
used as haptens. Again, adequate secondary antibodies labeled differ-
ently can be used *(14,17,32)*. A collection of protocols for multiple immunolabeling has been described by Beesley *(37)*.

4. ENHANCING SIGNALS
WITH ANTIBODY COMPLEXES

4.1. Enzyme–Antienzyme Methods

The PAP (*see* Fig. 1D) is composed of three peroxidase molecules and
two antiperoxidase antibodies. Each antibody has two identical binding
sites (antigen binding sites). One peroxidase molecule is bound to two
antibodies and the two other peroxidases are bound to the other binding
sites of the two antibodies. The PAP complex is used as a third layer, binding to a secondary layer of antibodies that has been applied in excess, so
that only one of its two identical binding sites bind to the primary antibody. This gives one free binding site for each secondary antibody to
bind the PAP complex. The APAAP system is a similar complex with
alkaline phosphatase. Silver-enhanced gold immunodetections are generally accepted to be more sensitive than the PAP technique *(32)*.

4.2. The Avidin–Biotin System

Avidin, a glycoprotein present in egg white, has a strong affinity for
biotin. Biotin is found in the liver. Avidin has four subunits each one with

a binding site for biotin. Advantage of this is taken by binding the low-molecular-weight biotin to antibodies (primary or secondary), binding avidin to this, and finally binding biotin coupled with, for example, phosphatase to the other three binding sites of avidin (*see* Fig. 1E). The actual order of these bindings vary from method to method *(32)*, but the end result is a significant improvement of sensitivity. Oligosaccharide residues on avidin can result in nonspecific binding to lectin-like proteins in the sections. Furthermore, avidin has an isoelectric point close to 10 and is positively charged at neutral pH and can, therefore, potentially bind to negatively charged groups in the sections. Therefore, avidin can be replaced by streptavidin, a similar protein extracted from the culture broth of *Streptomyces avidinii* without oligosaccharide residues and with an isoelectric point close to 7. Protocols for the above-mentioned methods can be found in *(14,32,37)*.

5. A PROTOCOL FOR IMMUNOVISUALIZATION OF PROTEINS WITH ALKALINE PHOSPHATASE

If you are about to try your first immunovisualization experiment, choose a simple method, consult your colleagues, ideally find a nearby laboratory (hospital), and try their favorite method if it is suitable for your work. The following method is simple and works very well with plant tissue and polyclonal antibodies. The procedure has been used to visualize antifungal proteins produced by sugar beet as a response to infections with *Cercospora beticola (39–43)*, to visualize the distribution of acetyl esterase and pectin methyl esterase in orange fruits *(44,45)*, and to show the distribution of uridine diphosphate-galactose epimerase in developing seeds of guar *(46)*.

5.1.1. IMMUNE PROCEDURE

5.1.1.1. Stocks

1. TBS (Tris buffer saline): Tris-HCl 0.05 M, NaCl 0.15 M, Triton X-100 1 mL/L, pH is adjusted to 7.6 with NaOH. (Can be made 10 times the foregoing concentration.)
2. Veronal-acetate buffer pH 9.2: Sodium acetate 30 mM, sodium barbitone or barbitone acid 30 mM, NaCl 100 mM, and MgCl$_2$ 50 mM. (Can be made five times the foregoing concentration.)
3. Swine serum/TBS: 20% Swine serum in TBS 0.05 M, pH 7.6, is made just before use. We use swine serum (normal) from DAKO A/S, code no. X0901.
4. Developer: Naphthol AS-BI phosphate (Sigma N4875) 2 mg, N,N-dimethylformamid 0.2 mL, veronal-acetate buffer 9.8 mL, pH 9.2, levamisol 1 M

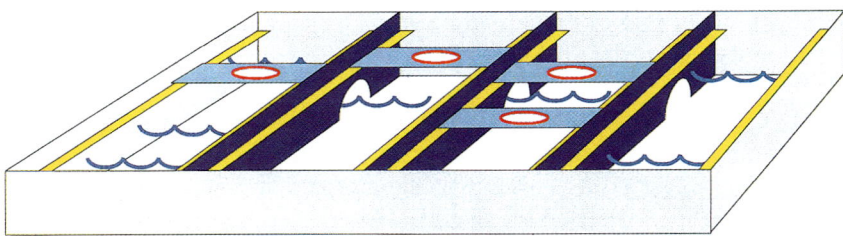

Fig. 2. Diagram of humidity chamber made from a plastic box. Water at the bottom of the chamber ensures a high humidity once the lid covers the chamber, this prevents significant evaporation of the slides placed on "shelves." The sections are encircled by the lipid content from a DAKO pen, limiting the amount of antibody used. The chamber is wrapped in tinfoil from the start when used for alkaline phosphatase experiments.

(Sigma L9756) 0.01 mL. The solution has to be kept dark; use tin foil. The substrates have to be added in the given order. This stock solution can be stored in the freezer for several months and for weeks at 4°C. Immediately before use, 10 mg Fast Red TR (Sigma F1500) is added to the solution and after dissolving, the solution is filtrated in the dark. The funnel and test tubes should be covered by tinfoil. The developer should be a clear solution. If it is colored (slightly red) a new developer has to be made. The 10-mL portions should be enough for 40–50 slides.

5.1.2. Immunoprocedure

1. Following deparaffination, the sections are treated with 0.05 M TBS, pH 7.6, for 5 min. Dry the slides with paper towels as close to the sectioned material as possible; encircle the sectioned material with a DAKO pen (code no. S2002).
2. Add 20% swine serum in TBS for 20 min in a humidity chamber (*see* Fig. 2) sealed with tinfoil. Pour off the swine serum and dry with paper towels around the circles made by the DAKO pen.
3. Incubate with primary antibody in swine serum/TBS for 60 min in a humidity chamber.
4. Wash three times for 5 min in TBS. Dry with paper towels after the last wash.
5. Incubate with secondary antibody 1:20 (Anti-Rabbit Immunoglobulins coupled with alkaline phosphatase, DAKO code no. D0306) in swine serum/TBS for 30 min. in a humidity chamber.
6. Wash three times for 5 min in TBS.
7. Wash in veronal-acetate buffer, pH 9.2, for 5 min. Dry with paper towels.
8. Develop with (naphthol AS-BI phosphate/Fast Red, fresh made) for 20 min in the dark (in tinfoil-sealed humidity chambers). For preparation of developer, *see* Subheading 5.1.1.1, item 4.

9. Washing 20 min in running tap water. Dry with paper towels.
10. Mount in glycerin/TBS 9:1.

5.1.2.1. Controls

1. Preimmune serum instead of the primary antibody. If positive, it indicates that unspecific binding has taken place. Try again with a higher dilution of preimmune serum and antibody.
2. Omission of primary antibody. If positive, it shows unspecific binding of secondary antibodies. Dilute or try another secondary antibody.
3. No antibodies, only application of developer. If positive, this shows endogenous enzyme activity. To solve this, raise levamisol concentration, with 20% acetic acid for 5 min, followed by intensive washing in water prior to the immunoprocedure (14).

What concentration of primary antibody should be chosen for the initial experiments with a new antibody? There is no clear answer to that, but if the optimal or suboptimal dilution is known from a Western blot, it often gives good results to start with this dilution, 10X and 100X less dilution in the initial experiment, and optimize further from there. Only in rare cases have we found that we should use less primary antibody than was used in Western blots.

5.1.3. RESULTS

In the immunocytochemical investigations performed on antifungal proteins in sugar beet leaves infected with *Cercospora*, we found that the antifungal proteins were localized both in the necroses and in the surrounding tissue as well (39–42), and in most cases also in uninfected leaf material (39–41). In the necroses, the antifungals were found both on the fungus and within the dead tissue. In the mesophyll surrounding the necroses, both intra- and extracellular localizations were found. Autolysis of single cells or groups was also observed (40–42). Xylem and stomata cells were found positive for most of the antifungal proteins in both infected and uninfected leaves. The stomata cells were either positive intracellularly, or the cell walls lining the substomatal lumen were positive. The findings of antifungal proteins in the stomata and xylem in uninfected leaves led us to investigate the stigma, as this structure shared the potential for fungal invasion. Stigma tissue was found to be positive for some of the antifungal proteins (39,40; see Figs. 3A and 3B).

As an example of the use of antibodies labeled with alkaline phosphatase for detection of *in situ* hybridization, an infection with BNYVV virus in sugar beet is shown in Fig. 3C. Lectins labeled with an avidin–biotin fluorescein conjugate was used to visualize α-galactosyl groups on the surface of *S. pombe* in Fig. 3D.

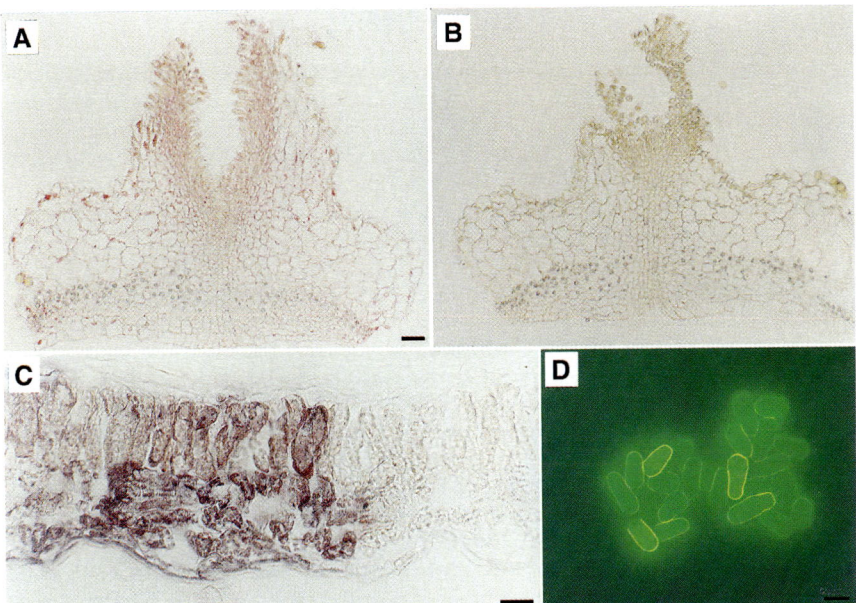

Fig. 3. (A) The antifungal Ax proteins visualized in stigma of sugar beet by and indirect immunodetection. Primary rabbit antibodies against Ax were applied as the primary antibody. Swine anti-rabbit coupled with alkaline phosphatase as the secondary antibody and the staining was accomplished with naphthol ASBI phosphate and Fast Red. **(B)** Preimmune serum control. **(C)** Visualization of the BNYVV (beet necrotic yellow vain virus) in sugar beet leaves by *in situ* hybridization. The visualization was accomplished by the hapten digoxigenin bound to oligonucleotides. After hybridization, antibodies coupled with alkaline phosphatase and directed against digoxgenin were applied and, finally, a phosphatase substrate with 5-bromo-4-chloro-3-indolyl phosphate and nitroblue tetrazolium was used for staining hybridized probe. **(D)** The Bandeiraea (Griffonia) simplicifolia lectin 1 binds to α-galactosyl groups on cell surface glycoconjugates. An avidin–biotin fluorescein conjugate of this lectin was used to visualize α-galactosyl groups on the surface of *Schizosaccharomyces pombe*. Bars = 50 μm in (**A**) and (**B**), 25 μm in (**C**), and 5 μm in (**D**).

6. CONCLUSIONS

Immunocytochemistry gives valuable information on the distribution of molecules of interest in tissue or at the cellular and subcellular levels, that could be difficult, or in most cases impossible, to obtain in another way. The detailed information obtained is often surprisingly high. It has a great potential in proving or rejecting an hypothesis and it gives further clues to new ideas. The illustrations obtained have a high value in educational research and innovation, in presentations and textbooks, and some of them are simply art.

REFERENCES

1. Adams AE, Pringle PR. Relationship of actin and tubulin distribution to bud growth in wild-type and morphogenetic-mutant *Saccharomyces cerevisiae*. *J Cell Biol* 1984; 98: 934–945.

2. Pringle JR, Adams AE, Drubin DG, Haarer BK. Immunofluorescent methods in yeast. *Methods Enzymol* 1991; 194: 565–602.

3. Hagan IM, Hyams JS. The use of cell division cycle mutants to investigate the control of micotubule distribution in the fission yeast *Schizosaccharomyces pombe*. *J Cell Sci* 1988; 89: 343–357.

4 Katembe WJ, Swatzell LJ, Makaroff CA, Kiss JZ. Immunolocalization of integrin-like proteins in *Arabidopsis* and *Chara*. *Physiologia Plantarum* 1997; 99: 7–14.

5. Singh MB, Taylor PE, Knox RB. Special preparation methods for immunocyto-chemistry of plant cells, in *Immunocytochemistry, A Practical Approach* (Beesley JE, ed.), Oxford University Press, Oxford, UK, 1993, pp. 77–100.

6. Johnstone A, Thorpe R. *Immunochemistry in Practice*, Blackwell Science Ltd, Oxford, UK, 1996.

7. Kaufmann A, Koenig R, Lesemann D-E. Tissue print-immunoblotting reveals an uneven distribution of beet necrotic yellow vein and beet soil-borne viruses in sugar beet. *Arch Virol* 1992; 126: 329–335.

8. Chan VT-W, McGee JO'D. Non-radioactive probes: preparation, characterization, and detection, in *In Situ Hybridization, Principles and Practice* (Polak JM, McGee JO'D, eds.), IRL Press at Oxford University Press, Oxford, UK, 1990, pp. 1–14.

9. Davies JT. *In situ* hybridization, in *Immunocytochemistry, A Practical Approach* (Beesley JE, ed.) Oxford University Press, Oxford, UK, 1993, pp. 176–205.

10. Dirks RW, van de Rijke FM, Fujishita S, van der Ploeg M, Raap AK. Methodologies for specific intron and exon RNA localization in cultured cells by hapterized and fluochromized probes. *J Cell Sci* 1993; 104: 1187–1197.

11. Gu J, ed. *In Situ Polymerase Chain Reaction and Related Technology*, Birchäuser, Boston, MA, 1995.

12. Johansen B. *In situ* PCR on plant material with sub-cellular resolution. *Ann Bot* 1997; 80: 697–700.

13. Yasuhiko M, Appels R. Direct chromosome mapping of plant genes by *in situ* polymerase chain reaction (*in situ* PCR). *Chromosome Res* 1996; 4: 401–404.

14. Ormerod MG, Imrie F. Immunohistochemistry, in *Light Microscopy in Biology. A Practical Approach* (Lacey AJ, ed.), IRL Press Ltd., Oxford, UK, 1989, pp. 103–136.

15. O'Brien TP, McCully ME. *The Study of Plant Structure Principles and Selected Methods*, Termarcarphi Pty. Ltd, Melbourne, Australia, 1981.

16. Harlow E, Lane D. *Using Antibodies. A Laboratory Manual*, Cold Spring Habor Laboratory Press, Cold Spring Harbor, NY, 1999.

17 Beltz BS, Burd GD. *Immunocytochemical Techniques*, Blackwell Scientific Publishers, Inc, Malden, MA, 1989.

18. Hjelm H, Hjelm K, Sjöquist J. Protein A from *Staphylococcus aureus*. Its isolation by affinity chromatography and its use as an immunosorbent for isolation of immunoglobulins. *FEBS Lett* 1972; 28: 73–76.

19. Patrick CC, Virella G. Isolation of normal human IgG3. Identical molecular weight for normal and monoclonal gamma-3 chains. *Immunochemistry* 1978; 15: 137–139.

20. Sonnewald U, Studer D, Rocha-Sosa M, Wilmitzer L. Immunocytochemical localization of patatin, the major glycoprotein in potato (*Solanum tuberosum*). *Planta* 1989; 178: 176–183.

21. Edge ABS, Faltynek CR, Hof L, Reichert LE Jr, Weber P. Deglycosylation of glycoproteins by trifluoromethanesulfonic acid. *Anal Biochem* 1981; 118: 131–137.

22. De Mey J, Moeremans M. Raising and testing polyclonal antibodies for immunocytochemistry, in *Immunocytochemistry, Modern Methods and Applications* (Polack JM, Van Noorden S, eds.), Wright, Bristol, UK, 1986, pp. 3–12.

23. Ritter MA. Raising and testing monoclonal antibodies for immunocytochemistry, in *Immunocytochemistry, Modern Methods and Applications* (Polack JM, Van Noorden S, eds.), Wright, Bristol, UK, 1986, pp. 13–25.

24. Kohler G, Milstein C. Continuous cultures of fused cells producing antibody of predefined specificity. *Nature* 1975; 256: 495–497.

25. Knox JP. The use of antibodies to study the architecture and developmental regulation of plant cell walls. *Int Rev Cytol* 1997; 171: 79–120.

26. Winter G, Griffiths AD, Hawkins RE, Hoogenboom HR. Making antibodies by phage display technology. *Annu Rev Immunol* 1994; 12: 433–455.

27. Rader C, Barbas CF. Phage display of combinatorial antibody libraries. *Curr Opin Biotechnol* 1997; 8: 503–508.

28. Griffeths AD, Duncan AR. Strategies for selection of antibodies by phage display. *Curr Opin Biotechnol* 1998; 9: 102–108.

29. Willats WGT, Gilmartin PM, Mikkelsen JD, Knox JP. Cell wall antibodies without immunization: generation and use of de-esterified homogalacturonan block-specific antibodies from a naive phage display library. *Plant J.* 1999; 18: 57–65.

30. Brooks SA, Leathem AJC, Schumacher U. *Lectin Histochemistry, A Concise Practical Handbook*, BIOS Scientific Publishers Ltd, Herndon, VA, 1997.

31. Knox RB, Clarke AE. Localization of proteins and glycoproteins by binding to labeled antibodies and lectins, in *Electron Microscopy and Cytochemistry of Plant Cells* (Hall JL, ed.), Elsevier-North Holland, Biomedical Press, Amsterdam, 1978, pp. 149–185.

32. Robinson G, Ellis IO, MacLannan KA. Immunochemistry, in *Theory and Practice of Histological Techniques* (Bancroft JD, Stevens A, eds.), Churchill Livingston, Edinburgh, New York, 1990.

33. Johnson GD, Nogueira Araujo GM de C. A simple method of reducing the fading of immunofluorescence during microscopy. *J Immunol Methods* 1981; 43: 349–350.

34. Verkleij AJ, Leunissen JLM. *Immuno-Gold Labeling in Cell Biology*, CRC Press, Boca Raton, FL, 1989.

35. Vandenbosch KA. Immunogold labeling, in *Electron Microscopy of Plant Cells* (Hall JL, Hawes C, eds.), Academic Press, San Diego, CA, 1991, pp. 181–218.

36. Monaghan P, Robertson D, Beesley JE. in *Immunocytochemistry, A Practical Approach* (Beesley JE, ed.) Oxford University Press, Oxford, UK, 1993, pp. 43–76.

37. Beesley JE. Multiple immunolabeling techniques, in *Immunocytochemistry, A Practical Approach* (Beesley JE, ed.), Oxford University Press, Oxford, UK, 1993, pp. 103–125.

38. Jackson P, Blythe D. Immunolabeling techniques for light microscopy, in *Immunocytochemistry, A Practical Approach* (Beesley JE, ed.), Oxford University Press, Oxford, UK, 1993, pp. 15–41.

39. Kragh KM, Nielsen JE, Nielsen KK, Drebolt S, Mikkelsen JD. Characterization and localization of new antifungal cysteine-rich proteins from *Beta vulgaris. Mol Plant-Microbe Interact* 1995; 8: 424–434.

40. Nielsen KK, Nielsen JE, Madrid S, Mikkelsen JD. New antifungal proteins from sugar beet (*Beta vulgaris* L.) showing homology to non-specific lipid transfer proteins. *Plant Mol Biol* 1996; 31: 539–552.

41. Nielsen JE, Nielsen KK, Mikkelsen JD. Immunohistological localization of a basic class IV chitinase in *Beta vulgaris* leaves after infection with *Cercospora beticola*. *Plant Sci* 1996; 119: 191–202.

42. Gottschalk TE, Mikkelsen JD, Nielsen JE, Nielsen KK, Brunstedt J. Immunolocalization and characterization of a β-1.3-glucanase in sugar beet, deduction of its primary structure and nucleotide sequence by cDNA and genomic cloning. *Plant Sci* 1998; 132: 153–167.

43. Kristensen AK, Brunstedt J, Nielsen JE, Mikkelsen JD, Roepsstorff P, Nielsen KK. Processing, disulfide pattern and biological activity of sugar beet defensin AX2, expression in *Pichia pastoris*. *Protein Expr Purif* 1999; 16: 377–387.

44. Christensen TMIE, Nielsen JE, Mikkelsen JD. Isolation, characterization and immunolocalization of orange fruit acetyl esterase, in *Progress in Biotechnology 14 Pectins and Pectinases* (Visser J, Voragen AGJ, eds.), Elsevier, Amsterdam, The Netherlands, 1996, pp. 723–730.

45. Christensen TMIE, Nielsen JE, Kreiberg JD, Rasmussen P, Mikkelsen JD. Pectin methylesterase from orange fruits: characterization and localization by *in situ* hybridization and immunohistochemistry. *Planta* 1998; 206: 493–503.

46. Joersboe M. Petersen SG, Nielsen JE, Marcussen J, Brunstedt J. Isolation and expression of two cDNA clones encoding UDP-galactose epimerase expressed in developing seeds of the endospermous legume guar. *Plant Sci* 1999; 142: 147–154.

7

The Fixation of Chemical Forms on Nitrocellulose Membranes

Rosannah Taylor

CONTENTS

1. OVERVIEW

Tissue printing is becoming a general procedure for the biochemical, molecular, anatomical, and physiological characterization of biological specimens and plant natural products. Tissue printing dates back to 1957, when Daoust *(1)* reported a substrate film printing. Tissue printing is the art and science of visualization of soluble materials and information that are transferred to membranes such as nitrocellulose membrane. In the 1960s, Daoust *(2)* used the technique to localize protease, amylase, RNase and DNase by placing cryostat sections of various animal organs (liver, kidney, pancreas, and intestine) on substrate films of gelatin, starch, or gelatin-nucleic acid, respectively. When the films were incubated for a few minutes and then stained for the substrate, negative images were obtained.

A breakthrough for tissue printing, at least for the botanical sciences, came when Cassab and Varner *(3)* combined the use of nitrocellulose and antibody technology. They placed sections of a freshly cut soybean seed on nitrocellulose membranes and probed the resulting imprints using specific antibodies. They were able to show that soluble extensin protein is primarily localized in the seed coat and vascular tissues.

From: *Methods in Plant Electron Microscopy and Cytochemistry*
Edited by: W. V. Dashek © Humana Press Inc., Totowa, NJ

The elegant simplicity of the technique has resulted in many new applications for both teaching and research, especially following the introduction of nitrocellulose membranes, agarose, and glue as physical printing substrates. The overall advantage of tissue printing is that it allows sensitive detection and localization of specific constituents. The binding properties of nitrocellulose membranes have made possible the cellular localization of proteins, nucleic acids, enzymatic activities, ions, and certain carbohydrate moieties in specific tissues. Although different substrates can be used, those producing an insoluble end product work best. Remarkable advancements have been made to overcome some of the earlier problems encountered with printing (e.g, soft tissues, large organs, etc.). Table 1 presents an abbreviated review of recent applications of tissue printing.

2. DETECTION AND LOCALIZATION OF AUXIN CONJUGATES IN OKRA

Actually doing tissue printing is the best way to understand its simplicity (Fig. 1). Wide variations in the technique are possible and can be developed easily and quickly according to need. The following protocol for using tissue prints to detect tissue distribution of IAA–protein conjuguates in okra fruit details the immunoblotting technique as follows.

2.1. How to Do Tissue Printing

2.1.1. TISSUE-PRINTING MATERIALS

1. Whatman No.1 filter paper
2. Blotting membrane (nitrocellulose, nylon, polyvinyldifluoridine, etc.)
3. Double-edged razor blades
4. Forceps
5. Appropriate plant material
6. Latex gloves
7. Paper to protect the membrane
8. Acrylic sheet (backing sheet for membrane)
9. Marking pen
10. Hand lens or microscope for viewing the specimen

2.1.2. PROCEDURE FOR TISSUE PRINTING

1. Place several layers of filter paper on a smooth, hard surface, and place a blotting membrane on top (*see* Fig.1). Depending on the size of the tissue, use a single- or double-edged razor blade to cut a tissue section 0.2–2.0 mm thick, depending on the particular tissue sample. It is necessary

Table 1
Summary of Some Recent Applications of Tissue Printing

Plant system investigated	Application	Reference
Winter rye in leaves, crowns, roots	Immunolocalization of antifreeze proteins	Antikainen et al. (4)
Virus-infected tobacco leaves	Spatial distribution of acidic chitinases and their messenger RNAs in tobacco plants infected with cherry leaf roll virus	Balsalobre et al. (5)
Soft tissue	Cryostat tissue printing	Conley and Hanson (6)
Deepwater rice	Expansion and internode growth	Cho and Kende (7)
Melon fruit	Evaluation and enhancement in cucurbit germplasm	Cohen et al. (8)
Garlic	Immunological detection of a Gar V-type virus	Helguera et al. (9)
Maize	Retention of maize auxin-binding protein in the endoplasmic reticulum	Henderson et al. (10)
Sporophytic maize tissues	Biochemical and tissue print analyses of hydroxyproline-rich glycoproteins	Hood et al. (11)
Developing soybean roots and hypocotyls	Localization of respiratory oxidase in meristematic and xylematic tissues	Hilal et al. (12)
Pineapple leaves and stems	Immunoassay to examine the distribution of pineapple *Clostero* virus	Hu et al. (13)
Transgenic tomato plants	Wound induced vascular bundle-specific expression of β-glucuronidase gene	Jacinto et al. (14)
Soybean	Low-molecular-weight heat shock protein	Jinn (15)
Wild-type mutant tomato fruit	Molecular cloning of a ripening-specific lipoxygenase and its expression	Kausch and Handa (16)
Soybean	Tissue-type-specific heat-shock response and immunolocalization	Key and Lin (17)

(continued)

Table 1 (Continued)

Plant system investigated	Application	Reference
Tobacco plants	Long-distance movement of cherry leaf roll virus in infected plants	Mas and Pallas (18)
Developing stem of tomato and popular	Lignification and cinnamyl alcohol dehydrogenase	Roth et al. (19)
Potato tubers	Tissue-specific distribution of glutamine synthetase	Pereira et al. (20)
Cynara cardunculus L	Cardosin A, an abundant aspartic proteinase, accumulates in protein storage vacuoles	Ramalho-Santo et al. (21)
Celery petiole	The role of calcium in cell-wall firmness	Taylor and Varner (22)
Celery petiole	Localization of ascorbic acid	Taylor and Varner (23)
Tobacco petiole and stem	Translocation of fungal protein in plants	Taylor et al. (24)
Soybean stem	Hydroxy-proline and glycine-rich protein and mRNA analysis	Varner and Ye (25)

to gently preblot the section on a separate piece of membrane or filter paper before printing to remove excess tissue exudate from cut cells and to help ensure a faithful print.

2. Using forceps, transfer the tissue section to the membrane, being careful not to move the section after it is in contact so as to avoid smearing. Several successive sections can be printed on the same piece of membrane.

3. Place a small piece of nonabsorbent paper (the paper separating the nitro-cellulose sheets works well) over the section to protect the membrane from fingerprints. In some instances, latex gloves may be required. When printing a thin section (200–300 µm), placing a piece of membrane on top of the section instead of the nonabsorbent paper frequently gives better results.

4. Apply the appropriate amount of pressure to the section for the type of print desired. A chemical print requires only light pressure, but a physical print requires several times as much. The proper pressure also varies with the tissue used.

5. Gently remove the protective paper and the section with forceps, air dry the print, and observe.

Prints may be illuminated from the top or from one side by white or ultraviolet light and may also be viewed with transmitted light.

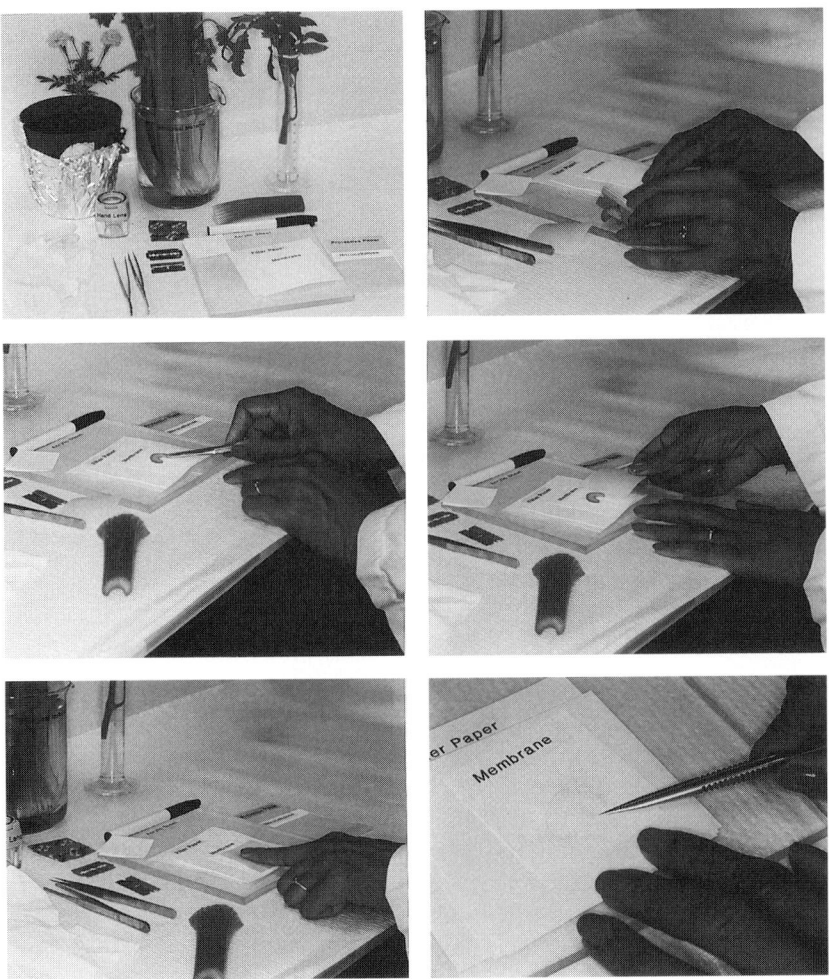

Fig. 1. Steps involved in tissue printing.

2.2. Detection and Localization
of Auxin Protein Conjugates in Okra

Okra, *Hibiscus escuentus* L. The immature pods of okra are popular as a vegetable and okra is also used for the ability to thicken soups and stews, and for fiber and oil. Not much is known about okra; most workers concentrate their research on the chemical composition *(26–29)* of the mucilaginous materials found in the pod *(30)* stalk, nutritional properties *(31)*. Research has also been conducted on the developmental properties of the okra seed *(32)*. This paper is the first report of the chemical composition of auxin conjugates in okra pods bound to nitrocellulose membrane.

2.2.1. MATERIALS FOR IMMUNOBLOTTING WITH CHEMLUMINESENCE DETECTION

1. Phosphate-buffered saline (PBS): 20 mM K$_2$HPO$_4$ and 0.15 M NaCl (pH 7.4)
2. Blocking solution: 5.0% Carnation powdered dry milk in
3. Tween-PBS (TPBS): PBS containing 0.5% Tween-20
4. Detection system:
   a. IAA-peptide specific antibody raised in rabbit *(33)* and preimmune serum
   b. Secondary antibodies goat anti-rabbit immunoglobulin conjugated with alkaline phosphatase commercially available
5. Tropix Chemluminesence kit (Western-Star Protein Detection Kit, #WL10RS Tropix, Bedford, MA).
6. Ponceau S protein stain *(34)*
7. Kodak T-max 100 film
8. Kodak XAR-5 X-ray film
9. Lightproof film cassette
10. X-ray film developer

2.2.2. METHODS

1. Tissue prints on nitrocellulose membranes (Schleicher & Schuell) (0.45 µm pore size) were first stained for protein using Ponceau S (Fig. 2A,B) and photographed immediately with T-max 100 black-and-white film. The membranes were then washed in PBS to remove the stain, and incubated for 2 h with shaking in blocking solution.
2. The blots were then incubated with shaking overnight at 4°C in fresh blocking solution containing a 1:1000 dilution of primary antibody (a polyclonal antibody made in rabbit to an IAA–peptide conjugate [Fig. 2D] purified from *Phaseolus vulgaris* seed; gift of Dr. J. D. Cohen).
3. As controls, replicate blots should be incubated with preimmune serum (Fig. 2C), or incubated directly in secondary antibody without incubation in primary antibody.
4. The blots were then washed five times with PBS (10 min each wash) with shaking and transferred to fresh blocking solution containing alkaline phosphatase conjugated secondary antibody (goat anti-rabbit; Sigma). The blots were shaken in secondary antibody for 1 h at room temperature before washing again five times with PBS.
5. After washing, the blots were processed for chemiluminescence detection according to the manufacturer of the kit.
6. Briefly, the membrane is incubated in the appropriate buffer for the chemiluminescence substrate then placed on a glass plate and a thin layer of substrate solution is pipetted onto the membrane to completely cover the surface and allowed to incubate for the required time (usually

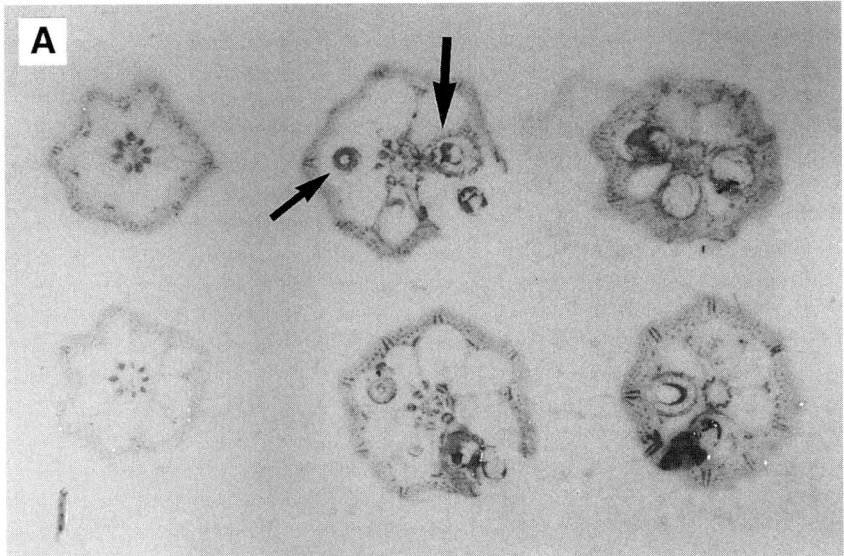

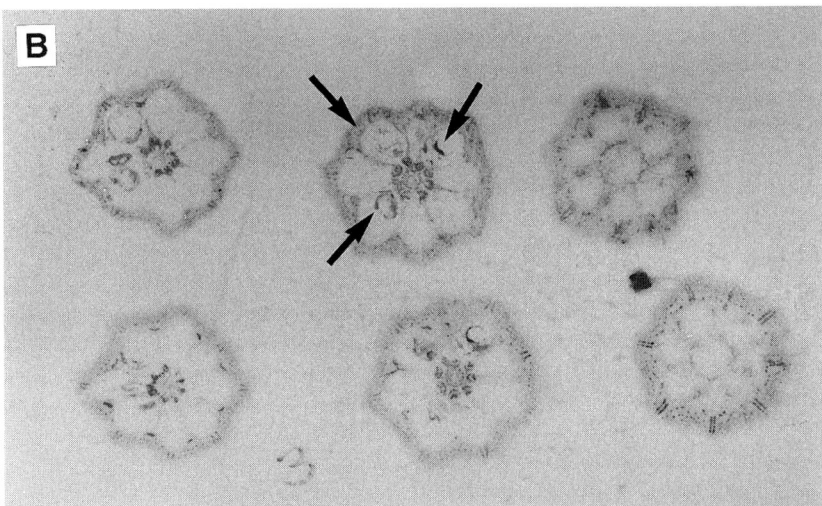

Fig. 2. Tissue-print immunoblots of cross sections from okra fruit pods. Ponceau S protein staining of blots before immunodetection reveals the anatomical detail of sections subsequently probed with control serum (**A**) or with *Phaseolus vulgaris* IAA-peptide antibody (**B**). Arrows indicate the position of seeds remaining within the cross-sectioned pods.

a few minutes at most). Excess substrate solution is allowed to drain off the membrane, the edge then blotted briefly on a laboratory wipe, and the membrane placed in a clear plastic envelope, which is then taken to a darkroom in a film cassette.

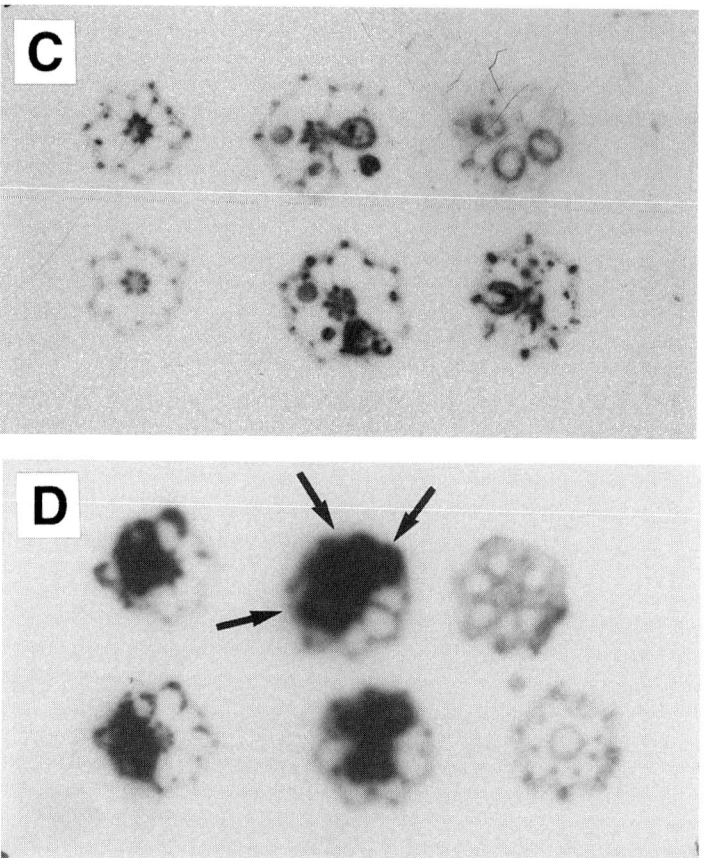

Fig. 2. (Continued). (**C**) Control protein tissue prints with antibody omitted. (**D**) Tissue-prints incubated with *Phaseolus vulgaris* IAA–peptide antibody. Cross-reaction visualized by chemiluminescence. Arrows indicate strong localization of the IAA-peptide in seeds.

7. In the dark, film is placed on top of the envelope and the cassette closed so that the film is pressed tightly against the plastic-covered membrane.
8. Usually, several exposures are made starting with a minute or less and the film developed immediately to determine if longer exposures are necessary. Very short exposures can be made by placing a sheet of paper on top of the membrane, followed by a sheet of film, then rapidly withdrawing the paper followed by the film. Frequently, the glow from the chemiluminescence signal can be seen by eye after several minutes in the dark if the antibody titer is high enough.

2.2.3. NOTES

1. The use of preimmune serum is recommended, as typically there is some nonspecific background binding of antibody to protein on the blot. Com-

paring the protein stained with Ponceau S and the blot incubated with preimmune serum with the signal obtained with primary antibody will differentiate background from "true" signal.

2. Large amounts of IAA-peptide in the seeds present in the tissue section caused "bleeding" of the chemiluminescence signal leading to a loss of anatomical detail. This may be alleviated by decreasing the exposure time of the film to a blot. To decrease the amount of antigen transferred to the blot, rinse the freshly cut tissue section in distilled water for 3–5 s. Alternatively, blot serial sections on the nitrocellulose.

The reader is referred to ref. *(35)* for an overview of tissue printing and ref. *(36)* for a tissue-printing protocol for celery.

ACKNOWLEDGMENTS

The author is grateful for the generosity of Dr. Jerry D. Cohen and Dr. Janet P. Slovin, for without them this work would not be possible, especially Dr. Alexander Walz for technical assistance. The author is also indebted to the late Joe Varner for being my best mentor and friend and, most of all, for teaching me how to make lasting impressions.

REFERENCES

1. Daoust R. Localization of deoxyribonuclease in tissue sections. A new approach to the histochemistry of enzymes. *Exp Cell Res* 1957; 12: 203–211.
2. Daoust R. Histochemical localization of enzyme activities by substrate film methods: Ribonucleases, proteases, amalyses and hyaluronidase. *Int Rev Cytol* 1965; 18: 191–221.
3. Cassab GI, Varner JE. Immunocytolocalization of extensin in developing soybean seed coats by immunogold-silver staining and by tissue printing on nitrocellulose paper. *J Cell Biol* 1987; 105: 2581–2588.
4. Antikainen M, Griffith M, Zhang J, Hon WC, Yang DSC, Pihakashi-Maunsbach K. Immunolocalization of antifreeze proteins in winter rye leaves, crowns, and roots by tissue printing. *Plant Physiol* 1996; 110: 845–857.
5. Balsalobre JM. Mas P, Sanchez-Pina MA, Pallas V. Spatial distribution of acidic chitinases and their messenger RNAs in tobacco plants infected with cherry leaf roll virus. *Mol Plant Microb Interact* 1997; 10: 784–788.
6. Conley CA, Hanson MR. Cryostat tissue printing: An improved method for histochemical and immunocytochemical localization in soft tissues. *Biotechniques* 1997; 22: 490–496.
7. Cho HT, Kende H. Expansions and internodal growth of deepwater rice. *Plant Physiol* 1997; 113: 1145–1151.
8. Cohen JD, Illic N, Taylor R, Dunlap JR, Slovin JP. Cucurbitaceaes: evaluation and enhancement in cucurbit germplasm, in *Cucurbitaceae* (Lester CE, Dunlop JR, eds.), Gateway, 1995, pp. 104–109.
9. Helguera MA. Immunological detection of a GarV-type virus in Argentine garlic cultivars. *Plant Dis* 1997; 81: 1005–1010.

10. Henderson J, Bauly JM, Ashford DA, Oliver SC, Hawes CR, Lazarus CM, Venis MA, Napier RM. Retention of maize auxin-binding protein in the endoplasmic reticulum: quantifying escape and the role of auxin. *Planta* 1997; 202: 313–323.

11. Hood KR, Baasiri RA, Fritz SE, Hood EE. Biochemical and tissue print analyses of hydroxyproline-rich glycoproteins in cell walls of sporophytic maize tissues. *Plant Physiol* 1991; 96: 1214–1219.

12. Hilal M, Castagnaro A, Moreno H, Massa EM. Specific localization of the respiratory alternative oxidase in meristematic and xylematic tissues from developing soybean roots and hypocotyls. *Plant Physiol* 1997; 115: 1499–1503.

13. Hu JS, Sether DM, Liu XP, Wang M, Zee F, Ullman DE. Use of a tissue blotting immunoassay to examine the distribution of pineapple *Clostero* virus in Hawaii. *Plant Dis* 1997; 81: 1150–1154.

14. Jacinto T, McGurl B, Franceschi V, Felano-Freier J, Ryan CA. Tomato prosystemin promoter confers would-inducible, vascular bundle-specific expression of the beta-glucuronidase gene in transgenic tomato plants. *Planta* 1997; 203: 406–412.

15. Jinn TL, Chang PFL, Chen YM, Key JL, Lin CY. Tissue-type-specific-heat-shock response and immunolocalization of class 1 low-molecular-weight heat-shock proteins in soybean. *Plant Physiol* 1997; 14: 429–438.

16. Kausch KD, Handa AK. Molecular cloning of a ripening-specific lipoxygenase and its expression during wild-type and mutant tomato fruit development. *Plant Physiol* 1997; 113: 1041–1050.

17. Key JL, Lin CY. Tissue-type-specific heat-shock response and immunolocalization of class 1 low-molecular-weight heat-shock proteins in soybean. *Plant Physiol* 1997; 114: 429–438.

18. Mas P, Pallas V. Long-distance movement of cherry leaf roll virus in infected tobacco plants. *J Gen Virol* 1996; 77: 531–540.

19. Roth R, Boudet AM, Pont-Lezica R. Lignification and cinnamyl alcohol dehydrogenase activity in developing stems of tomato and popular: a spatial and kinetic study through tissue printing. *J Exp Bot* 1997; 48: 247–254.

20. Pereira S, Pissarra J, Sunkel C, Salema R. Tissue-specific distribution of glutamine synthetase in potato tubers. *Ann Bot* 1996; 77: 429–432.

21. Ramalho-Santos M, Pissarra J, Verissimo P, Pereira S, Salema R, Pires E, Faro CJ. Cardosin A, an abundant aspartic proteinase, accumulates in protein storage vacuoles in the stigmatic papillae of *Cynara cardunculus* I. *Planta* 1997; 203: 204–212.

22. Taylor R, Varner JE. The role of calcium in maintaining cell-wall firmness studied by physical printing on nitrocellulose membranes, in *Tissue Printing: Tools for the Study of Anatomy, Histochemistry, and Gene Expression* (Reid P, Pont-Lezica R, del Campillo E, Taylor R, eds.), Academic Press, New York, 1992, pp. 15–17.

23. Taylor R, Varner JE. Ascorbic acid location by tissue printing on nitrocellulose, in *Tissue Printing: Tools for the Study of Anatomy, Histochemistry, and Gene Expression* (Reid P, Pont-Lezica R, del Campillo E, Taylor R, eds.), Academic Press, New York, 1992, pp. 163–164.

24. Taylor R, Bailey BA, Dean JFD, Anderson JD. Translocation of fungal protein in plants, in *Tissue Printing: Tools for the Study of Anatomy, Histochemistry, and Gene Expression* (Reid P, Pont-Lezica R, del Campillo E, Taylor R, eds.), Academic Press, New York, 1992, pp. 43–58.

25. Varner JE, Ye Z. Tissue printing. *FASEB J* 1994; 8: 378–384.

26. Al-Wandawi H. Chemical composition of seeds of two okra cultivars [Emerald, Ibtaria, *Abelmoschus esculentus*]. *J Agric Food Chem* 1983; 31: 1355–1358.

27. Ames JM, MacLeod G. Volatile components of okra, in *Phytochemistry* 1990; 29: 1201–1207.

28. Baxter L, Waters L Jr. Chemical changes in okra stored in air and controlled atmosphere. *J Am Soc Hortic Sci* 1990; 115: 452–454.

29. Ndjouenkeu R, Goycoolea FM, Morris ER, Akingbala JO. Rheology of okra (*Hibiscus esculentus* L.) and dika nut (*Irvingia gabonensis*) polysaccharides. *Carbohydr Polym* 1996; 29: 263–269.

30. Bryant LA, Montecalvo J Jr, Moren KS, Loy, B. Processing, functional, and nutritional properties of okra seed product. *J Food Sci Off Publ Inst Food Technol* 1988; 53: 810–816.

31. Moawas FG, Wahab EI, Shehata FW, Abdel-Naiem FM. Preparation of some chemical components from okra stalks. *Ann Agric Sci Moshtohor* 1984; 21: 603–612.

32. Cook DA, Brown A III. Somatic embryogenesis and organogenesis in okra (*Abelmoschus esculentus* L. Moench.) *Somatic Embryogenesis and Synthetic Seed* 1995; 31: 164–169.

33. Bialek K, Cohen JD. Isolation and partial characterization of the major amide-linked conjugate of indole-3-acetic acid from *Phaseolus vulgaris. Plant Physiol* 1986; 80: 99–104.

34. Sambrook J, Fritsch EF, Mantiatis T. *Molecular Cloning: A Laboratory Manual*, 2nd ed., Cold Spring Harbor Laboratory Press, Cold Spring Harbor, New York, 1989, pp. 1867–1868.

35. Taylor R. Demonstration of tissue printing; an overview, in *Tissue Printing: Tools for the Study of Anatomy, Histochemistry and Gene Expression* (Reid P, Pont-Lezica R, del Campillo E, Taylor R, eds.), Academic Press, New York, 1992, pp. 5–7.

36. Taylor R, Inamine G, Anderson JD. Tissue printing as a tool for observing immunological and protein profiles in young and mature celery petioles. *Plant Physiol* 1993; 102: 1027–1031.

8 Dark-Field Microscopy and Its Application to Pollen Tube Culture

Tetsuya Higashiyama

1. DARK-FIELD MICROSCOPY

1.1. Principle of Dark-Field Microscopy

Although there are several methods to construct dark-field illumination, it is universal in dark-field microscopy that no light for illumination directly enters the light path of the objective lens (*see* Fig. 1A). Only the scattered light by the specimen enters the objective lens (*see* Fig. 1C). Specimen features thus appear bright on a dark background. When a powerful light source such as a mercury lamp is used for illumination, small objects can be detected even far below the resolution of light microscopy, as in fluorescence microscopy.

Dark-field illumination is classified into three types. The first one is for a microscope equipped with low numerical aperture (NA) objective lenses (*see* Fig. 1). To cast a "shadow" at the objective lens, a ring-slit as shown in Fig. 1B is inserted into the light path. The second is for high NA (>0.5) objective lenses. Special, ready-made dark-field condensers or lenses are used for dark-field illumination. The third is independent

From: *Methods in Plant Electron Microscopy and Cytochemistry*
Edited by: W. V. Dashek © Humana Press Inc., Totowa, NJ

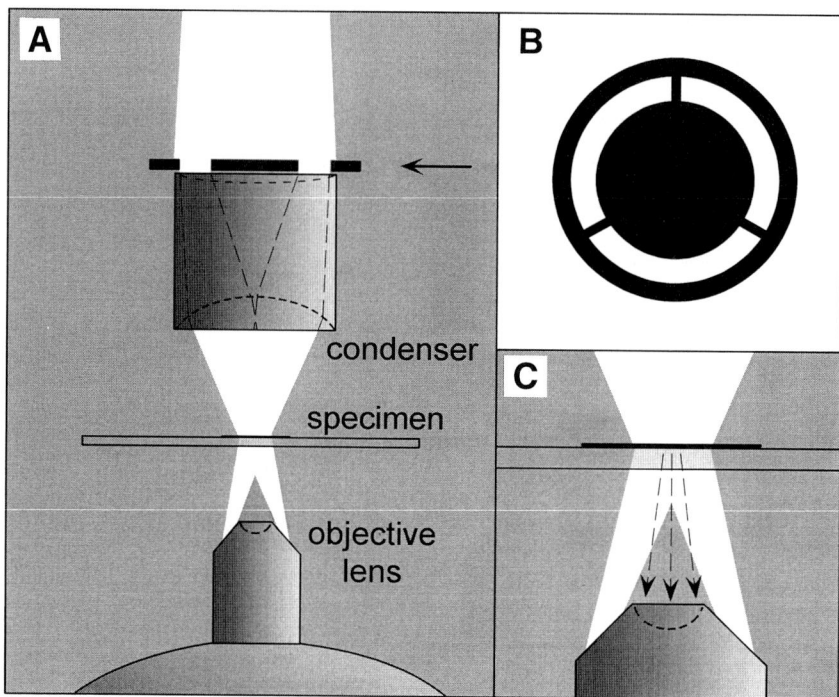

Fig. 1. Dark-field microscopy. (**A**) Schematic representation of a simple inverted dark-field microscope. An arrow indicates a ring-slit for dark-field illumination. In dark-field microscopy, no light (white region) directly enters the objective lens. (**B**) A ring-slit for dark-field microscopy. By simply placing a handmade slit on the condenser of an inverted microscope as in (**A**), the bright-field illumination is converted to the dark-field illumination. (**C**) High-magnification view of the dark-field microscope in (**A**). The surface of the objective lens is shaded by the ring-slit within the light path, and only scattered light (arrows with a broken line) from the sample enters the objective lens.

of the NA of objective lens; the specimen is illuminated vertically. Neither a specific condenser nor lenses are necessary when a specimen is illuminated vertically against the axis of the objective lens, because the illumination light docs not directly enter the objective lens.

1.2. Instrument Setup for Simple Dark-Field Microscopy

As described before, special built-in equipment is not always necessary for dark-field microscopy, and one can readily convert bright-field illumination to dark-field illumination. An inverted microscope equipped with long-distance (low NA) lenses is suitable for setup of a dark-field microscope. A handmade ring-slit is available when objective lenses such as ×4 lens with NA 0.13, ×10 lens with NA 0.30, and ×20 lens with NA 0.40 are used in combination with a long-distance condenser with NA 0.55. Remember that the condenser NA must always be higher than that of the objective lens. The long-distance condenser is suitable for observation of the specimen in the Petri dish. This simple dark-field microscopy by using an inverted microscope has been used for observation of pollen tubes *(1)* and *in situ* hybridization *(2)*.

2. APPLICATION OF DARK-FIELD MICROSCOPY TO POLLEN TUBE CULTURE

The pollen tube is a slender structure extruded from a pollen grain (haploid male gametophyte of flowering plants) after pollination. Pollen tubes grow down through the stigma, style, and ovary of the pistil to deliver nonmotile gametes into the female gametophyte within the ovule. Because it is difficult to investigate the physiology of the pollen tube in vivo (in the pistil tissue), pollen tubes are often cultured on the medium. In vitro analysis makes it possible to observe growing pollen tubes directly and, moreover, makes it possible to investigate physiological mechanism of the pollen tube for a definite culture condition. Dark-field microscopy is very useful especially when pollen tubes are observed at a low magnification to evaluate their growth *(1)*.

3. THE PROTOCOL

3.1. Instrument Setup for Simple Dark-Field Microscopy

1. Make a ring-slit as shown in Fig. 1B from a cardboard, and place it onto the condenser (*see* Fig. 1A). Suitable inner and outer diameters of the ring-slit are, for example, 24 mm and 33 mm, respectively, when a condenser IMT2-LWDNC by Olympus (Tokyo, Japan) is used. A built-in phase contrast slit is sometimes available for dark-field illumination. Use the largest slit (a slit for the most high-magnification objective lens).
2. In order to adjust the dark-field illumination, project the illumination light onto a piece of paper placed on the microscopic stage or on the objective lens. Adjust the position of the slit and the height of condenser so that the transmission light illuminates the sample uniformly and brightly, and that the diameter of the shaded area produced on the surface of the objective lens is larger than that of the objective lens (*see* Fig 1C).

3.2. Preparing the Culture Medium for Pollen Tubes

1. Make the medium shown in Table 1 (200 mL for 10 Petri dishes). For the growth of pollen tubes in vitro, calcium and borate are generally required. Sucrose is also added to the medium, mainly for the purpose to adjust the osmotic pressure; it is unclear whether the external sucrose plays a role

Table 1
Medium for Culture
of Pollen Tubes of Flowering Plants

Component	mg/L	% (w/v)
$Ca(NO_3)_2 4H_2O$	500	0.05
H_3BO_3	100	0.01
Sucrose [a]	50,000	5

[a] Concentration of sucrose should be changed according to the appropriate osmotic pressure of each plant species.

There is no need to adjust pH. Solidify this medium with 0.15% gellan gum.

as a carbon source. Appropriate osmotic pressure differs in plant species, e.g., 15% in tobacco, 15–20% in *Arabidopsis*, and 5% in *Torenia*. Thus, an appropriate concentration of sucrose should be determined at the onset of experiments. Other minor components such as magnesium, zinc, and copper are sometimes used for culture of pollen tubes. These components are not, however, essential for the growth of pollen tubes in most cases. At least these components are not the factors that should be examined at first.

2. Examine the pH after preparing the medium. When the pH is around 5.8, it is preferable that it is not adjusted. The pH of the medium shown in Table 1 is approximately 5.6. If there is a need to adjust the pH, use KOH and H_2SO_4; sodium (e.g., NaOH) must not be used because it often inhibits the growth of pollen tubes considerably.

3. After 0.15 % (w/v) gellan gum (gelrite) is added, autoclave the medium for 20 min at 120°C.

4. Solidify less than 20-mL aliquots in 9-cm Petri dishes. Thinner medium is preferable for observation. Liquid media generally support the growth of pollen tubes in vitro well; however, liquid media are not suitable for observation of pollen tubes. Agarose or gellan gum (gelrite) is usually used for solidification of media. The concentration of these compounds is usually low (e.g., 0.8% of agarose), thus the media is soft. Gellan gum is preferable for dark-field microscopy because the medium becomes more transparent. Both plastic and glass dishes are available.

3.3. Cultivation and Observation of Pollen Tubes

5. Dust pollen grains of a plant of your interest on the medium by using forceps, and seal the dish. Pollen grains soon hydrate and begin to extrude a tube. Pollen tubes usually grow well between 20°C and 30°C. Pollen tubes of *Torenia* grow at a maximum rate of approximately 600 μm/h in vitro.

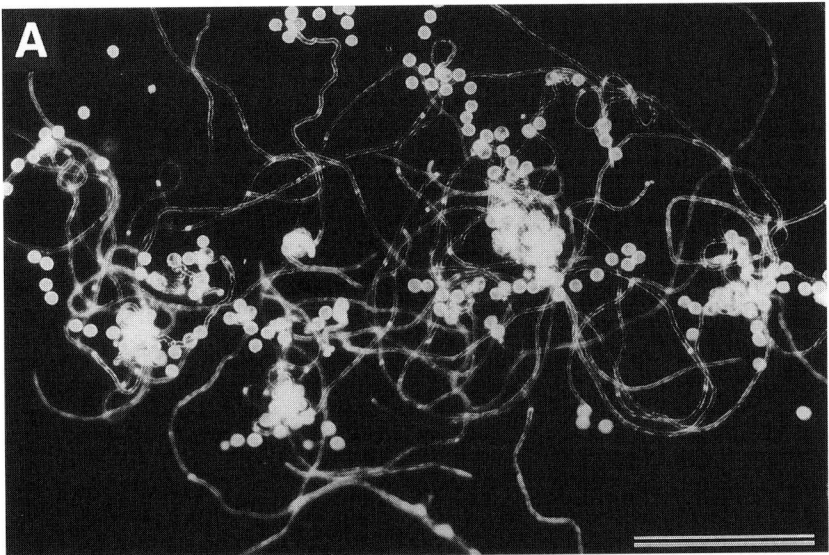

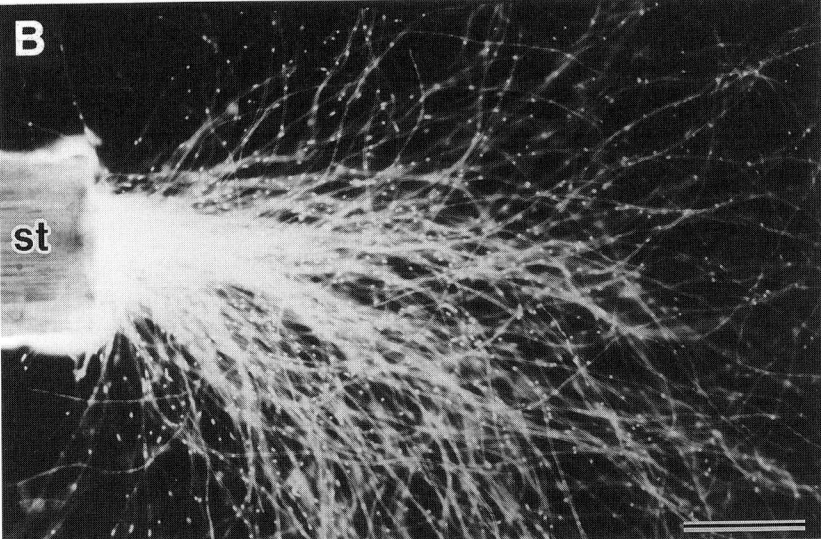

Fig. 2. Dark-field images of cultured pollen tubes of *Torenia*. (**A**) Pollen tubes germinated in vitro. Pollen grains are directly dusted onto the medium. *Torenia* is the only plant in which in vitro guidance system of the pollen tube to the embryo sac has been established *(1)*. (**B**) Semi-in vitro growth of pollen tubes. Pollen tubes grow through the cut style (st) and continue to grow on the medium after emergence from the cut end of the style. Semi-in vitro pollen tubes show different characteristics from those germinated in vitro. In *Torenia*, only pollen tubes growing semi-*in vitro* are guided to the embryo sac. Bars = 500 μm.

6. Observe pollen tubes by dark-field microscopy. Figure 2A shows dark-field images of pollen tubes of *Torenia*. Growth of pollen tubes can be evaluated readily under dark-field microscopy. To obtain high-contrast image, the illumination must be intensified. Scratches, fingerprints, dirt, dust, and collected moisture on the top of the dish destroy a good dark-field image.

3.4. Semi-In Vitro Culture of Pollen Tubes

7. While the culture medium supports the growth of pollen tubes germinated in vitro, semi-in vitro culture as shown in Fig. 2B is also possible. Cut the style after pollination, and place it onto the medium. Note that the cut end of the style is always attached to the surface of the medium. If the examined plant has self-incompatibility, use pollen grains and the style from different strains. Semi-in vitro pollen tubes show different characteristics from those germinated in vitro; compared with pollen tubes germinated in vitro, pollen tubes under semi-in vitro conditions elongated in straight lines, at higher growth rates, and with smaller diameters. Accidental bursting of pollen tubes occurred less frequently and, thus, pollen tubes became longer under semi-in vitro conditions. In *Torenia*, only pollen tubes growing semi-in vitro are guided to the embryo sac in vitro *(1)*.

4. GENERAL COMMENTS
FOR POLLEN TUBE CULTURE

When pollen tubes grow poorly or do not germinate, it might be difficult to judge whether the culture condition is inappropriate or the pollen of the examined plant species does not have an ability to extrude a tube in vitro. As described previously, an appropriate osmotic pressure is critical for the cultivation of pollen tubes. The concentration of sucrose should be investigated at first and at wide range. The phase of the medium is also critical for the growth of pollen tubes; liquid medium seems preferable for the growth of the pollen tube in most plant species. In some plant species, the medium solidified with gelatin supports the growth of pollen tubes, but the liquid medium and the agarose medium do not.

There are two types of pollen in flowering plants—bicellular and tricellular. Bicellular pollen consists of the vegetative cell and the generative cell. The generative cell divides into two sperm cells within growing pollen tubes after germination of the pollen. In tricellular pollen, on the other hand, the generative cell divides within the pollen grain during maturation of the pollen and, thus, each pollen grain already consists of the vegetative cell and the two sperm cells. Approximately seventy percent of flowering plants have the bicellular pollen *(3)*. The bicellualr pollen tends to grow in vitro (e.g., tobacco, lily, *Torenia*), whereas the tricellular ones do not (e.g.,*Arabidopsis*). Because tricellular pollen emerges at a phylogenetically later stage during evolution and the nearly all aquatic species with submersed flowers have tricellular pollen, it has been proposed that most of the tricellular pollen have acquired the ability to avoid accidental hydration *(3)*.

Two other additives have also been proposed to promote growth of pollen tubes. One is nontoxic buffer MES-KOH (e.g., 25 mM, pH 5.0), and the other is polyethylene glycol 4000 (PEG-4000). PEG-4000 is sometimes added to the medium and drastically promotes the growth of the pollen tube, whereas the mechanism of physiological action is largely unknown *(4–6)*. Note that osmotic pressure should be adjusted, when high concentration of PEG-4000 (e.g. 15% (w/v)) is added to the medium (15% PEG-4000 is similar in osmolarity to that of 6% sucrose). Unfortunately, a high concentration of PEG-4000 is not suitable for dark-field microscopy because it precipitates in agarose and gellan gum media (liquid medium is available).

REFERENCES

1. Higashiyama T, Kuroiwa H, Kawano S, Kuroiwa T. Guidance *in vitro* of the pollen tube to the naked embryo sac of *Torenia fournieri*. *Plant Cell* 1998; 10: 2019–2031.
2. Matsunaga S, Kawano S, Higashiyama T, Inada N, Kuroiwa T. Clear visualization of the products of nonradioactive in situ hybridization in plant tissue by simple dark-field microscopy. *Micron* 1997; 28: 185–187.
3. Brewbaker JL. The distribution and phylogenetic significance of binucleate and trinucleate pollen grains in the angiosperms. *Am J Bot* 1967; 54: 1069–1083.
4. Jahnen W, Lush WM, Clark AE. Inhibition of *in vitro* pollen tube growth by isolated S-glycoproteins of *Nicotiana alata*. *Plant Cell* 1989; 1: 501–510.
5. Taylor LP, Hepler PK. Pollen germination and tube growth. *Annu Rev Plant Physiol Plant Mol Biol* 1997; 48: 461–491.
6. Higashiyama T, Kuroiwa H, Kawano S, Kuroiwa T. Explosive discharge of pollen tube contents in *Torenia fournieri*. *Plant Physiol* 2000; 122: 11–13.

9

Computer-Assisted Microphotometry

Howard J. Swatland

1. INTRODUCTION

For almost any experiment that involves a light microscope, quantitative data to test a hypothesis may be obtained by microphotometry. Add a photometer and a few accessories to a light microscope, and it may be possible to quantify a cytochemical reaction product, measure the spectrum of a pigment, quantify natural or induced fluorescence, quantify birefringence, or map and analyze an image. But manually collecting thousands of numbers is unbelievably tedious. The solution is obvious. Use a personal computer (PC) to collect the numbers.

From: *Methods in Plant Electron Microscopy and Cytochemistry*
Edited by: W. V. Dashek © Humana Press Inc., Totowa, NJ

This chapter provides an overview of the computer-assisted light microscope (CAM). The prerequisites for getting started are competence in light microscopy and in programming a PC using a simple high-level language such as BASIC. Neither skill is particularly difficult, but the details will depend on the hardware available to the reader. One reader may be poised, soldering iron in hand, ready to connect a familiar PC to a well-used microscope, whereas another reader may be cowering before the stack of an expensive, commercial CAM. Hopefully, a general overview will give them both some idea of what to do next. Suggested reference works for the prerequisites are Piller *(1)* for the fundamentals of light microscopy and microphotometry, Pluta *(2)* for microscopy theory and accessories, Hartshorne and Stuart *(3)* for polarized light microscopy, Harris and Bashford *(4)* for practical spectrophotometry and spectrofluorometry, Russ *(5)* for image analysis, and Nickalls and Ramasubramanian *(6)* for PC programming and interfacing.

2. MAIN COMPONENTS OF THE CAM

Finding your way around a CAM may require a little detective work (*see* Fig. 1). Start at the top, where most likely you will find the photometer (Fig. 1, 1), with high-voltage cables if it is a photomultiplier. Somewhere below that may or may not be a filter wheel (Fig. 1, 2) for stray light from a monochromator (Fig. 1, 3). Sometimes the monochromator is a rectangular box with manual settings for the bandpass (slit-width), or it may be an elongated continuous interference filter (which will not need stray-light filters or have a variable bandpass). Somewhere below these components for spectrophotometry will be a unit containing a shutter (to protect the photometer from bright light while you are using the microscope in a normal mode to look for a suitable specimen) plus different sized apertures (Fig. 1, 4). Behind the eyepiece you look through (Fig. 1, 20) will be some type of beam splitter (Fig. 1, 5). A simple splitter might have just a flappable mirror (like a single-lens reflex camera) enabling you to look at a specimen and then measure it (it will disappear while you measure it). A more sophisticated splitter might use a hemispherical mirror with a half-silvered aperture in the optical axis. Thus, the sample will always be visible, and measurements will be taken through the dim circle in the center of the field. Below this, if the microscope is fitted with accessories for polarized light, may be found a rotary analyzer, possibly with some type of compensator (a thin slider with a quartz window) below it (Fig. 1, 6). If the microscope is set up for epifluorescence, there will be a dichroic beam splitter (Fig. 1, 7) above the objective lens (Fig. 1, 8). Epifluorescence accessories may include an arc lamp (Fig. 1, 16), a shutter (Fig. 1, 17), a monochromator (Fig. 1, 18),

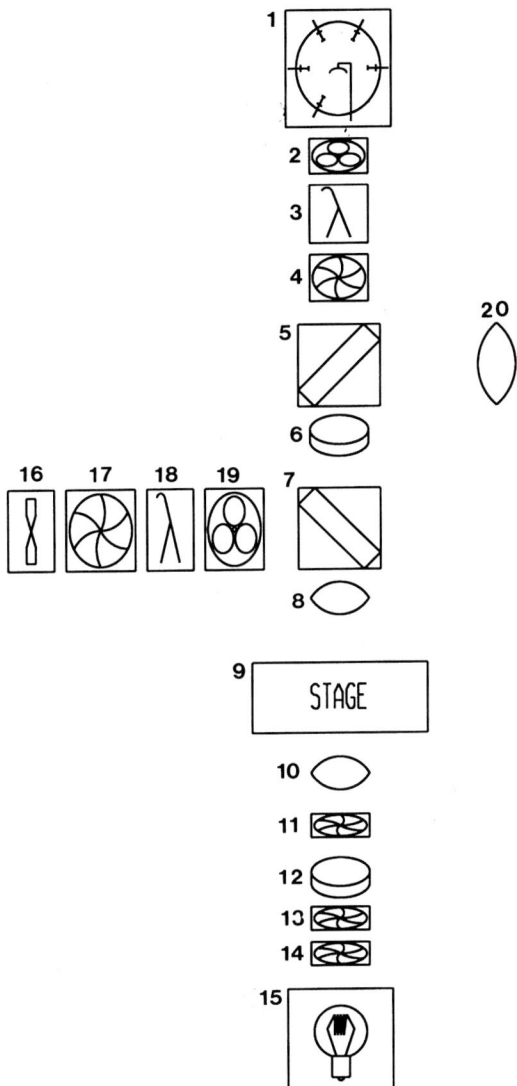

Fig. 1. Typical locations for CAM components, showing the photometer, 1; filter wheel, 2; monochromator, 3; shutter and aperture unit, 4; beam splitter, 5; accessories for polarized light such as a rotary analyzer and a compensator, 6; beam splitter for epi-excitation fluorescence, 7; objective lens, 8; stage, 9; substage condenser, 10; condenser aperture, 11; polarizer, 12; field aperture for photometry, 13; shutter, 14; primary illuminator, 15; arc lamp, 16; shutter, 17; monochromator, 18; filter wheel, 19; and ocular, 20.

and stray-light filters (Fig. 1, 19). If the fluorescence pathway is not equipped for computer-assisted scanning, the monochromator may be replaced by heat and low-pass excitation filters, and there will be a barrier filter in position 6 of Fig. 1. The stage (Fig. 1, 9) could be an ordinary square stage with knurled knobs to move the slide, a rotary stage if it is a polarizing microscope, or a scanning stage complete with two stepper motors and a joystick for you to move the specimen. Beneath the stage is the substage condenser (Figs. 1 and 10) which you must learn to operate properly in order to obtain Köhler illumination. Accessories for phase contrast or differential interference contrast (DIC) may be found in the substage region, but most of the time they will be offline and we will be using just the simple iris diaphragm for the condenser (Fig. 1, 11). In this region, too, you may find a polarizer (Fig. 1, 12), if it is a polarizing microscope (or if the microscope has differential interference contrast [DIC]), plus apertures and a shutter for photometry (Fig. 1, 13 and 14). The shutter will block the primary illuminator (Fig. 1, 15), whereas the aperture will block stray light around the edge of the measured field. Thus, the aperture at Fig. 1, 13, should appear larger in the field of view than the measuring aperture higher up (Fig. 1, 4). If the microscope is set up for diaexcitation fluorescence, components analogous to those at positions 7, 16, 17, 18, and 19 in Fig. 1 may be much lower, on the laboratory bench, and may replace or alternate with the primary illuminator and shutter (Fig. 1, 14, and 15).

3. INTERFACING

For building a CAM from scratch, the key point is to choose a suitable interface. The least expensive and easiest to improvise is RS-232, but remember that microphotometry is more than simply feeding the output of a photometer to an analog to digital (A:D) board. At least one shutter is needed. Other actuators will depend on the experiments planned, but probably will involve stepper motors or servomotors. Thus, with both talkers and listeners on the bus, the extra cost and complexity of a general purpose interface bus (GPIB or IEEE-488) is worthwhile. At the high end of cost and performance, VXI is the way to go.

If these terms mean nothing, then briefly: the RS-232 (Recommended Standard 232) is used for serial, asynchronous data transfer through two wires (one in each direction, and both relative to ground) and usually requires a handshaking protocol to make the talker wait for the listener. There are many different handshaking protocols, so you must find which method is used by the peripheral device (as well as baud rate, byte size, parity, and stop bits), then program the PC to deal with it. RS-232 handshaking may be hardwired (using lines 4, 5, 6, 8, or 20) or may be programmed using XON/XOFF. Decimal 17 (coded as DC1, for device control 1, with a hexadecimal value of 11h) is XON, while decimal 19

(DC2, hex 13h) is XOFF. The controller sends XOFF to stop a peripheral sending data, and XON to resume.

The more sophisticated GPIB hardware consists of a 24-pin connector with knurled knobs for tightening by hand and a screwdriver slot for loosening. It has 16 parallel data lines accessed by OUTPUT and ENTER statements in software, plus 3 handshake control lines, and 13 management lines accessed by SEND statements. With odd problematic exceptions, everyone uses the same handshaking protocol that can become invisible to the programmer. The GPIB may interconnect up to 15 devices, but one of them must be designated as the active controller. The peripherals may be listeners, talkers, or both, with their addresses set from hardware switches. Remember that some of the peripheral devices used in a CAM may be operable in local mode from their own front panels. Control of each or all of them may be assumed from software with the REMOTE statement and released by a LOCAL statement. The GPIB may be operated with various types of maskable interrupts (temporarily ignorable) and nonmaskable interrupts (top-priority emergencies) using the service request (SRQ) line (pin 10) of the GPIB. Interrupts enable the active controller to respond to situations arising peripherally, such as a photomultiplier becoming saturated (after which it may be too noisy to use for hours or even days). The active controller can solicit a status byte response from all peripheral devices with a parallel poll (PPOLL) statement and, if necessary, obtain further information from an individual device by means of a serial poll (SPOLL) statement.

VXI is a system used by test and measurement engineers. VXI is a noiseproof version of the standard VME bus (Versa Module European) found in PCs (VXI = Vme bus eXtension for Instrumentation). The cards to operate the CAM are located in a mainframe separate from the PC. It is expensive and sophisticated, but easy to operate.

To summarize, interfacing may involve cards in the PC connected to the CAM, perhaps using RS-232; or standalone boxes (like a photometer) connected to the PC using GPIB; or a PC plus a VXI mainframe. For our reader unfamiliar with a commercial CAM, get someone to demonstrate the system, then label the cables. The most common cause of failure is an improperly connected cable.

4. SOFTWARE

When programming a homemade system or replacing the restrictive software of a commercial CAM, the goal is to write general purpose software to collect and analyze data for any conceivable experiment or configuration of the CAM. This is more difficult than writing software for a single situation, but soon will minimize delays between experiments, maximize the return on time and effort, and clarify the logical design of experiments. First, find the similarities of a wide variety of experimental situations. For example, data collection generally involves a series of step-and-measure operations, whereby a device such as a grating monochromator or polarized light analyzer is set to a start position and a measurement is taken. Then the remainder of a spectrum or data set is collected by incremental stepping and re-measuring. Thus, the same software may be used for radically different operations, ranging from measuring a spectrum to scanning an image to form a matrix.

Second, decide on a common data format that will (1) identify the data, (2) record the operating conditions of the CAM, (3) be flexible on the x-axis to take wavelength for spectra, angular rotation for polarization, or distance for spatial scanning, and (4) allow multidimensionality. Measuring one spectrum is easy enough, but multidimensionality soon occurs in normal experiments. In fluorometry, for example, we may have excitation wavelength versus emission wavelength versus time. For an image, we may have x-axis position versus y-axis position versus absorbance versus sample pH. Do not assume that data always form an ascending series ($0°$ may come after $360°$ when rotating a polarized light analyzer).

In subprograms for operating scanning devices, the primary algorithm may include a "do and wait" operation (for a stepper motor) or a "do and look" operation (for a servo system). Try to avoid problems that could arise if you move your software to a faster, newer PC. With a new PC running at a faster clock speed than an old one, the faster "do" operation will require a longer "wait" if the scanner operation is matched to the duration of "do + wait." Thus, the CAM should be treated as three separate units: the microscope, the interface, and the PC. The PC will become obsolete before the microscope.

For a commercial CAM, having worked through the instruction manual, if things go wrong, look for a feature on the microscope altered by a previous user but invisible to the PC. This may involve a step-by-step trace of the light path, making sure that light reaches the photometer but, equally important, also making sure that a shutter stops the light when the software reads the dark-field output of the photometer.

5. ILLUMINATION

It is essential to use Köhler illumination, which is a method for adjusting the illumination pathway to achieve maximum resolution and minimum chromatic error. The information is in the documentation for your

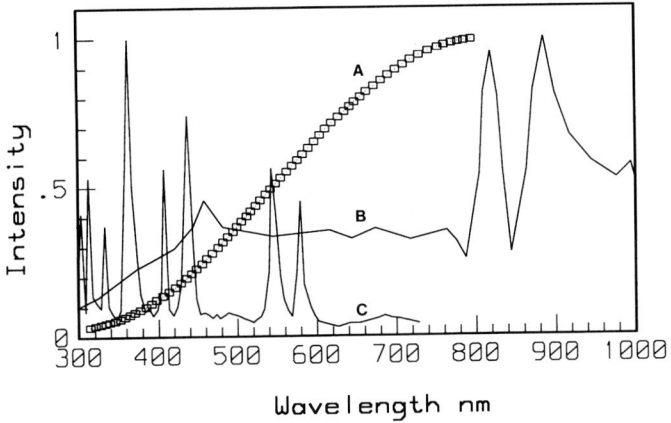

Fig. 2. Relative radiant intensity of halogen (**A**), xenon (**B**), and mercury sources (**C**).

microscope and is essential reading. The only exception is when the specimen is in an unavoidably thick sample chamber (e.g., to keep it alive or alter its environment), in which case, do the best you can, but be aware that you have lost resolution and spectra may change with depth of focus.

Different types of microscope photometry require different types of illuminators. The halogen lamps found in most research microscopes are excellent for spectrophotometry from about 420–700 nm (visible light), and even above 700 nm (into the near-infrared), but they seldom emit enough light for the lower wavelengths of visible light (down to 400 nm) or for UV (below 400 nm). It is important to know the emission spectrum of the illuminator. If it has sharp peaks, it will emphasize any small errors from the monochromator stopping at a particular wavelength on the side of a peak. A halogen source (*see* Fig. 2, A) causes the least problem in this regard, but the spectral range is limited for practical purposes. The spectrum of a xenon source (often termed XBO) has a couple of nuisance peaks, but is otherwise fairly flat (*see* Fig. 2, B). A mercury source (often termed HBO) is terrible for scanning (*see* Fig. 2, C), but provides high levels of UV at certain peaks and so is excellent for exciting fluorescence (but not for measuring excitation spectra). Some examples are shown in Fig. 2. Precise information is best obtained from the manufacturers of the sources you are using. To switch from one source to another, it is convenient to buy or build a mounting rail with a rotating mirror. Xenon lamps tend to produce ozone, so good ventilation or an extractor fan above the lamp may be required.

The illuminator must have a stabilized DC power supply, otherwise fluctuations in intensity will appear in the signal from the photometer.

If you have doubts about stabilization, examine the power output with an oscilloscope while it is operating. Remember to turn the system on early and let the illumination system get hot before making any measurements, especially if there are nonsoldered connections in the wiring (resistance may change) or supplementary filters in the light path (they may move). Modern programmable DC power supplies may be excellent (such as Hewlett-Packard 6642A), but remember that the emission spectrum will change with the current and remember to wait for the spectrum to stabilize if the current is changed. On a low budget, try a battery charger working through a 12-V automobile battery. As in color photomicrography, changes in illumination intensity are best made using neutral density filters (stopping equally all wavelengths in use). Also, try fine wire-mesh in the illumination pathway.

6. SHUTTERS

At least one shutter is needed to find the dark-field output of the photometer, but it can be a fake shutter (a prompt to the operator to block the light path). The dark-field reading is subtracted from all measurements, both of the standard (e.g., an empty space next to a specimen used to set 100% transmittance) and the specimen. Other shutters are useful for (1) protecting the photometer from bright light when the operator searches for a specimen, (2) protecting the specimen from continuous illumination between measurements, (3) switching in a known source of white light when adjusting fluorescence emission spectra, and (4) checking the level of ambient illumination. Shutter closures or openings are not instantaneous, and a pause may be required in software. Solid, swing-in solenoid shutters may become erratic as they age, and gravity may affect speed of closure or opening (depending on how the shutter is oriented in the light path).

7. TYPES OF PHOTOMETERS

With bright illumination, a large measuring aperture and a low-power objective, quite reasonable measurements may be made with an inexpensive photoresistor costing a couple of dollars. In a photoresistor, light energy removes electrons from their parent atoms so that the free electrons can carry a current and reduce overall resistance. However, with dim light, a small measuring aperture or a high-power objective, a photomultiplier may be needed. In a photomultiplier, photons strike a photocathode immediately inside a window (usually located on the side of a vacuum tube). This causes the photoemission of electrons, which are then focused by an electrostatic field and accelerated toward the first dynode. Each high-energy electron causes the emission of two or more secondary electrons, which, in turn, accelerate toward the next dynode, which is at a higher positive voltage than the last. An anode collects the current, in front of the last dynode with the highest voltage in the series. Side-window photomultipliers in a CAM are normally used to a maximum of approximately 1.2 kV (so be very careful with the electrical connections!). The dark-field current from thermionic emission at the cathode may become quite substantial at high voltage, which is why a shutter is needed to measure it. The most widely used type of photomultiplier is the Hamamatsu R928HA.

Photomultipliers are rather old fashioned when compared with the charge-coupled devices (CCDs) now in widespread use for high-tech scientific apparatus as well as domestic video applications, but CCDs have not yet had much impact in the sort of simple microscope photometry we are reviewing here. In a photodiode array (PDA) formed from a matrix of photodiodes across a chip, exposure of each photodiode generates a current. The photovoltaic current discharges a capacitor, and the extent of the discharge is determined when the photodiode is read and refreshed. Photodiodes are scanned in sequence, and the output is fed to a video line. One advantage of a PDA is that, with numerous photoreceptors, it is possible to measure a whole spectrum almost instantaneously, instead of stepping through all the wavelengths mechanically with a monochromator. When a monochromator is coupled directly to a PDA for this purpose the whole assembly is called a spectrograph. So why are we still using vacuum tube technology?

The answer lies in the precise alignments required in microscopy. A microscope spectrophotometer is a very precise optical bench. Either it must be built in such a way that a near-perfect factory alignment can be maintained during shipping and everyday use, or the user must be given alignment possibilities and appropriate instruction. The correct alignment of all the elements leading to a side-window photomultiplier is relatively easy (although it requires knowledge, patience and common sense) because the side window is relatively large (≈ 2 cm^2). Coupling the optical axis of the microscope to the optical axis of the spectrograph, however, is far more difficult (although one can cheat by using optical fibers if the light intensity is high). Thus, the main problem is commercial. Right now, not enough CAMs are sold to amortize the cost of solving the problem. If microphotometry was popular technology with a mass market, the whole spectrograph would be mounted in the microscope ocular giving real-time spectra from the field of view through the ocular. Until this happens (if ever), most CAM programmers will be using something like the following protocol.

1. Create an appropriate blank (usually a clear area next to the specimen) and find the wavelength that gives the maximum photometer response.
2. Adjust the photometer to give almost a maximum response (say 90%).
3. Temporarily close a shutter in front of the light source and measure the photometer response (dark-field measurement).
4. Go to each wavelength and measure the photometer response.
5. At each wavelength, subtract the dark-field measurement from the photometer response measured in step 4 to get a blank measurement.
6. Place a sample in position and go to each wavelength to measure the photometer response.
7. At each wavelength, subtract the dark-field measurement from the photometer response measured in step 6 to get the sample measurement.
8. At each wavelength, express the sample measurement as a function of the blank measurement.

The main concepts in powering up a photomultiplier are as follows. Always avoid saturating the photomultiplier. At best, quit any operation before the photomultiplier reaches its maximum output (which you must find gently by trial and error one Friday afternoon, so that you can continue on Monday morning). At worst, create a nonmaskable interrupt to crash the program in the event of some unexpected event (like someone foolishly using a large measuring aperture and low magnification combined with an arc lamp, or manually taking the monochromator offline). A safe, but slow, method is to select the lowest amplifier gain, then slowly crank up the high voltage until you reach 90% output. If light intensity is too low, increment the amplifier gain and start from scratch with the voltage. Short cuts will soon occur to you once you become familiar with the operation of your particular hardware. As you gain experience, you may find it worthwhile to create an acceptance window for the photomultiplier output (e.g., $x\%$ saturation $\pm y\%$ range) to deal programmatically with large ranges in light intensity and noisy photomultipliers.

The spectral response curves of various types of photometers are usually fairly flat or smoothly rounded and do not normally create programming problems, but they must be known so that you do not stray beyond the usable dynamic range.

8. MONOCHROMATORS

There are three types of monochromators in common use, depending on the age and cost of the CAM. Least expensive and easiest to use is the continuous interference filter composed of very thin, alternating layers of strong-reflecting and weakly reflecting materials. These create interference effects so that only monochromatic light is transmitted. For an elongated wedge of interference layers, thin at one end and thick (approx 1 mm) at the other, the wavelength of transmitted light changes along the wedge. Thus, the whole wedge is moved across the light path to change the wavelength without altering the tube length (the distance from objective to eyepiece, normally fixed at 160 mm in older microscopes). A disadvantage is that, because the bandpass is determined by the photometric aperture, there may be a spectral shift across a low-magnification field of view (low wavelengths on one side of the field of view and higher wavelengths on the other). However, some advantages with a wedge are its neutrality to polarized light and a uniform transmittance (typically 20–25% from 400–800 nm).

A prism monochromator may be found on an older CAM. Prism monochromators tended to be very heavy, so they could only be used sitting on the laboratory bench in the illumination pathway. This is a disadvantage because ambient light may create a serious error when imposed on the weak monochromatic light passing through the monochromator. The only real use for prism monochromators now is for continuous fluorescence excitation with a strong mercury source (which sooner or later burns out a more fragile grating monochromator).

A third type of monochromator, using a diffraction grating, allows the monochromator to be placed above the microscope in the measuring pathway. This is the best location because it avoids problems from ambient light (which now is trivial relative to that of the microscope illuminator). Holographic interference diffraction gratings for a CAM are small and lightweight, consisting of straight, parallel, evenly spaced grooves in a layer of aluminum or gold. The angle of the grating is programmed to give the required wavelength, usually with a stepper motor as the actuator. The maximum spectral efficiency is given by the blaze wavelength specified for the grating, which relates to the spacing and cutting angle of the grooves (the grooves are triangular in cross section). Different gratings may be available for the same box, so it is important to know what is inside. Figure 3 shows smoothed curves for the grating efficiencies of two gratings, one for UV absorbance measurements or excitation (Fig. 3, A), and the other for general use from 300–1000 nm (Fig. 3, B). UV absorbance measurements are particularly useful in botanical studies for the identification of aromatic compounds in cell walls (7). Smoothing is needed for grating efficiency curves because they may show numerous irregularities some of which, such as Wood's anomalies, may be polarization dependent. A depolarizer may be needed at an appropriate point in the light path if spectral measurements are to be made of polarized light.

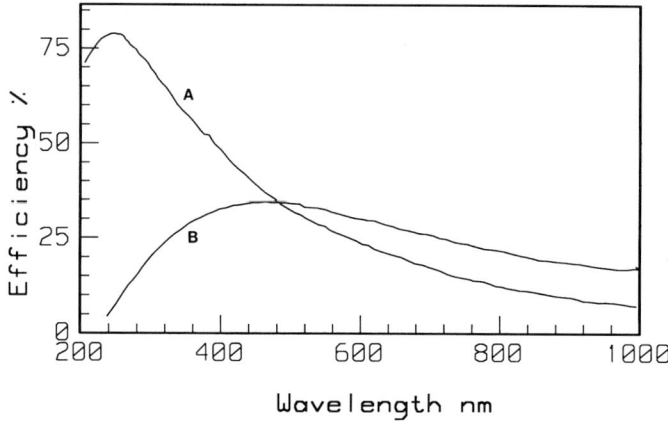

Fig. 3. Monochromator grating efficiencies for UV (**A**) and general-purpose use (**B**).

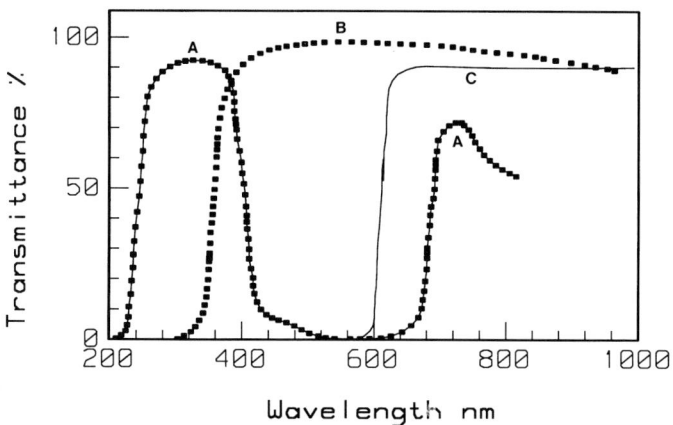

Fig. 4. Transmittance spectra of filters with a bandpass for UV (**A**), visible (**B**), and red to near-infrared (**C**).

Another important programming point is that a grating monochromator does not produce a single spectral peak output but a series of harmonic peaks. For example, with a blaze wavelength of 300 nm, a second-order peak also occurs at 600 nm, and a third-order peak at 900 nm. Thus, a grating monochromator on a CAM should always be used with filters to remove the higher-order harmonics, which are regarded as stray light. The programmer must decide at which wavelength to swing in a particular stray light filter. Thus, for the examples shown in Fig. 4, ascending in wavelength from 200 nm to 1000 nm, the change from filters A to B would be placed before 400 nm, probably at 380 nm.

The change from filters B to C at 700 nm is less critical. The curves in Fig. 4 may be different from those in the CAM you are using. Bandpass spectra must be obtained from the manufacturer's documentation or by direct measurement, especially if you change on a shoulder (i.e., where a small change in wavelength will produce a large change in transmittance).

9. DISTRIBUTIONAL ERROR

If a rectangular cuvette is filled with an evenly dispersed chromophore in solution, the photometric laws of Bouguer, Lambert, and Beer may be used to measure the concentration of the chromophore from the cuvette dimensions and optical absorbance. But suppose the chromophore has precipitated in the middle of the light path, then the photometric laws will fail, because of unabsorbed white light passing around the precipitated chromophore. Yet the same amount of chromophore is in the light path. This distributional error may be corrected using the two-wavelength method (8–10). The absorbance spectrum of the stained structure is measured in a situation where distribution is homogeneous and there is minimal light scattering. Choose two wavelengths, such that absorbance at one wavelength is half that at the other. When measuring a heterogeneous sample, measurements at these two wavelengths then are examined for their deviation from the photometric laws and corrected accordingly (11). Corrections also may be required for other deviations from the photometric laws, such as those caused by uneven sample thickness (12). These correction methods date back to a time before the PC when, for example, it was required to find the DNA content of a nucleus with minimum data collection through a large aperture. A more reliable solution with the CAM is to scan the nucleus (either by video or with a mechanical scanning stage) to integrate the results of a large number of measurements (pixels) in each of which there is minimal distribution error (because the pixel is very small relative to the pattern of heterogeneity of the sample). Both the two-wavelength method and pixel integration will require empirical calibration and constant checking, unless all that is needed is a method to separate haploid from diploid nuclei.

10. FLUOROMETRY

Fluorescence occurs when radiant energy is absorbed and then, almost instantly, some of the energy is re-emitted, usually at a longer wavelength. Primary fluorescence (autofluorescence) occurs in flavoproteins (13), plant cell wall materials such as lignin (7), and in flagella (14). Secondary fluorescence is when a material binds a fluorescent dye or fluorochrome *(15)*, or a fluorescent-labeled antibody (immunofluorescence) or antibiotic *(16)*. Other applications include measurement of intracellular pH from the emission spectrum of an indicator dye, such as fluorescein *(17)*.

It is important to distinguish between a fluorescence excitation spectrum (scanning through excitation wavelengths to find the peak response) and a fluorescence emission spectrum (scanning through the emitted fluorescence to find its peak). Fluorometry of an unknown fluorochrome requires the determination of a response surface (excitation wavelength versus emission wavelength). For most fluorochromes there is a single peak, which may be found quickly using the simplex algorithm.

Diaexcitation (with the excitation passing upward through the microscope slide) will require a slightly different protocol to epiexcitation (with the excitation directed downward through the objective lens onto the top of the specimen, which need not be sectioned). A useful principle when programming is to ask the operator what type of measurement is required, give the operator a plan of the components required, and ask for verification. The key point is that computer assistance is to facilitate data collection, not to totally automate the microscope. Movement of the parts that move frequently (shutters, monochromators, etc.) is automated, but infrequent movement of components and adjustment of the optics is left to the operator. Thus, the programmer should provide copious, helpful instructions for the operator.

Programming a CAM for fluorometry is far more complex than for spectrophotometry. Spectrophotometry is simple because it is based on the ratio of light in to light out. But fluorometry creates many of the problems associated with true radiometry—measuring the emission spectrum of an unknown source. The logic may become circular. Radiometry to determine an emission spectrum requires the relative spectral sensitivity of the photometer to be known, but how can this be determined without a source with a known emission spectrum? Fortunately, physicists in our national standardization organizations provide us with calibrated sources and photometers.

For the CAM programmer, most of the fluorescence emission spectra required are in the visible spectrum, which enables us to use a halogen source with a known emission spectrum to "show" the PMT what white light looks like. Thus, the spectrophotometer scans from 400–700 nm, usually in steps of 10 nm. The photometer reading at each wavelength is weighted by a factor corresponding to the emission spectrum of the source. For example, comparing only two wavelengths for the sake of simplicity, if emission at the first is only one tenth that at the second, the photometer response for the first wavelength is multiplied by 10, whereas the photometer response for the second wavelength is multiplied by 1. Thus, if this system were to be exposed to a hypothetical source with equal intensity at all wavelengths, the PMT should read the same for both wavelengths. Test your software by measuring the fluorescence emission spectrum of uranyl glass and comparing it to the manufacturer's technical data. Other standards are optical fibers doped with uranium or europium *(18)* or fluorescent pigments in a mounting medium *(19)*.

This short-cut protocol does not allow fluorescence spectra to be expressed in radiometric units or as quantum yields. Thus, the spectra are only relative spectra, and are often normalized (scaled to a maximum value of 1). Alternatively, split the excitation beam to direct 15% to a quartz cuvet containing Rhodamine B in ethylene glycol. This emits red light in proportion to the intensity of the excitation so that fluorescence may be expressed ratiometrically *(20,21)*.

11. VIDEO AND SPATIAL SCANNING

Many laboratories are equipped with commercial video frame grabbers and video image analysis systems, the operation of which is nicely explained by Russ *(5)*. Many of the concepts involved apply both to video analysis and to the analysis of matrices obtained with a mechanical scanning stage driven by a stepper motor. Video scanning is very rapid, but slow mechanical scanning can be very precise and accurate.

Programming instructions for scanning-stage stepper motors (or monochromator stepper motor, etc.) may contain a bit pattern that reflects the structure of the motor. For example, to step in one direction the POKE instructions to the motor address may be 3, 6, 12, and 9, which may look illogical until you consider the respective bit patterns, 0011, 0110, 1100, and 1001. Sending in reverse order will step the motor in the opposite direction.

The stator is the outer nonrotating part of the motor and the rotor is the inner rotating part. In a simple four-phase stepper motor, the stator is composed of two pairs of electromagnets on teeth facing each other across the axis, and the rotor is a permanent magnet. The activation of individual electromagnets in single-phase mode pulls the rotor into alignment.

Single-phase operation moves the rotor from 0°, to 90°, to 180°, to 270°, and to 360°, whereas the two-phase mode uses adjacent pairs of electromagnets so that rotor comes to a balance point between them, pulling the rotor from 45°, to 135°, to 225°, and to 315°. With pairs of electromagnets activated, the torque is increased. The half-step mode is a combination of both single phase and two phase, pulling the rotor from

0° to 360° in steps of 45°. For microstepping in a scanning stage, the current to electromagnets is varied so that the rotor reaches balance points other than the half-way position between pairs of electromagnets. This increases resolution but increases the cost and complexity of control circuits.

For stepper motors and more sophisticated actuators such as variable reluctance motors, there are typical operational problems in CAM programming, such as gear backlash. Small movements require the motor to be geared down. When rotating in one direction, all the gear teeth are pressing in one direction but, if the motion is reversed, some of the motor rotation may be used to take up the slack between the teeth. Thus, for accurate placement, it may be necessary to approach the end point of a programmed movement from a constant direction, sometimes going past the end point, then reversing back to it. If control of the scanning stage is passed to the operator for manual location of the sample, a slip clutch may be used to override the stepper motor. But excessive wear or loss of alignment may occur on delicate gears, or the clutch may slip so that rotor movements do not result in stage movement. When programming for open-loop control systems, there is no feedback on the stage or rotor position, and it is difficult to check programmatically that an operation has been properly performed. The motor must be capable of rapid rotation, otherwise stage movements are too slow. But if a stationary motor is instantaneously set to maximum speed, the mass of the microscope slide and friction cause a slow start and some of the first steps are missed. This is why it is preferable to hold the microscope slide in a separate jig rather than dragging it over the stage, why an operator joystick is preferable to a mechanical override of the stepper motor, and why automatic circuits for smooth acceleration to maximum velocity avoid a lot of extra programming.

With mechanical scanning, the measuring aperture is usually round so that the matrix matches the image. But with video, it is important to match the width:height aspect ratio of the video camera (such as 4:3) to the aspect ratio of screen and printer pixels, in order to maintain isotropic scaling (so that a circle does not become an ellipse when displayed or plotted). For the following figures of xylem vessels in a corn stalk (*see* Fig. 5), the frame-grabber had 256 gray levels in an area of interest (AOI) of 50×64 pixels. The objective was to create artificially large pixels to show clearly what is happening as we transform an image in the PC. To create graphical displays on a Laserjet printer, the gray level of each pixel was converted to black-and-white stippling, with the relative numbers of dots in the stippling pattern giving the gray level. In a mono-

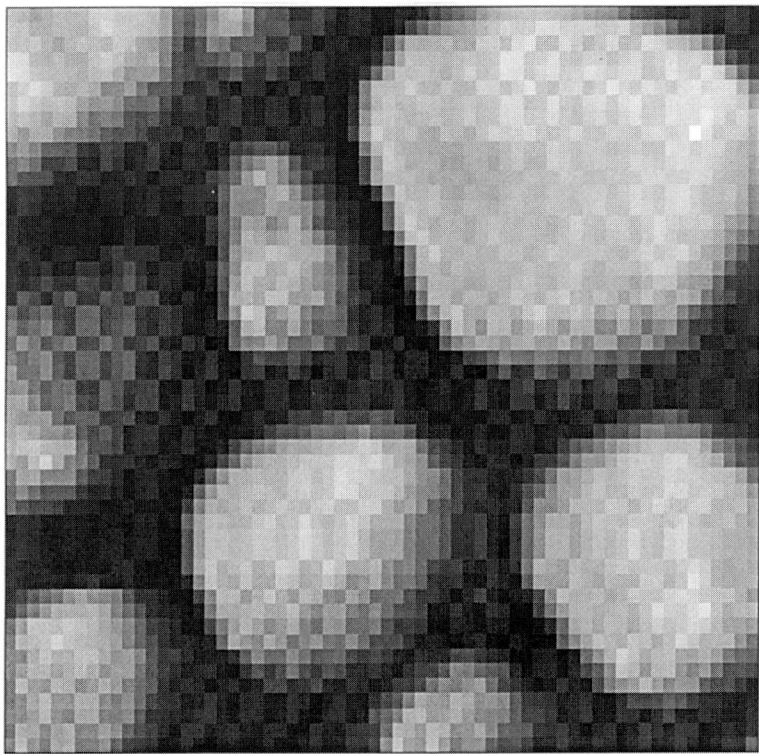

Fig. 5. Part of a video image of xylem vessels, exaggerating the size of the pixels and plotting them on a Laserjet printer.

chrome image composed of closely related gray levels, the pixels typically are grouped over a relatively small fraction of the range from 0–255, with none completely black (gray level 0) or white (gray level 255). By subtracting from the lower gray levels and adding to the higher gray levels, pixel intensity can be spread over the whole dynamic range from 0–255. Thus, when pixels are returned to their original positions in the image, the contrast is enhanced from black to white.

The noise in an image is really obvious if the data are plotted as a three-dimensional view, using pixel intensity in the z-axis (*see* Fig. 6), or by mapping contour lines drawn around topographical features (*see* Fig. 7). A video matrix may be smoothed with a kernel, which is a small matrix moved systematically over the image matrix, transforming the image by weightings specified in the kernel. For example, each pixel of the image may be averaged with the four adjacent pixels with which it shares a side. This simple method of neighborhood averaging has a kernel with the structure,

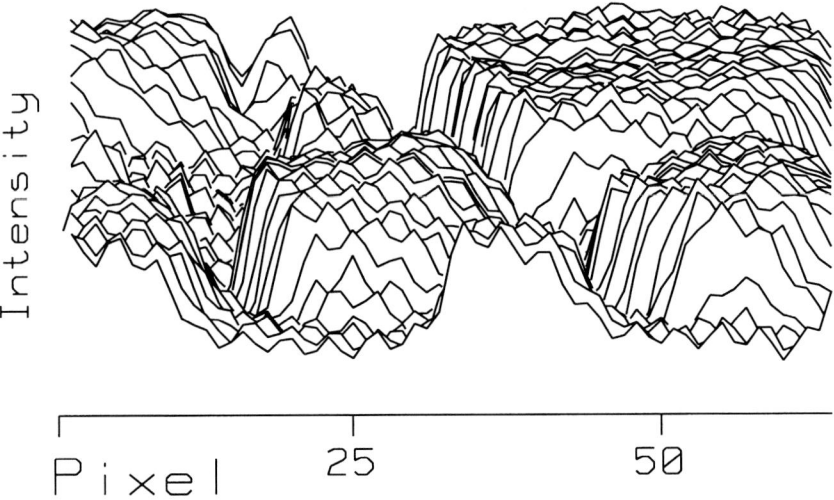

Intensity

Pixel   25   50

Fig. 6. A high level of interpixel noise seen in a three-dimensional plot of Fig. 5.

$$0\ 1\ 0$$
$$1\ 1\ 1$$
$$0\ 1\ 0$$

where the center point is the pixel being averaged with its four immediate neighbors: above, below, left, and right of center. Corner positions are unused (0 weighting), and the others have equal weighting (weighting = 1). The kernel moves across the image matrix so that each pixel, in turn, lies at the center of the kernel. The value at the center point is replaced by the average of the five pixels set to 1 in the kernel. However, on the outermost rows and columns of an image matrix, not all the kernel positions are available for averaging, so less pixels are averaged than for internal rows and columns. Working row by row down an image matrix it is important to use pixel intensities from the original or preceding image, and pixels that already have been averaged should not be reused. Thus, to reuse the existing matrix to store the newly processed image, a few extra rows (the height of the kernel) must be held in a memory buffer to avoid averaging data that have already been averaged, unless a recursive filter is being applied *(22)*. Compare Figs. 8–10, which were processed by a single pass of the kernel shown earlier, with the originals in Figs. 5–7, and you will see how useful this simple technique can be.

The weighting of the center pixel may be increased to reduce the chance of losing details, for example, giving the center pixel a weighting of 4, the major neighbors 2, and the farthest neighbors 1. The weighting of the center pixel may be increased for even greater preservation of a

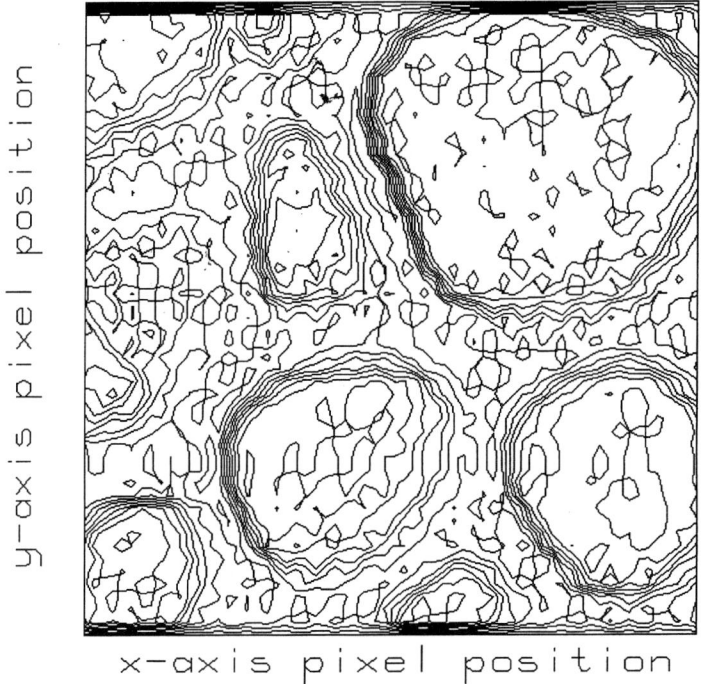

Fig. 7. A high level of interpixel noise seen in a contour plot of Fig. 5.

single pixel detail, but then it is better to increase the size of the whole kernel to maintain the averaging effect of canceling noise between pixels. Much larger kernels than these are found in commercial software, and the weighting of the center point may be Gaussian. With a median filter, instead of replacing the center pixel by the arithmetic mean, the pixels of the kernel are ranked in order, then the median value is used to replace the center pixel. Sometimes median filters may be better than averaging filters because they maintain the brightness differences between pixels and cause less movement of boundaries. If spots and small areas are the main subject of measurement, then a top-hat filter may be used to preserve information at the center of a kernel *(23)*.

Kernels may be adapted for sharpening boundaries or increasing the contrast between structures in an image, as in a Laplacian kernel.

$$-1\ -1\ -1$$
$$-1\ +8\ -1$$
$$-1\ -1\ -1$$

Moving the kernel across relatively smooth areas produces little effect, but a strong effect occurs at boundaries, giving sharper images for the

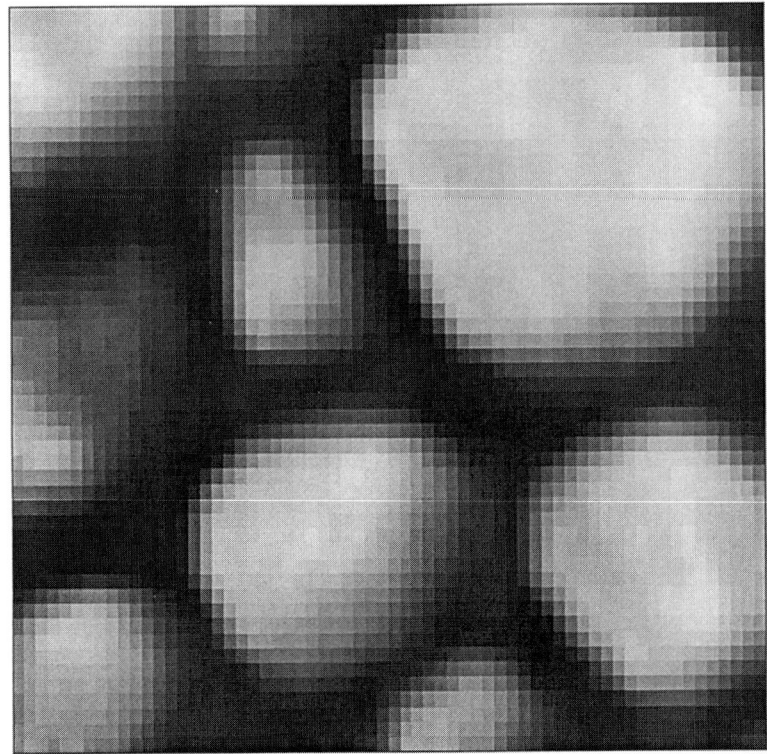

Fig. 8. Improvement of Fig. 5 by a single pass of a simple kernel.

human eye. However, this is only because the eye is very responsive to boundaries and there may be no improvement for computational analysis. Averaging kernels tend to act as low-pass filters, letting through broad shapes but filtering out noise, whereas Laplacian kernels tend to act as high-pass filters, missing broad shapes but emphasizing regional differences, which are easily influenced by noise.

12. POLARIZED LIGHT

Polarized light is useful for investigating birefringent structures such as plant cell walls, chloroplasts, and polysaccharide gums *(24–25)*. In birefringent structures, a transmitted light ray splits into two components traveling at different velocities, the ordinary (o) and extraordinary (e) rays, with $o \perp e$. Birefringence is measured as the refractive index (n) of the extraordinary ray minus that of the ordinary ray, and may be positive or negative in sign.

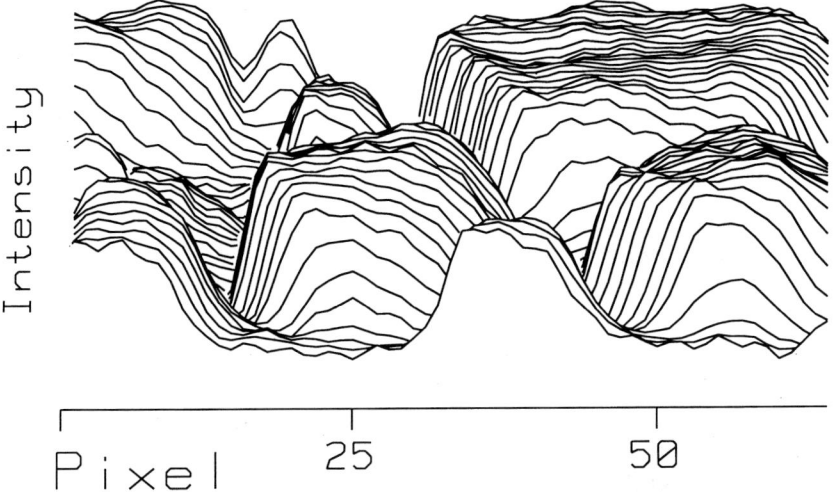

Fig. 9. Improvement of Fig. 6 after a single pass of a simple kernel.

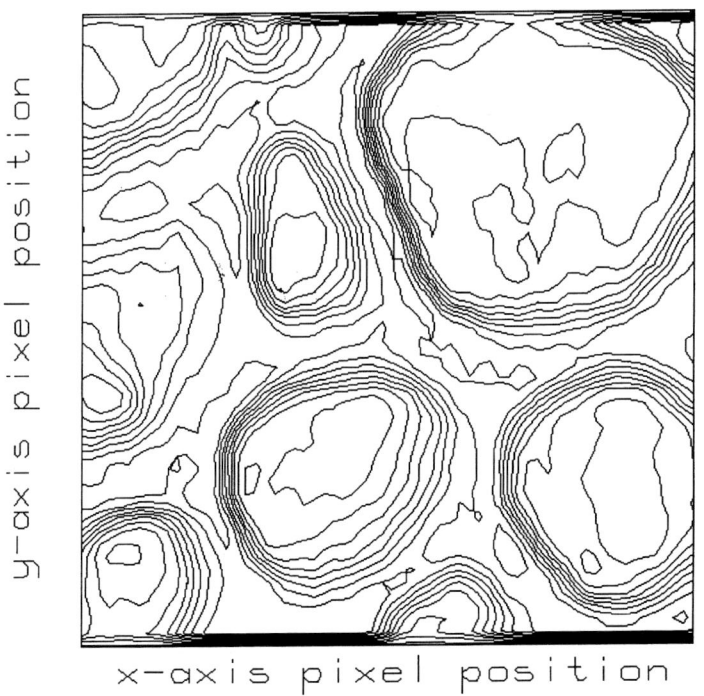

Fig. 10. Improvement of Fig. 7 after a single pass of a simple kernel.

In microscopy, an azimuth is an angle measured relative to a north–south axis of the microscope tube. Normally, the primary north–south axis divides the visible field into left and right sides and corresponds to a position of 0° on the first polarizer, usually below the substage condenser. Be careful if the orientation of the visible field has been altered by microscope accessories (such as cameras), which is why Bennett *(26)* defined the 0° axis relative to the stand of the microscope. From the 0° position, we follow the convention used in the mathematics of polar coordinates, moving counterclockwise to increment the angles. Points of the compass also are used to describe the orientations of components used for polarized light microscopy, and are abbreviated to N, S, E, and W.

Relative to specimens examined under the microscope, the α or fast axis corresponds to the direction of the minimum refractive index, the minimum dielectric constant, and the maximum velocity. The γ or slow axis corresponds to the maximum refractive index, the maximum dielectric constant, and the minimum velocity. Occasionally, a β axis is recognized with intermediate properties between α and γ. When working with elongated birefringent structures, birefringence usually is taken as positive when the γ axis is parallel to the longitudinal axis.

Retardation is a decrease in the velocity of light caused by an interaction with the medium through which the light is passing. Phase retardation is an interference caused by ordinary and extraordinary diverging and taking different paths through the specimen (one path longer than the other). When the rays recombine after passing through the specimen they are out of phase by an amount equal to the path difference (between long and short paths). Thus, the path difference depends on both the degree of birefringence (divergence of paths) and the thickness of the specimen.

$$\text{phase retardation} = (n_e - n_o) \times \text{thickness}$$

A polarizing microscope for transmitted light usually has a polarizer at a fixed azimuth beneath the substage condenser, and an analyzer in the microscope tube above the objective. Typically, a compensator is inserted at 45° beneath the analyzer. Some compensators (such as the de Sénarmont) are fixed in azimuth, and measurements are made with a rotary analyzer with a variable azimuth. For other compensators, the analyzer is fixed and the compensator is rotated or tilted to make measurements in the axis of the microscope.

Adding accessories for polarized light to a CAM enables us to use the photometer for an objective measurement of extinction positions (the darkest point when the analyzer is rotated). All that is required is to adapt the actuator driving the monochromator to drive the analyzer instead. For those researchers with generous funding and access to a commercial system, there are exciting new developments in this field. The Pol-Scope uses circularly polarized light so that birefringent structures appear equally bright at all orientations (27). With electro-optical modulators as compensators, and path differences being found electrically rather than electromechanically, ellipsometry can be undertaken on whole video fields, rather than on part of the field defined by an aperture.

13. COMMON CAM PROBLEMS

1. The primary illuminator is on, but you can see nothing down the microscope. Check for:
   a. A mirror sending the light in the wrong direction (rotate it)
   b. A solenoid shutter wrongly closed (you may have an untrapped error in the interface, clear the bus or turn off all the digital components and start again)
   c. The optical pathway not properly adjusted so that one or more apertures is off field (slide out all photometry apertures and check the Köhler illumination)
   d. A manual shutter to exclude ambient light reaching the photometer through the oculars (open it)
   e. A beam splitter sending all the light to the photometer or to a camera (change it)
   f. Crossed polarizers (rotate the analyzer)
   g. A monochromator in the illumination pathway set to UV or near-infrared (take it off line or change the wavelength manually)
   h. A monochromator in the illumination pathway with a very low or zero bandpass (increase the slit width)
   i. Inappropriate accessories for phase contrast or DIC in the light path (remove them and check Köhler illumination)
2. The CAM appears to be working and you can see the sample, but there are no data or the sample spectrum is flat. Check for:
   a. An inactive dark-field shutter so that the dark field is bright and subtracted from all data (check shutters or dark-field data)
   b. Beam-splitter not letting light to the photometer (change it)
   c. Monochromator off line (put on line, or check for spectral colors transmitted through it)
   d. Photometer is saturated, giving a maximum reading for all measurements (reset photometer and watch for excessive noise)
   e. With fluorometry, strong intrinsic fluorescence from dust or an inappropriate mounting medium is being subtracted from the sample measurements (check sample is considerably brighter than the background)

3. The data are unacceptably noisy or irregular. Check for:
   a. A ripple in the power supply to the illuminator (test against a battery-operated source or lengthen photometer integration time)
   b. A loose electrical connection to the illuminator (burnish and tighten all connections—sometimes they must be soldered, even to the base of the lamp)
   c. A dangling filament in the lamp bulb (replace lamps regularly)
   d. External interference in a homemade system (twist and shield all exposed wiring)
   e. Not enough light reaching the photometer (increase the size of the measuring aperture, use stronger illumination, decrease the magnification, use an objective with a higher numerical aperture, use immersion oil beneath the slide to increase the numerical aperture of the condenser, swing out any frosted glass in the light path, enlarge the aperture of the condenser, use quartz optics if working with UV, check any diaphragms or centering screws in high-quality objectives, increase monochromator bandpass, or use a thinner specimen)
   f. Noisy photomultiplier (give it a rest in the dark, use a standby, cool it down but beware of condensation on high-voltage wires)
4. Damage to the CAM may arise from:
   a. Touching illuminator bulbs with bare fingers
   b. Excessive exposure of gratings and filters to a strong arc lamp
   c. Positioning collector lenses too close to the source in an illuminator
   d. Movement or vibration of hot filaments
   e. Mounting arc lamps in a tilted position
   f. Not cleaning glass chips from microscope slides off a scanning stage
   g. Programming a device not protected by limit switches past its normal range
5. Damage to you could occur from:
   a. Retinal damage caused by removing heat or barrier filters when using an arc lamp
   b. Electrocution from wet, damaged, or worn-out high voltage wires to arc lamps or photomultipliers
   c. Changing the high-pressure burner of an arc lamp without using gloves and a full-face mask
   d. Continuous exposure to ozone from an arc lamp

14. CONCLUSIONS

A PC interfaced with a microscope photometer provides a powerful analytical tool working down to cellular and subcellular levels (with polarized light). With the relatively low cost of modern PCs and the wide availability of surplus components for light microscopy, the methodology is available to anyone who can combine microscopy with computer programming. For those with minimal funding, this is a way to have some fun and do good science. The main thing to remember is that most of the components you are dealing with have a spectral response curve. Look at them all together, and do not stray too far up or down the spectrum where energy levels are too low to make reliable measurements. For further information on wires to solder and programming concepts to develop, refer to Swatland *(28)*.

REFERENCES

1. Piller H. *Microscope Photometry*, Springer-Verlag, Berlin, 1977.
2. Pluta M. *Advanced Light Microscopy*, Vol. 1, *Principles and Basic Properties*, Elsevier, Amsterdam, Polish Scientific Publishers, Warsaw, 1988.
3. Hartshorne NH, Stuart A. *Crystals and the Polarising Microscope*, 4th ed., Edward Arnold, London. 1970, pp. 308–318.
4. Harris DA, Bashford CL. *Spectrophotometry and Spectrofluorimetry*, IRL Press, Oxford, UK, 1987.
5. Russ JC. *Computer-Assisted Microscopy. The Measurement and Analysis of Images*, Plenum Press, New York, 1990.
6. Nickalls RWD, Ramasubramanian R. *Interfacing the IBM-PC to Medical Equipment: the Art of Serial Communication*, Cambridge University Press, Cambridge, UK, 1995.
7. Ames NP, Hartley RD, Akin DE. Distribution of aromatic compounds in coastal Bermudagrass cell walls using ultraviolet absorption scanning microdensitometry. *Food Struct* 1992; 11: 25–32.
8. Swift H, Rasch E. Microphotometry with visible light, in *Physical Techniques in Biological Research*, Vol. III (Oster G, Pollister AW, eds.), Academic Press, New York, 1956, pp. 353–400.
9. Mendelsohn ML. Absorption cytophotometry: comparative methodology for heterogeneous objects, and the two-wavelength method, in *Introduction to Quantitative Cytochemistry* (Wied GL, ed.), Academic Press, New York, 1966, pp. 201-214.
10. Oostveldt P van, Boeken G. Absorption cytophotometry: evaluation of three methods for the determination of DNA in Feulgen stained nuclei. *Histochemistry* 1976; 50: 147–159.
11. Galassi L, Della Vecchia B. Selection of optimal transmittances with the cytophotometric two-wavelength method. *Basic Appl Histochem* 1988; 32: 279–291.
12. Chikamori K, Yamada M. Determination of dehydrogenase activity in tissue sections by tridensitometry and its applications. *Acta Histochem Cytochem* 1986; 19: 41–50.
13. Kunz WS, Gellerich FN. Quantification of the content of fluorescent flavoproteins in mitochondria from liver, kidney cortex, skeletal muscle, and brain. *Biochem Med Metab Biol* 1993; 50: 103–110.
14. Kawai H, Nakamura S, Mimuro M, Furuya M, Watanabe M. Microspectrophotometry of the autofluorescent flagellum in phototactic brown algal zoids. *Protoplasma* 1996; 191: 172–177.
15. Rost FWD. *Fluorescence Microscopy*, Cambridge University Press, Cambridge, UK, 1995.
16. Crissman HA, Oka MS, Steinkamp JA. Rapid staining methods for analysis of deoxyribonucleic acid and protein in mammalian cells. *J Histochem Cytochem* 1976; 24: 64–71.

17. Visser JWM, Jongeling AAM, Tanke HJ. Intracellular pH-determination by fluorescence measurements. *J Histochem Cytochem* 1979; 27: 32–35.

18. Velapoldi RA, Travis JC, Cassatt WK, Yap WT. Inorganic ion-doped glass fibres as microspectrofluorimetric standards. *J Microsc* 1974; 103: 293–303.

19. Walker PJ, Watts JMA. Permanent fluorescent test slides. *J Microsc* 1970; 92: 63–65.

20. Mayer RT, Novacek VM. A direct recording corrected microspectrofluorometer. *J Microsc* 1974; 102: 165–177.

21. Wreford NGM, Schofield GC. A microspectrofluorometer with on line real time correction of spectra. *J Microsc* 1975; 103: 127–130.

22. Zimmer H-G. Digital picture analysis in microphotometry of biological materials. *J Microsc* 1979; 116: 365–372.

23. Bright DS, Steel EB. Two dimensional top hat filter for extracting spots and spheres from digital images. *J Microsc* 1987; 146: 191–200.

24. Ruch F. Birefringence and dichroism of cells and tissues, in *Physical Techniques in Biological Research*, Vol. III, Part A (Pollister AW, ed.), Academic Press, New York, 1966, pp. 57–86.

25. Schorsch C, Garnier C, Doublier JL. Microscopy of xanthan/galactomannan mixtures. *Carbohydr Polym* 1995; 28: 319–323.

26. Bennett HS. Methods applicable to the study of both fresh and fixed material. The microscopical investigation of biological materials with polarized light, in *McClung's Handbook of Microscopical Technique*, 3rd ed., (McClung Jones R, ed.), Paul Hoeber, New York, 1950, pp. 591–677.

27. Oldenbourg R. A new view on polarization microscopy. *Nature* 1996; 381: 811–812.

28. Swatland HJ. *Computer Operation for Microscope Photometry*, CRC Press, Boca Raton, FL, 1998.

10 Isolation and Characterization of Endoplasmic Reticulum from Mulberry Cortical Parenchyma Cells

Norifumi Ukaji, Seizo Fujikawa, and Shizuo Yoshida

CONTENTS

1. INTRODUCTION

1.1. Endoplasmic Reticulum in the Overwintering Mulberry Tree

The endoplasmic reticulum (ER) is the largest organelle of the endomembrane system and it has many physiological functions, such as the synthesis of proteins and lipids, addition of oligosaccharide chains, transport of proteins and membrane materials, and the regulation of cytosolic calcium concentrations. Classical literature distinguishes three ER subcompartments: rough ER, smooth ER, and the nuclear envelope. However, it has been suggested that the ER has a multifunctional nature and that specialized subregions in addition to the three classical domains exist in plant ER *(1,2)*.

It is known that ER changes from flattened cisternae to small vesicles during seasonal cold acclimation in mulberry cortical parenchyma cells

From: *Methods in Plant Electron Microscopy and Cytochemistry*
Edited by: W. V. Dashek © Humana Press Inc., Totowa, NJ

(3). This vesiculated ER is mainly localized in the cortical cytoplasm. When the mulberry twigs were subjected to slow freezing at about −5°C, the vesiculated ER was changed into large flattened multiple lamellae *(4)*. The multiple lamellae were localized near the plasma membrane and did not fuse on further freezing. In onion cells, it was also shown that these morphological changes in the ER were induced by low temperature or cytosolic calcium levels *(5)*. On the basis of these observations, it is suggested that morphological changes in the ER are due to environmental stresses such as freezing and play an important role in protecting the cells against such stresses *(4)*. However, the precise function of the ER is not yet clear.

To identify the specialized features of the ER, we isolated the ER from mulberry cortical parenchyma cells every month from August to June. In this chapter, we described the method of isolation of ER from mulberry cortical parenchyma cells. At the end of this chapter, we show the results of our characterization of the protein component in ER during cold acclimation in mulberry cortical parenchyma cells.

1.2. Organelle Isolation from Plant Cells

The isolation of organelles or membranes is an important technique for the study of the physiological and biochemical mechanisms of the organelles themselves. Plant cells have characteristic organelles, such as vacuoles and chloroplasts. In addition, strengthened homogenization may be necessary for plants materials as compared with mammalian cells, as plant cells are surrounded by stacked cell wall. Consequently, the isolation of plant-cell membrane requires more steps. In mammalian cells, the "microsome" fraction, which is obtained by the ultracentrifugation of a postmitochondrial supernatant, mainly consists of the ER. However, in plant cells, the "microsome" fraction contains various organelle membrane vesicles such as the ER, the Golgi apparatus, tonoplasts, chloroplasts, and plasma membranes. To obtain an ER-enriched fraction from the microsome fraction, density gradient centrifugation is needed.

There are two density gradients for the separation of the microsome fraction: the linear density gradient and the step gradient. In cases in which the density of object organelle is already evident, step gradient centrifugation is available. In general, the step gradient centrifugation is faster and easier than the linear density gradient centrifugation. If the density of the object organelle density is not clear, linear density gradient centrifugation is effective in distributing the subcellular components according to their densities. Our purpose was to isolate an ER-enriched fraction from mulberry cortical parenchyma cells every month from

August to June. Because the density of the ER differs with the season, we collected ER by the "linear sucrose density gradient" method.

Most organelle membranes, such as the tonoplast *(6)* and the Golgi apparatus *(7)*, can be separated by density gradient ultracentrifugation of plant cell homogenates. However, other effective methods for the isolation of the plasma membrane *(8,9)* have been described. Moreover, another method that uses an aqueous two-phase system for the isolation of ER is also described *(10)*. Those interested in these details for these methods should consult the original articles.

2. PROTOCOLS

2.1. Linear Density Gradient Centrifugation and Fractionation

2.1.1. PLANT MATERIALS

We used one-year-old twigs of mulberry (*Morus bombycis* Koidz.) grown under field conditions as our experimental material. Mulberry trees are pruned before their leaves open in spring. As a result of pruning, many year-old twigs develop by August.

2.1.2. SOLUTIONS FOR HOMOGENIZATION AND MICROSOME PREPARATION

The compositions of the solutions for the isolation of mulberry ER are described as follows. The compositions of the solutions (in particular, the homogenizing solution) can be changed to fit the conditions of each plant material. For some plant species, it is necessary to modify the composition of the homogenizing solution, so a similar approach has been successfully utilized to isolate ER from a range of plant tissues. The basic components are (1) an osmotic balance component (sucrose fulfills this purpose), (2) a buffer condition (MOPS-KOH, pH 7.6), (3) a reducing agent (potassium metabisulfite, BTH), and (4) a protease inhibitor (PMSF). Some plant tissues can be homogenized in a solution containing only these four components. However, we added a further component to the solution. PVP (soluble) and PVPP (insoluble) are added to remove polyphenolic compounds, large amounts of which are present in woody plants. In the case of herbaceous plants, 1% PVP or PVPP is enough.

2.1.2.1. Homogenizing Buffer Solution

1. 300 mM sucrose
2. 75 mM MOPS-KOH (pH 7.6)
3. 5 mM EGTA-KOH (pH 8.0)
4. 2 mM EDTA-KOH (pH 8.0)
5. 3% (w/v) PVP (mol wt 24,500)

6. 10 μg solution mL^{-1} 2,6-di-t-butyl-p-cresol (BHT) *†
7. 5 mM potassium metabisulfite†
8. 1.5% (w/v) PVPP (polyvinyl-polypyrrolidone)
9. 2 mM PMSF†

2.1.2.2. Resuspending Solution

1. 10% (w/w) sucrose
2. 15 mM HEPES-BTP (pH 7.3)
3. 1 mM EGTA-KOH (pH 8.0)
4. 0.1 mM EDTA-KOH (pH 8.0)
5. 2 mM DTT†

2.1.3. METHODS OF HOMOGENIZATION AND PREPARATION
OF MICROSOMAL FRACTION

2.1.3.1. Preparation of Crude Microsomal Fraction. It is important to keep the temperature of the samples between 0°C and 4°C throughout homogenization, fractionation, and analysis.

1. Cool the bark tissues (25 g FW [fresh weight]) on ice and cut into small pieces (about 5 × 5 mm) with a razor blade.
2. Place small pieces of the tissue into 180 mL of the homogenizing solution.
3. Homogenize with Polytron PT20 (Kinematica, Lucerne, Switzerland) at medium speed for 90–120 s.
4. Pass the homogenate through three layers of gauze.
5. Pour the filtrate into centrifugation tubes and then centrifuge at 10,000g for 15 min. The cell debris and organelles such as nuclei, mitochondria, chloroplasts, and PVPP are precipitated in this step.
6. Collect the supernatant and then ultracentrifuge at 200,000g for 20 min to precipitate the crude microsomal fraction that contains the ER, Golgi complex, tonoplast, and plasma membrane.
7. Suspend the pellet in 15 mL of resuspending solution using a Teflon homogenizer.
8. Reultracentrifuge the resuspended crude microsome fraction at 200,000g for 20 min to exchange the buffer system.
9. Suspend the pellet in the resuspending solution and adjust the volume of the suspension to a final volume of 4.5 mL.

2.1.3.2. Comments

1. Because the water content of woody plant tissues is generally less than that of herbaceous plant tissues, 6–8 mL of homogenizing solution is

*Makes a 100 mg mL^{-1} stock solution in isopropanol
† BHT, potassium metabisulfite, PMSF and DTT are added just before homogenization, as these reagents readily inactivate.

needed for 1g FW of bark tissue for homogenization in this experiment. When herbaceous plants are used as the experimental material, the volume of homogenization solution should be reduced; e.g., 3 mL g FW^{-1} for wheat leaves, and 1 mL g FW^{-1} for mung bean hypocotyls.

2. The period required for homogenization also varies with the plant material. In general, 90 s may be enough.
3. To suspend a pellet of microsomal fraction, we used a glass pipet with a bent tip and a Teflon/glass homogenizer.
4. Most of the intact mitochondria and chloroplasts may be removed at the first centrifugation. However, a large amount of these organelles still existed in the microsomal fraction as estimated from the results of marker enzyme assays.

2.1.3.3. Preparation of Linear Sucrose Density Gradient

A. Solution I: 15% (w/w) sucrose in resuspending solution
B. Solution II: 50% (w/w) sucrose in resuspending solution

1. Place solution I and II in the gradient former (ISCO Model 570, Lincoln, NE, or Advantech GR-40, Tokyo, Japan).
2. Add 30 mL of 15–50% (w/w) gradient solution to the ultracentrifuge tubes of the swinging bucket rotor (max vol 40 mL). Sucrose concentrations increase from the top to the bottom of the tube.

2.1.3.4. Comments

1. The gradient tubes should be handled as gently as possible.
2. Linear density gradient centrifugation involves two methods: isopycnic density gradient centrifugation and sedimenting velocity density gradient centrifugation. In the isopycnic linear density gradient centrifugation adopted in this examination, ultracentrifugation is carried out until organelle sedimentation stops.
3. When discontinuous sucrose gradient centrifugation is used, two different density sucrose solutions should be piled up into the tube. If 20% (w/w) and 30% (w/w) density solutions are piled up, the ER-enriched fraction should be layered on the interphase between 20% (w/w) and 30% (w/w) sucrose solutions.

2.1.3.5. Centrifugation and Fractionation

1. Overlay 4 mL of crude microsomal fraction onto the linear sucrose density gradient solution.
2. Ultracentrifuge the tubes at 84,000g for 17 h in a swing rotor (Hitachi, RPS27-2, Tokyo, Japan). Note that at the beginning and end of centrifugation, acceleration and deceleration rates should be low so as not to disturb the gradient of the tubes.
3. After centrifugation, pick up the tubes from the bucket and place them on ice.

4. Fractionate the gradient solution into 1.2 mL aliquots by a gradient fractionator (ISCO, Model 640).
5. Place the fractionated samples on ice.
6. Assay the sucrose concentration in each fraction in a perfectometer.

2.1.3.6. Comments

1. Discontinuous step gradient is performed with a fixed angle rotor (Hitachi, RP50T-2). In the case of mulberry cortical parenchyma cells, ER-enriched fraction can be sufficiently separated enough by ultracentrifugation of 20% (w/w) and 30% (w/w) discontinuous sucrose gradients at 200,000g for 1.5 h.
2. The fractionator is not necessary. Place a hole in the bottom of the centrifuge tube, and the solution that dripped from the bottom will be collected into a test tube (same volume each tube).

2.1.4. SUGGESTIONS

1. In this protocol, the ER was collected as smooth ER, which is free from the attached ribosome, as a chelating agent is contained in the homogenizing medium.
2. To separate smooth ER and intact rough ER, an Mg^{2+}-induced density shift is used. The density shift can separate smooth ER and rough ER at different fractions for a linear gradient centrifugation. To perform an Mg^{2+}-induced density shift, 3 mM $MgCl_2$ is added to and EGTA is removed from the homogenizing solution. Furthermore, only 0.5 mM of EDTA is added to the solution. However, each organelle's membranes sometimes diffuses in the Mg^{2+}-induced density shift.

2.2. Marker Enzymes Assay

It is well-established that marker enzyme activity is an index of organelle (or organelle-related component) existence. In this experiment, assays of marker enzymes, which are localized in organelle membranes, are carried out as described below.

The marker enzymes used in this experiment are as follows: vanadate-sensitive H^+-ATPase (plasma membrane), nitrate-sensitive H^+-ATPase or pyrophosphatase (tonoplast), Triton X-100 stimulated-UDPase or IDPase (Golgi complex), antimycin A-insensitive NADPH cytochrome c reductase (ER), and cytochrome c oxidase (mitochondria inner membrane). NADH cytochrome c reductase activity is found to be 10 times higher than NADPH cytochrome c reductase activity. Chlorophyll content can be measured as the chloroplast marker. The chlorophyll content is calculated by the following equation. Before measurement, auto zero is performed at 750 nm.

$$\text{Chlorophyll (mg mL}^{-1}) = 8.05\text{Å} \times \text{ABS}_{663} + 20.29\text{Å} \times \text{ABS}_{645}$$

However, simple evaluation of choloplast contamination in the fractions can be performed visually from the color of the fraction.

2.2.1. SUBSTRATE SOLUTIONS FOR MARKER ENZYME ASSAY

2.2.1.1. Assay Reagents for Released Pi

**Stock Solution of H⁺-ATPase Substrate Mixture
for Vanadate- or Nitrate-Sensitive H⁺-ATPase***

40 mL	Final concentration in a working solution
91 mg ATP-2Na	3 mM
37 mg MgSO$_4$	3 mM
297.8 mg HEPES-BTP, pH 7.0	25 mM
186.3 mg KCl	50 mM
0.0375% (v/v) Triton X-100	0.03% v/v

* When nitrate-sensitive H⁺-ATPase (tonoplast) activity is measured, 250 mM (final concentration, 200 mM) KNO$_3$ is added to the stock solution of the substrate mixture. When vanadate-sensitive H⁺-ATPase (plasma membrane) activity is measured, 125 μM Na$_3$VO$_4$ (final concentration, 100 μM) is added to the stock solution of the substrate mixture.

Stock Solution of H⁺-Pyrophosphatase (PPase) Substrate Mixture*

40 mL	Final concentration in a working solution
33.4 mg sodium pyrophosphatase	1.5 mM
18.5 mg MgSO$_4$	1.5 mM
297.8 mg HEPES-BTP, pH 7.4	25 mM
180.3 mg KCl	50 mM
0.0375% (v/v) Triton X-100	0.03% v/v

* To prevent the precipitation of Mg^{2+} salts, MgSO$_4$ should be added to the PPase substrate mixture just before use.

Stock Solution of Triton-Stimulated UDPase Substrate Mixture

40 mL	Final concentration in a working solution
80 mg UDP-2Na	3 mM
36.2 mg MnSO$_4$	3 mM
297.8 mg HEPES-BTP, pH 7.4	25 mM
186.3 mg KCl	50 mM
0.0375% (v/v) Triton X-100.	0.03% v/v

Pi Assay Reagent 1 (reagent for termination of enzyme reaction)

200 mL
0.5 g (NH$_4$)$_6$Mo$_7$O$_{24}$
2.6 g sodium dodecyl sulfate (SDS)
2.8 mL H$_2$SO$_4$

Pi Assay Reagent 2 (reagent for color development)

100 mL

5.7 g NaHSO$_4$
0.2 g Na$_2$SO$_4$
0.1 g 1-Amino-2 naphthol sulfonic acid (ANS)

2.2.1.2. Other Assay Reagents for Marker Enzymes

Antimycin A-Insensitive NADH Cytochrome c Reductase

Assay Mixture

50 mL

13.5 mg cytochrome c
50 mM Potassium phosphate buffer, pH 7.4
8.8 mg sodium azide
75 µL antimycin A *, **

> * Stock solution of antimycin A (20 mM) is prepared in ethanol and stored at −20°C.
> ** Antimycin A should be added to the assay mixture just before use.

NADH Stock Solution

7.1 mg mL^{-1} NADH   10 mM
(10 mM NADPH solution is also available)

Cytochrome c Oxidase

Solution 1: 50 mM potassium phosphate buffer, pH 7.4
Solution 2: 1% (v/v) Triton X-100
Substrate solution: cytochrome c (reduced form)*

> * To obtain reduced cytochrome c, oxidative form of cytochrome c is reduced by
> ascorbic acid described as follows.

1. Dissolve 10 mg of cytochrome c in 0.5 mL of 50 mM potassium phosphate buffer, pH 7.4.
2. Add about 5–20 mg ascorbic acid to the cytochrome c solution and mix well.
3. Remove excess ascorbic acid by gel filtration (Sephadex G25, Pharmacia, Uppsala, Sweden) and collect the reduced cytochrome c fraction.
4. Bring the reduced cytochrome c solution to a final volume of 4.5 mL with 50 mM potassium phosphate buffer.

2.2.2. ASSAY OF MARKER ENZYME

2.2.2.1. H$^+$-ATPase, H$^+$-PPase, and UDPase.
Enzyme activities of vanadate- or nitrate-sensitive H$^+$-ATPase, H$^+$-PPase, and UDPase are measured by the quantitation of released inorganic phosphate from the substrate by enzymatic hydrolyzation. When H$^+$-ATPase activities are assayed, at least three substrate mixtures should be prepared: (1) a substrate mixture containing nitrate, (2) a substrate mixture containing vanadate, and

(3) a substrate mixture without any inhibitor. Vanadate-sensitive H^+-ATPase activity is calculated by the subtraction of the enzyme activity observed in (2) from that observed in (3). Nitrate-sensitive H^+-ATPase activity is calculated by the subtraction of the enzyme activity observed in (1) from that observed in (3).

1. Place 50 μL of each fraction into test tubes on ice water at 0°C. As a control, pour 50 μL of resuspending solution into test tube.
2. Add 200 μL of the assay mixture of each enzyme in the test tubes.
3. Transfer the test tubes to water bath at 30°C, mix well, and then incubate for 10 min
4. Immediately after incubation, transfer the test tubes to ice water, and then add 2.5 mL of Pi assay reagent 1.
5. After mixing well, add 0.5 mL of Pi assay reagent 2 to the test tubes and mix well.
6. Incubate the samples at 30°C for 10 min.
7. Pour the control sample solution into a cuvet, place into the spectrophotometer, and zero at 700 nm.
8. Measure the absorbance of each samples with a spectrophotometer at 700 nm.
9. Enzyme activity (μmol 10 min^{-1}) is calculated by the following equation.

$$0.714 \times OD_{700} = Pi\ mol$$

2.2.2.2. Antimycin A-Insensitive NADH Cytochrome *c* Reductase

1. Pour 1.5 mL of the assay mixture (25°C) into a cuvet.
2. Add 50 μL of sample to the cuvet and mix well.
3. Add 50 μL of 10 m*M* NADH stock solution and mix well.
4. Immediately, follow the reduction of cytochrome *c* by spectrophotometry as an absorbance increase at 550 nm for 1 min.
5. Enzyme activity is calculated by the following equation:

$$Enzyme\ activity\ (\mu mol/min) = reduction\ rate$$
$$of\ OD_{550}\ min^{-1} \times 1/18.5 \times 1.60$$

2.2.2.3. Cytochrome *c* Oxidase

1. Add 25 μL of sample to 15 μL of 1% Triton X-100 solution and mix well.
2. After incubation for 5 min, add 1.5 mL of potassium buffer to the sample solution and mix well.
3. Pour the solution into the cuvet.
4. Add 100 μL of reduced cytochrome *c* solution to the sample solution, and mix well.
5. Immediately, follow the oxidation of cytochrome *c* by spectrophotometry as an absorbance decrease at 550 nm for 1 min.
6. Enzyme activity is calculated by the following equation:

$$\text{Enzyme activity (μmol/min)} = \text{oxidation rate}$$
$$\text{of OD}_{550} \text{ min}^{-1} \times 1/18.5 \times 1.65$$

2.2.2.4. Assay of Protein Content. Protein content is measured by a protein assay kit (Bio-Rad, Hercules, CA). Assay is performed according to the manufacturer's instructions.

2.2.2.5. Comments

1. Substrate concentration is quite limited in each enzyme assay mixture. Enzyme assay should be performed under unsaturated conditions.*

$$\text{ATPase and UDPase; OD}_{700} < 0.90$$

$$\text{PPase; OD}_{700} < 0.45$$

   Antimycin A-insensitive NADH cytochrome c reductase and cytochrome c oxidase; Absorbance plots of OD_{550} for 1 min need to follow a straight line.
2. A hydrogensulfite gas is produced during Pi assay reaction. This gas is an irritant.

2.3. Analysis of an ER Fraction from the Mulberry Tree

The microsome fractions (*see* Fig. 1) that were prepared from mulberry cortical parenchyma cells were fractionated to 24 or 25 fractions using the 15–50% sucrose linear density gradient centrifugation (*see* Fig. 2). Profiles of the marker enzymes and the protein content are described in Fig. 3. In general, the antimycin A-insensitive cytochrome c reductase activity is exhibited at a lower density than are those of the marker enzymes. The fraction that exhibited the highest antimycin A-insensitive cytochrome c reductase activity for each month was used as the ER-enriched fraction.

In this experiment, we analyzed the seasonal change in protein composition of the ER-enriched fraction. Sodium dodecyl sulfate-polyacrylamide gel electrophoresis (SDS-PAGE) analysis indicated that the levels of two major proteins, WAP20 and WAP27, increased during seasonal cold acclimation (*11*). We confirmed using the antibodies raised against these two proteins that the proteins might be specifically localized in the ER (Fig. 4). This result suggested that the fraction we collected by sucrose linear density gradient was ER enriched.

3. CONCLUSIONS

Organelle isolation from plant cells is an important technique for the investigation of the physiological functions of the organelle itself. Due to

* The maximum limits.

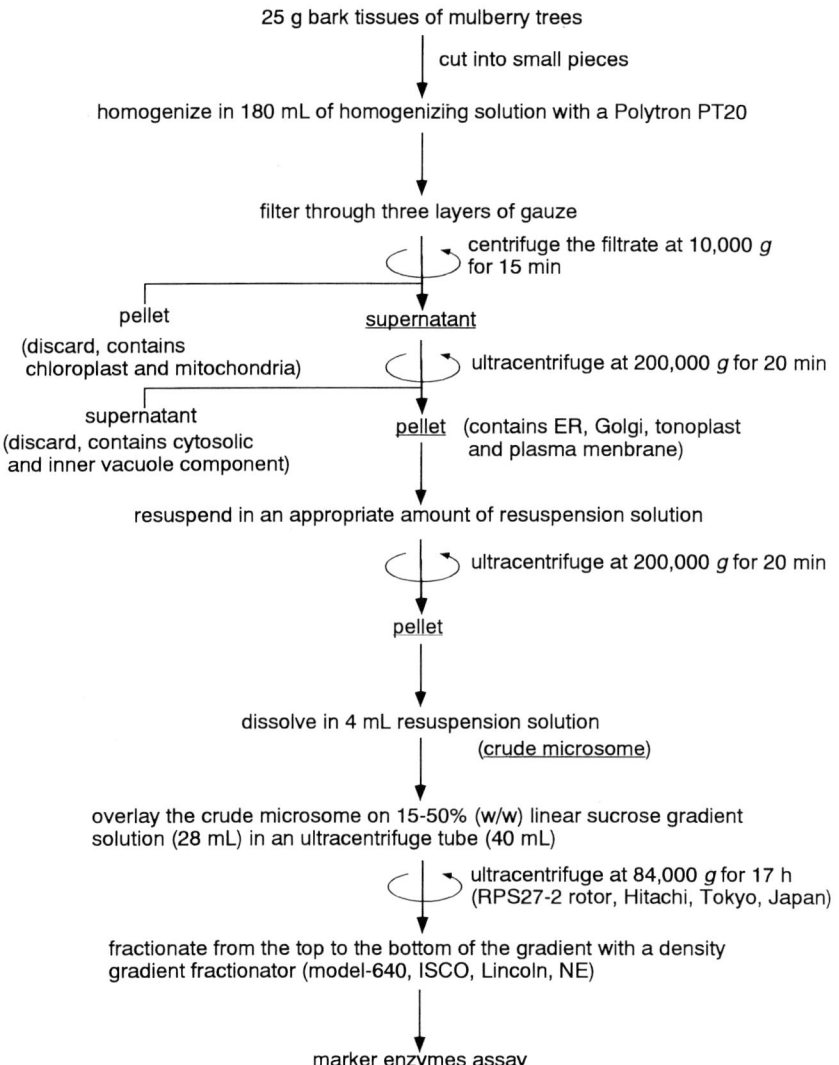

Fig. 1. Flow chart for isolation of endoplasmic reticulum from mulberry cortical parenchyma cells. All procedures are performed at 4°C. Centrifugation times are given as the time at full speed, not included the acceleration and braking time. Resuspension of the microsomal fraction pellet is accomplished using a Teflon/glass homogenizer.

the work of previous investigators, and to recent technical advances, high-purity membrane isolation has become possible. Sucrose density gradient centrifugation is one of the most effective techniques for separating cellular components. As shown in this chapter, the ER-enriched fraction was separated from other cellular membranes of cortical parenchyma

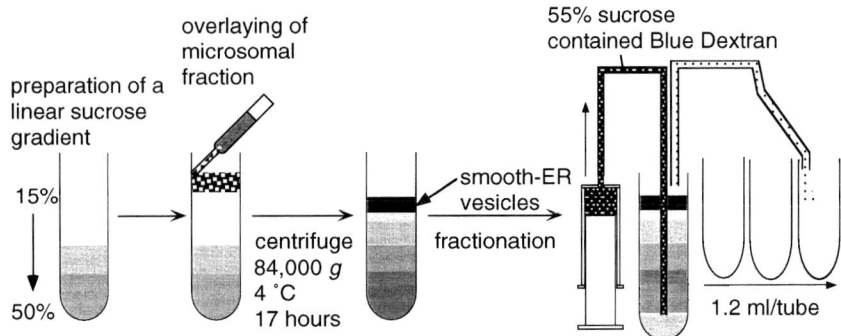

Fig. 2. Preparation of linear sucrose density gradient, centrifugation, and fractionation. Fractionation is performed using an ISCO model 640 density gradient fractionator in this experiment.

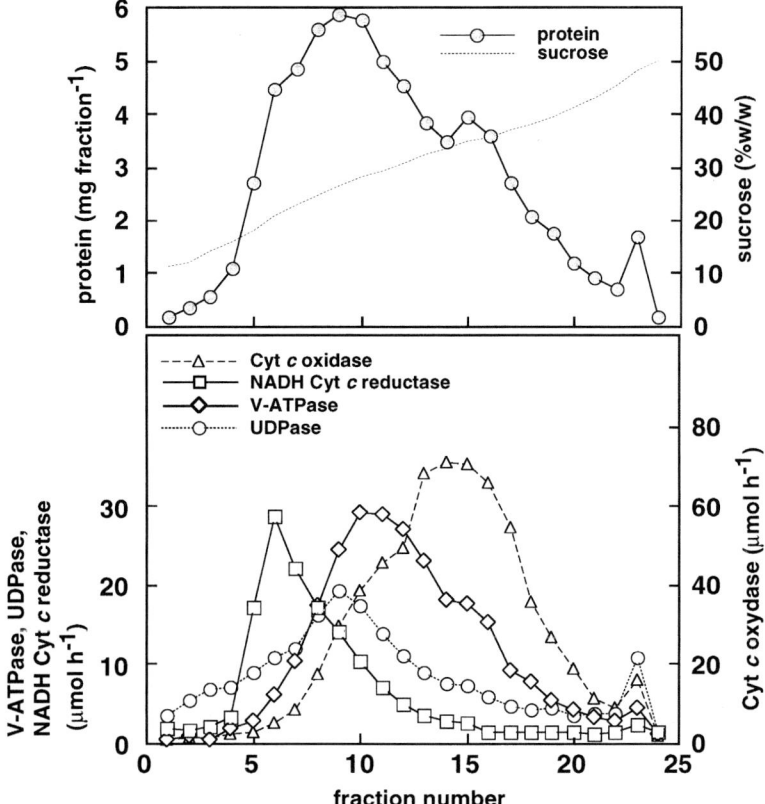

Fig. 3. Distribution of various membrane markers (ER, tonoplast, Golgi complex, and mitochondrion) in the fractions of linear sucrose density gradient fractionation of mulberry cortical parenchyma cells in February.

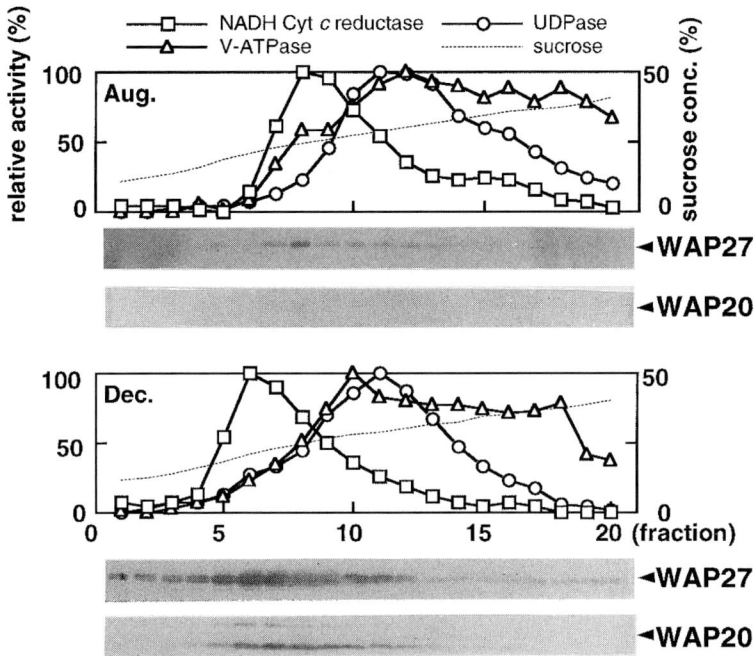

Fig. 4. Localization of WAP27 and WAP20 in the crude microsome fractions and the relation with marker–enzyme activities in three organelles (ER, tonoplast, and Golgi). SDS-PAGE of fractionated proteins by isopycnic linear sucrose density gradient centrifugation of microsome fraction of mulberry cortical parenchyma cells was performed using 6-µL samples in each fraction. Immunoblot analysis was performed with anti-WAP27 and anti-WAP20 antibodies. (From ref. *[1]*, with permission from the American Society of Plant Physiologists.)

cells of mulberry trees. However, the separated cellular organelle or membranes may not be homogeneous. In particular, isolation of the ER fraction may be difficult, as the ER is known to form a continuous membrane system with others such as the nuclear or Golgi complex. In our experiment, the ER-enriched fractions contained a small amount of tonoplast and Golgi membranes, as judged from the distribution of the marker enzymes (Fig. 3). A consideration of contamination is sometimes necessary for the accurate evaluation of the obtained results. This approach must lead to an elucidation of the nature of plant ER.

REFERENCES

1. Hepler PK, Palevitz BA, Lancelle SA, McCauley MM, Lichtschleidl I. Cortical endoplasmic reticulum in plant. *J Cell Sci* 1990; 96: 355–373.
2. Staehelin LA. The plant ER: a dynamic organelle composed of a large number of discrete functional domains. *Plant J* 1997; 11: 1151–1165.

3. Niki T, Sakai A Ultrastructural changes related to frost hardiness in the cortical parenchyma cells from mulberry twigs. *Plant Cell Physiol* 1981; 22: 171–183.

4. Fujikawa S, Takabe K. Formation of multiplex lamellae by equilibrium slow freezing of cortical parenchyma cells of mulberry and its possible relationship to freezing tolerance. *Protoplasma* 1996; 190: 189–203.

5. Quader H. Formation and disintegration of cisternae of the endoplasmic reticulum visualized in live cells by conventional fluorescence and confocal laser scanning microscopy: evidence for the involvement of calcium and the cytoskeleton. *Protoplasma* 1990; 155: 166–175.

6. Maeshima M, Yoshida S. Purification and properties of vacuolar membrane proton-translocating inorganic pyrophosphatase from mung bean. *J Biol Chem* 1989; 264: 20,068–20,073.

7. Ali MS, Nishimura M, Mitsui T, Akazawa T. Isolation and characterization of Golgi membrane from suspension-cultured cells of sycamore (*Acer pseudoplatanus* L.). *Plant Cell Physiol* 1985; 26: 1119–1133.

8. Yoshida S, Uemura M, Niki T, Sakai A, Gusta LV. Partition of membrane particles in aqueous two-polymer system and its partial use for purification of plasma membranes from plants. *Plant Physiol* 1983; 72: 105–114.

9. Yoshida S. Chemical and biophysical changes in the plasma membrane during cold acclimation of mulberry bark cells (*Morus bombycis* Koidz. cv Goroji). *Plant Physiol* 1984; 76: 257–265.

10. Yoshida S. Isolation of smooth endoplasmic reticulum and tonoplast from etiolated mung bean hypocotyls. *in Methods in Enzymology*, Vol. 228 (Walter H, Johansson G, ed.), Academic Press, New York, 1994, pp. 482–489.

11. Ukaji N, Kuwabara C, Takezawa D, Arakawa K, Yoshida S, Fujikawa S. Accumulation of small heat shock protein homologs in the endoplasmic reticulum of cortical parenchyma cells in mulberry in association with seasonal cold acclimation. *Plant Physiol* 1999; 120: 481–490.

11 Methods for the Identification of Isolated Plant Cell Organelles

William V. Dashek

1. INTRODUCTION

1.1. Methods for Identifying Isolated Organelles

Organelles within cellular homogenates can be identified with some degree of confidence through a combination of transmission electron microscope (TEM) (Chapter 14) and marker enzyme analysis *(1–3)* for fractions prepared by differential and/or gradient centrifugations *(4,5)*. For example, this author *(6)* employed TEM to locate Golgi bodies (*see* Fig. 1) in discontinuous, sucrose-gradient centrifugations of homogenates of sycamore–maple cell suspension cultures. Later, this author and his students utilized marker enzyme analysis to tentatively identify organellar fractions obtained by centrifugation of homogenates of excised soybean roots. More than a decade of research regarding the isolation of organelles and their subsequent identifications led this author and others to conclude that the identification of organelles in subcellular fractions of homogenates requires both TEM and marker enzyme analysis, as even gradient-enriched organellar fractions can be cross-contaminated.

From: *Methods in Plant Electron Microscopy and Cytochemistry*
Edited by: W. V. Dashek © Humana Press Inc., Totowa, NJ

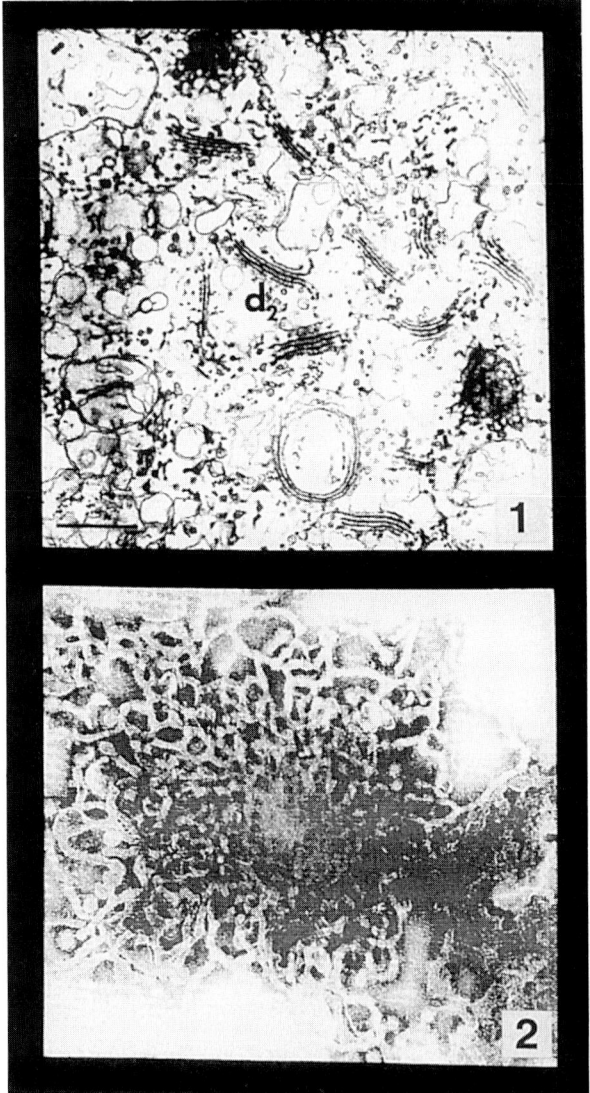

Fig. 1 and 2. TEM of positively stained isolated Golgi bodies. TEM of negatively stained isolated Golgi bodies. From Morré and Buckhout *(16)* with permission.

Yet, inspite of advanced technology, certain investigators still report the results of only one of the two approaches. This outlook is analogous to biochemists identifying chemicals via paper, thin-layer, or high-pressure-liquid chromatographies rather than by more rigorous analysis such as gas chromatography-mass spectrometry (GC-MS).

Table 1
Marker Enzymes Useful in the Identification of Plant Organelles [a]

| Membrane | Equilibrium density in sucrose | | Marker enzyme/substance |
	g/cm^3	%w/w	
Clathrin-coated vesicles	1.18–1.26	40–55	as-ni-ATPase polypeptides; 190 kDa clathrin polypeptide
Golgi apparatus	1.12–1.15	27–34	IDPase; GS 1
Intact vacuoles	Various, depending on content		AP; α-M
Microbodies	1.21–1.25	46–53	CAT
Nucleus	1.30–1.32	60–65	RNA polymerase; CCR
(nuclear envelope)	1.13–1.18	30–40	
Plasma membrane	1.13–1.18	30–40	cis-vi-ATPase; GS 11
Rough ER	1.11–1.13	26–30	CCR; RNA
Secretory vesicles	1.08–1.12	21–27	IDPase
Smooth ER	1.07–1.09	18–22	CCR
Tonoplast (from mesophyll and root storage tissue)	1.17–1.19	37–42	as-ni-ATPase; cs-PPase
Tonoplast (from seed storage tissue and maize root and coleoptile)			as-ni-ATPase; cs-PPase TP-25 (seeds only)

[a]Abbreviations: α-M, α-mannosidase; AP, acid phosphatase; as-ni-ATPase, anion-stimulated, nitrate-inhibitable ATPase; CCR, NAD(P)H-dependent cytochrome c-reductase; cs-vi-ATPase, cation-stimulated, vanadate-inhibitable ATPase, CAT, catalase; GS 1/11, glucan synthase 1 or 11; IDPase, inosine diphosphatase; cs-PPase, cation-stimulated pyrophosphatase; RNA polymerase, DNA-dependent RNA polymerase; TP-25, 25 kDa tonoplast integral protein.
Adapted from Robinson et al. (5). Examples of other marker enzymes can be found in Bowles et al. (1).

1.2. Transmission Electron Microscopy

With regard to TEM of isolated organelles, both conventional TEM and negative staining have been beneficial. The latter technique has yielded information regarding the ultrastructural morphology of Golgi bodies (see Fig. 2) in general and mitochondria in particular.

1.2.1 MARKER ENZYMES

The enzymes that appear to be associated with certain plant organelles are listed in Table 1. In contrast, Table 2 offers references for the assay conditions of certain of the enzymes. As an example of an assay, protocol 1 summarizes the assay of inosine diphosphatase, a presumed marker for

Table 2
References for Assay Conditions of Significant Marker Enzymes

Marker enzyme	EC number	Reference for conditions
Antimycin A insensitive	1.6.99.3	Sauer and Robinson (9)
NADH: cytochrome c reductase	3.6.1.3	Halliwell (10)
Anion-stimulated,nitrate-inhibitable H⁺-ATPase		
Cation-stimulated vanadate inhibitable H⁺-ATPase	3.6.1.35	Lanzetta et al. (11)
Catalase	1.11.16	Stegink et al. (12)
Glucan synthase I	2.4.1.12	Ray (13)
Glucan synthase II	2.4.1.34	Kause and Jeblich (14)
Cs P Pase inorganic pyrophosphatase	3.6.1.1	Walker and Leigh (15)

the Golgi apparatus, as the author possesses experience in the isolation of this organelle (6). Since that time, a wealth of papers has appeared regarding the isolation of the Golgi apparatus (see reviews of Morré and Buckhout [16]; Morré and Keenan [17]). The reader is urged to consult the papers cited in Table 2 prior to assaying other purported marker enzymes.

1.2.2. Assays Other Than Marker Enzymes

In addition to marker enzymes, other cellular chemicals can obviously serve to tentatively identify the organelles in fractions prepared by centrifugations of cellular homogenates. For example, the differences in the buoyant densities of nuclear, chloroplast, and mitochondrial DNA following $CsCl_2$–density gradient are well-known. Indeed, the current information explosion resulting from research in molecular biology has yielded comparisons in organelle genes and genomes (18,19).

2. ASSAY OF INOSINE-5-DIPHOSPHATASE (IDPASE) EC3.6.1.6, PRESUMED MARKER ENZYME FOR GOLGI APPARATUS

2.1. Reagents

1. IDPmix: 50 mM Tris-HC1, pH 7.2 containing 10 mM KC1, 2–4 mM IDP (inosine diphosphate), + and -0.2% (v/v) Triton X-100
2. Trichloroacetic acid (TCA): 10% w/v
3. Color reagent 3 M H_2SO_4, 2.5% w/v ammonium molybdate. 4 H_2O, 10% w/v-ascorbic acid and H_2O in proportions of 1:1:1:6 by volume
4. Bovine serum albumin (BSA) (140 µg mL^{-1} for standard curve construction)
5. KH_2PO_4

2.2. Supplies

Beakers	1, 5, and 10 mL pipets
Erlenmeyer flasks	Spatulas
Ice bucket	Water wash bottle
Kimwipes (Kimberly-Clark)	Volumetric flasks

2.3. Equipment

Analytical balance	Top-loading balance
Eppendorf pipets and μL tips	Water bath
Glass and quartz cuvets	Weighing paper and boats
Preparative centrifuge	UV-visible spectrophotometer

Perform the necessary controls, i.e., withhold the substrate in the reaction mixture, boil the membrane preparation, and then carry out the enzyme reaction. Adapted in part from Morré et al. *(22)*.

1. Add 50 mL of membrane preparation to 50 μL IDP mix in an Eppendorf tube.
2. Incubate with shaking at 37°C for 10, 20, 40, and 60 min (perform a 0-min determination on ice).
3. Terminate reaction by adding 500 μL cold 10% (w/v) TCA.
4. Maintain at 4°C for at least 2 h.
5. Centrifuge at 9000g, 10 min (use a nomogram).
6. To the supernatant
   a. Add 200 μL–1.8 mL color reagent (*see* Subheading 2.1.)
   b. Incubate 37°C for 1 h.
   c. Make sure that there are no particles. Determine the absorbance at 820 nm with a spectrophotometer.
   d. Quantify the amount of phosphate released from a KH_2PO_4 calibration curve, i.e., A_{820} nm y axis vs nmol KH_2PO_4 (0–250) on the x axis.
   e. Express specific activity as nmol phosphate released min^{-1} mg^{-1} protein (protein content of membrane preparation can be quantified by UV spectroscopy and a BSA standard curve [20,21]).
7. Subtract −Triton values from + Triton values for measure of detergent-stimulated latency.

3. CONCLUSIONS

Perhaps, the future of identifying organelles within fractions derived from cellular homogenates resides in the immunocytochemical localization of possible unique organellar antigens *(23)*. This would, of course, require time-consuming screening of organelles for such antigens. As an interim measure, monoclonal antibodies to the marker enzymes could be prepared. This would permit transmission immunoelectron microscopy, thereby avoiding dividing the organellar fraction in two for TEM and biochemistry.

For those laboratories routinely performing marker enzyme analysis, it is hoped that future investigators will be more attentive to the preparation of balance sheets (24), i.e., comparing the specific activities of marker enzymes in each organellar fraction relative to those in the total cellular homogenate. Unfortunately, some investigators assay for the presence or absence of marker enzymes only in the fraction(s) that he/ she is interested in.

REFERENCES

1. Bowles DJ, Quail PH, Morré DJ, Hartnann GC. Use of markers in plant cell fractionation, in *Plant Organelles* (Reid E, ed.), Ellis Horwood, Chichester, UK, 1979, pp. 207–227.
2. Quail PH. Plant cell fractionation. *Ann Rev Plant Physiol* 1979; 30: 425–484.
3. Robinson DG. *Plant Membranes. Endo and Plasma Membranes of Plant Cells*, Wiley-Interscience, New York, 1985.
4. Reid E. *Plant Organelles*, Ellis Horwood, Chichester, UK, 1979.
5. Robinson DG, Hinz G, Oberbeck K. Isolation of endo- and plasma membranes, in *Plant Cell Biology* (Harris GN, Oparka KJ, eds.), IRL Press, Oxford University, Oxford, UK, 1994.
6. Dashek WV. Synthesis and transport of hydroxyproline-rich components in suspension cultures of sycamore-maple cells. *Plant Physiol* 1970; 46: 831–838.
7. Danley JM, Staggers S, Varner A, Llewellyn GC, Dashek WV. A combined biochemical and ultrastructural analysis of aflatoxin action on the endomembrane system of plant cells. *Actual Bot* 1982; 129: 5–13.
8. Walker SJ, Llewellyn GC, Lillehoj EB, Dashek WV. Uptake and distribution of aflatoxin B_1 in excised, soybean roots and toxin effects on root elongation. *J Environ Exper Bot* 1984; 24: 113–122.
9. Sauer A, Robinson DG. Intracellular localization of posttranslational modifications in the synthesis of hydroxyproline-rich glycoproteins. Peptidyl hydroxylation in maize roots. *Planta* 1985; 164: 287–294.
10. Halliwell B. Marker enzymes of plant cell organelles, in *Methodological Developments in Biochemistry. 4. Subcellular Studies* (Reid E, ed.), Longman, London, pp. 357–366.
11. Lanzetta PA, Alvarez LJ, Reinach PS, Condia OA. An improved assay for normal amounts of inorganic phosphate. *Anal Biochem* 1979; 100: 95–97.
12. Stegink, SJ, Vaughn KC, Kunce CM, Trelease RN. Biochemical, electrophoretic and immunological characterization of peroxisomal enzymes in three soybean organs. *Plant Physiol* 1987; 69: 211–220.
13. Ray PM. Maize coleoptile cellular membranes bearing different types of glucan synthetase activity, in *Plant Organelles* (Reid E, ed.), Ellis Horwood, Chichester, UK, 1979, pp. 135–158.
14. Kause HL, Jeblich W. Synergistic activation of 1,3-β-D-glucan synthase by Ca^{2+} acid polyomers. *Plant Sci* 1986; 43: 103–107.
15. Walker RR, Leigh RA. Mg^{2+}-dependent, cation-stimulated inorganic pyrophosphatase associated with vacuoles isolated from storage roots of red beet (*Beta vulgaris* L.). *Planta* 1981; 133: 150–155.

16. Morré DJ, Buckhout TJ. Isolation of Golgi apparatus, in *Plant Organelles* (Reid E, ed.), Ellis Horwood, Chichester, UK, 1979, pp. 117–134.

17. Mooré DJ, Keenan TW. Membrane flow revisited. *Bioscience* 1997; 47: 489.

18. Tager JM. *Organelles in Eukaryotic Cells. Molecular Structure and Interactions*, Plenum Press, New York, 1990.

19. Gillham NW. *Organelle Genes and Genomes*, Oxford University Press, Oxford, UK, 1994.

20. Zeidan H, Dashek WV. *Experimental Approaches in Biochemistry and Molecular Biology*, Wm. C. Brown Co, Dubuque, IA, 1996.

21. Dashek WV, Micales JM. Purification of enzymes: oxalate decarboxylase, in *Methods in Plant Biochemistry and Molecular Biology* (Dashek WV, ed.), CRC Press, Boca Raton, FL, 1997, pp. 49–71.

22. Morré DJ, Lembi CA, Van Der Woude W. Latent inosine-5^1-diphosphatase associated with Golgi apparatus-rich fractions from onion stem. *Cytobiologie* 1977; 16: 72–81.

23. Hasegawa Y, Nakamura S, Kakizoe S, Sato M, Nakamura N. Immunocytochemical and chemical analyses of Golgi vesicles isolated from the germinated pollen of *Camellia japonica. J Plant Res* 1998; 11: 421–429.

24. Laisson IC, Moller IM. (eds.). *The Plant Plasma Membrane: Structure, Function and Molecular Biology*, Springer Verlag, Berlin, 1990.

12 Methods for the Detergent Release of Particle-Bound Plant Proteins

William V. Dashek

1. MEMBRANE PROTEINS

There are three major classes of membrane proteins, i.e., peripheral or extrinsic proteins, integral or intrinsic proteins and transport proteins *(1)*. The peripheral proteins (*see* Fig. 1) appear to be restricted to either the outer or inner layers of the membrane bilayer. These proteins, which are bound chiefly by ionic forces to the polar heads of phospholipids (electrostatic bonding), can be isolated through alteration of the pH or ionic strength of the suspension medium. Integral or transmembrane proteins are linked either electrostatically or by means of biophysical lipophilicity to the inner domains of the membrane bilayer *(1)*. Integral proteins appear to be anchored in the membrane bilayer via polar amino acid sequences *(2)*. There are two types of integral proteins, simple and complex. Whereas the former possess γ-helical structure *(3)*, the latter are globular and comprised of several γ-helical loops that may traverse the membrane more than once (multiple hairpin bending). The transport proteins may be either of the peripheral or integral type consisting of

From: *Methods in Plant Electron Microscopy and Cytochemistry*
Edited by: W. V. Dashek © Humana Press Inc., Totowa, NJ

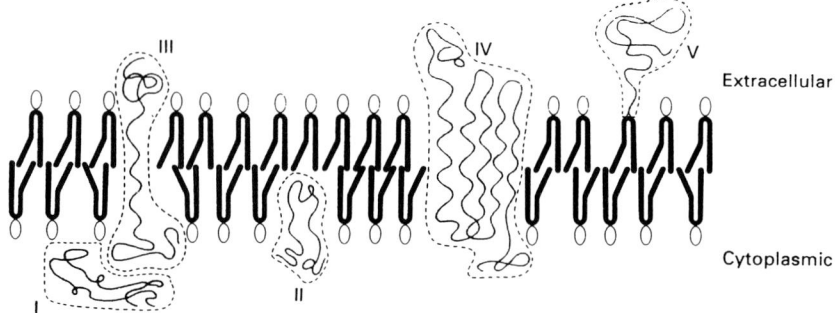

Fig. 1. Classes of membrane proteins. From Evans and Callow *(2)*, with permission.

pumps, carriers, and channels based on energy input *(1)*. Thorough treatments of various aspects of membrane proteins can be found in Martanosi *(4)*, Azzi *(5)*, Gennis *(6)*, Turner *(7)*, Shinitzky *(8)*, Lee *(9)*, Papa and Tager *(10)*, Findlay *(11)*, and Van Heijne *(12)*.

2. SOLUBILIZATION OF MEMBRANE-BOUND PROTEINS BY DETERGENTS

Following the isolation *(13–17)* and identification of membrane fractions, integral membrane proteins can be solubilized by detergents *(18–27)*. The chemistry of certain detergents (*see* Table 1) is presented in Fig. 2. Detergents are classified as either ionic or nonionic. The ionic detergents are anionic (bile salts-cholate, deoxycholate), cationic (alkytrimethyl-ammonium salts), or zwitteronic; [(3-cholamidopropyl)-dimethylammonio-1-propane sulfate (CHAPS)]. Examples of the nonionic detergents include octylglucoside and polyoxyethylene derivatives such as Triton X-100, Lubrol PX and the Tween reagents. To obtain successful solubilization of membrane proteins, i.e., preservation of biological activity, detergent selection is critical. The criteria for detergent selection are presented in Table 2.

Subsequent to possible solubilization of membrane-bound proteins, solubilization must be verified. The criteria listed in Table 2 are relevant in assessing whether solubilization has been accomplished. To ascertain whether the solubilized protein has retained biological activity, membrane reconstitution *(28)* is attempted subsequent to detergent removal *(24)*. Reconstitution is often visualized by electron microscopy employing either negative staining or freeze fracture.

Table 3, which summarizes some recent applications of detergents to solubilizing particulate plant proteins, provides an indication of the diversity of proteins that can be solubilized by detergent action.

Table 1
Chemical and Physical Properties of Commonly Used Detergents

Detergent[a]	m.p. °C	Mol. wt.[c] Monomer	Micelle	CMC[e] % (w/v)	M
SDS	206	288	18,000	0.23	8.0×10^{-3}
Cholate	201[b]	430	4300	0.60	1.4×10^{-2}
Deoxycholate	175[b]	432	4200	0.21	5.0×10^{-3}
C_{16}TAB	230[c]	365	62,000	0.04	1.0×10^{-3}
Lyso PC (C_{16})	—	495	92,000	0.0004	7.0×10^{-6}
CHAPS	157[c]	615	6150	0.49	1.4×10^{-3}
Zwittergent 3–14	—	364	30,000	0.011	3.0×10^{-4}
Octyl glucoside	105[c]	292	8000	0.73	2.3×10^{-2}
Digitonin	235[c]	1229	70,000	—	—
$C_{12}E_8$	[d]	542	65,000	0.005	8.7×10^{-5}
Lubrol PX	[d]	582	64,000	0.006	1.0×10^{-4}
Triton X–100	[d]	650	90,000	0.021	3.0×10^{-4}
Tween-80	[d]	1310	76,000	0.002	1.2×10^{-5}

From Jones (21) with permission.
[a] See Fig. 2 for structures.
[b] Based on free acid.
[c] Decomposes.
[d] Viscous liquid at room temperature.
[e] Determined at 20–25°C.

3. PURIFICATION OF SOLUBILIZED MEMBRANE PROTEINS

The purification of solubilized proteins involves standardized protein purification protocols. Because the literature already contains a copious supply of available reviews and monographs regarding protein purification (35–43), the topic will not be discussed here. The reader is directed to ref. (38) for techniques specifically applicable to membrane proteins.

Once solubilization of the membrane protein has been achieved, a reliable assay for it must exist. If the protein is an enzyme, then one must quantify the specific activity (spc. act.) of the enzyme, i.e., μmol product formed or substrate disappeared per min per mg protein. Thus, not only must the activity of the enzyme be assayed (44) but also the protein content of the enzymatic preparation. In this connection, Dashek and Micales (45) have discussed the factors that must be considered when assaying enzyme activity. In addition, they review protein quantification.

4. PROTOCOLS

The utilization of three different detergents for the solubilization of particulate enzymes is presented to provide the reader with the scope of

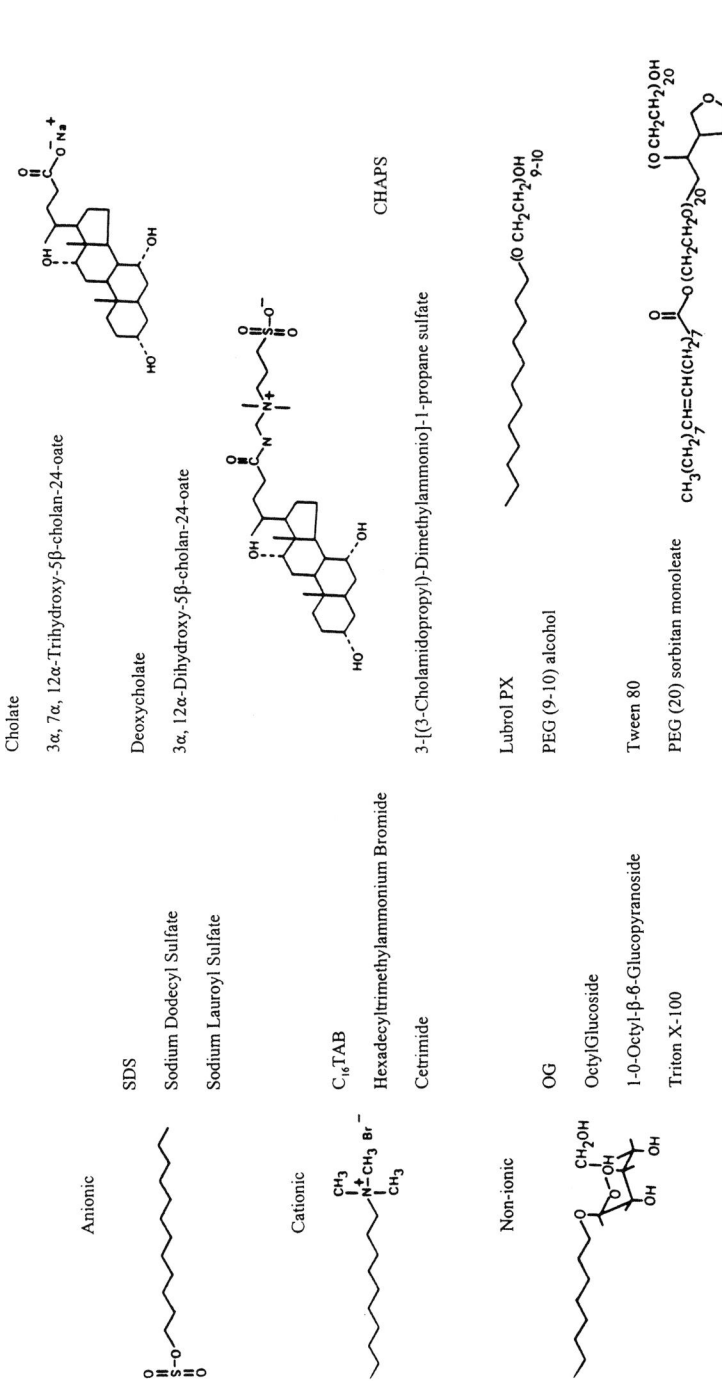

Fig. 2. Chemical composition of some detergents.

Anionic

SDS — Sodium Dodecyl Sulfate
Sodium Lauroyl Sulfate

Cholate — 3α, 7α, 12α-Trihydroxy-5β-cholan-24-oate

Deoxycholate — 3α, 12α-Dihydroxy-5β-cholan-24-oate

Cationic

C₁₆TAB — Hexadecyltrimethylammonium Bromide
Cetrimide

CHAPS — 3-[(3-Cholamidopropyl)-Dimethylammonio]-1-propane sulfate

Non-ionic

OG — OctylGlucoside
1-0-Octyl-β-6-Glucopyranoside
Triton X-100

Lubrol PX — PEG (9-10) alcohol

Tween 80 — PEG (20) sorbitan monoleate

172

Table 2
Summary of Criteria for Detergent Selection and Protein Solubilization

Criteria for detergent selection	*Criteria for protein solubilization*
Detergent should solubilize but not denature the protein	Retention of a protein or protein activity after 105,000g, 1 hr
Detergent must be readily available in pure form	Chromatography on gel filtration columns with large pore sizes
Detergent must not be very expensive	Electron microscopy—however, sample preparation may partially reconstitute membranes
Detergent should be readily removable after treating membrane fraction	
Detergent must not interfere with assays such as lipids, protein colorimetric determinations and enzymatic activity; many of the nonionic detergents contain high phosphate levels that interfere with certain lipid analyses and certain of the polyoxyethylene derivatives affect protein colorimetric and enzyme assays	Decrease in solution turbidity, which may be detected by a diminution in light scattering or an enhancement in light transmission
Anionic and cationic surfactants are more effective in membrane solubilization than nonionic surfactants.	Diffusion of membrane lipids as assayed by nuclear magnetic resonance and electron spin resonance
Anionic surfactants (SDS and deoxycholate) are more effective solubilizing agents at higher concentration than nonionic surfactants (Lubrol, Triton)	

Criteria for detergent selection
Detergent should solubilize but not denature the protein
Detergent should be readily available in pure form
Detergent should be inexpensive
Detergent should be easily removable
Detergent should not interfere with assays such as lipid and protein colorimetric assays and enzyme assays

detergent action. One of these enzymes, oxalate decarboxylase, is fungal-derived but is included because of its role in wood decay. The two other enzymes, polyphenol oxidase and fatty acid desaturase, are found in higher plants.

Table 3
Some Recent Examples of the Use of Detergents
to Solubilize Particulate Plant-Bound Plant Proteins

Protein	Detergent	Reference
Polyphenol oxidase	Triton X-114	Sajo et al. *(29)*
Oligopeptide elicitor receptor		Nennstiel et al. *(30)*
δ-12 and δ-6 fatty acid desaturases		Galle et al. *(31)*
Binding site for a glycopeptide elicitor		Fath and Boller *(32)*
PS 11-S pigment-binding protein		Funk et al. *(33)*
Sulphite oxidase		Jolivet et al. *(34)*

4.1. Triton X-114-Mediated Solubilization of Vicia Faba *Chloroplast Polyphenol Oxidase*

Polyphenol oxidase occurs within certain mammalian tissues as well as both lower *(46,47)* and higher *(48–55)* plants. In mammalian systems, the enzyme as tyrosinase *(56)* plays a significant role in melanin synthesis. The PPO complex of higher plants consists of a cresolase, a catecholase and a laccase. These copper metalloproteins catalyze the one and two electron oxidations of phenols to quinones at the expense of O_2. Polyphenol oxidase also occurs in certain fungi where it is involved in the metabolism of certain tree-synthesized phenolic compounds that have been implicated in disease resistance, wound healing, and antinutrative modification of plant proteins to discourage herbivory *(53,55)*. This protocol presents the Triton X-114-mediated solubilization of *Vicia faba* chloroplast polyphenol oxidase as performed by Hutcheson and Buchanan *(57)*.

4.1.1. GREENHOUSE GROWN VICIA FABA (BROAD WINDSOR HORSEBEAN) UNDER NATURAL CONDITIONS

1. Isolate chloroplasts from 1 kg chilled leaflets in 0.35 *M* sucrose and 25 m*M* Tris-HCl, pH 7.9, by blending, filtering, and centrifuging.
   a. Blend: 1 kg/lamina (minus midrib) in 122 mL of 0.35 *M* sucrose and 25 m*M* Tris-HCl, pH 7.9
   b. Filter: strain through cheesecloth
   c. Centrifuge: 4000*g*, 15 min
2. Release membrane-bound PPO by sonicating 10 mL aliquot of chloroplast membrane suspension on ice for 105 s (30-s pulse followed by a 10-s rest-repeat) using a Bronson Sonifier (Model S125), setting #3 or an alternative.
3. Centrifuge 105,000*g*, 90 min, pellet, chloroplast, membrane.

4. Supernatant: Apply to a 2×30 cm DEAE cellulose column pre-equilibrated with 50 mM Tris-HCl, pH 7.9.
5. Elute column with 600 mL of a 0.0–0.4 M linear NaCl gradient added to 50 mM Tris-HCl, pH 7.9.
6. Collect 6.6 mL fractions and assay for PPO.
7. Pool fractions containing PPO and concentrate to 5 mL using YM 10 Amicon ultrafiltration filters.
8. Centrifuge YM-retained fraction at 105,000g, 30 min.
9. Apply 5 mL to a 2×90 cm Sephadex G-100 column equilibrated with 50 mM Tris-HCl, pH 7.9.
10. Collect 2.5 mL fractions at a flow rate of 16 mL/h and assay for PPO.
11. Pool PPO containing concentrate to 5 mL and apply to a 2×5 cm hydroxylapatite column Bio-Gel HTP equilibrated with 50 mM Tris-HCl, pH 7.9.
12. Elute with 200 mL buffer supplemented with a linear gradient of 0.0 to 0.35 M pH 7.7 potassium phosphate buffer.
13. Collect 2.5 mL fractions.
14. Assay for PPO.
15. Concentrate and store at −10°C.

The reader may consult Sanchez-Ferrer et al. *(58)*, Nunez-Delicádo et al. *(59)*, and Sajo et al. *(29)* for updates concerning the use of Triton surfactants to solubilize membrane-bound polyphenol oxidase. In addition, Kieselbach et al. *(60)* offers a contemporary view regarding the isolation and characterization of chloroplast thylakoids. Wilhelnova *(61)* compared the abilities of different detergents to fragment thylakoid membranes.

4.2. Triton X-100-Promoted Solubilization of Postia Placenta *Hyphal Sheath Oxalate Decarboxylase*

Oxalate decarboxylase (ODC) (EC4.1.2) is one of a group of wood decay enzymes (e.g., cellulose, lignase, manganese-dependent peroxidase) isolated from various wood-rotting basidiomycetes. This enzyme appears to regulate the accumulation of, or lack thereof, of oxalic acid during decay. There are two main types of basidiomycetes, i.e., white-rot and brown-rot fungi. Although the former degrades both wood cellulose and lignin, the latter degrades the cellulose, leaving the lignin undegraded but chemically modified *(62)*. Until recently, oxalate decarboxylase was associated with white-rot fungi, but recently Micales *(63)* detected it in *Postia placenta*, a brown-rot fungus. Dashek and Micales *(45)* have reviewed the occurrences and assay procedures *(71)* for ODC as well as the reported attempts at its purification.

Recently, Dashek and Micales *(64)* attempted to further purify *P. placenta's* ODC beyond that initially reported by Micales. Although they were not able to purify ODC to homogeneity, the specific activity of the enzyme could be enhanced through the insertion of dialysis, ammonium sulfate fractionation, Sephadex gel filtration, and DEAE Sepharose chromatography into the procedure employed by Micales. Most recently, Micales *(64,65)* has used differential centrifugation and Triton X-100 to determine that the enzyme is primarily extreacellular and weakly associated with the hyphal sheath (*see* Table 4). Figure 3 is a flowchart showing how to use Triton X-100 to solubilize hyphal ODC.

4.3. CHAPS Solubilization
of Microsome-Bound Desaturases

1. Prepare microsomes from borage seeds, lipid extraction, and fatty acid analysis according to Galle et al. *(31)*. Protocol adapted from ref. *(31)*.
2. Resuspend microsomes buffer to a final concentration of 18 mg total protein mL^{-1} 70 mM, pH 7.2 buffer.
3. Solubilize desaturases with 1% CHAPS at a detergent–membrane protein ratio of equal to one, 5°C for 30 min with gentle stirring.
4. Centrifuge at 100,000g for 190 min to sediment unsolubilized membranes.
5. Assay supernatant for Δ^{12}–Δ^6-desaturase activites according to Galle et al. *(31)*.
6. Dilute the 100,000g supernatant with Tris-HCl, 20 mM pH 7.5, 10% glycerol, 3 mM sodium azide, 10 mM mercaptoethanol to achieve partial purification of the enzymes.
7. Load diluted supernatant onto a column of Pharmacia DEAE Trisacryl.
8. Elute column with 10 M NaCl or dilution buffer.
9. Assay fractions.

5. CONCLUSIONS

A recent review of the detergent literature indicated the following trends: (1) improvement in dodecyl maltoside surfactants *(73)*, (2) expansion of the use of Triton detergents in biochemistry *(74)*, and (3) the use of detergents for solution nuclear magnetic resonance studies *(75)*.

REFERENCES

1. Leshem YY with the participation of Shewfelt RL, Willmer CM, Pantoja O. *Plant Membranes: A Biological Approach to Structure, Development and Senescence,* Kluwer Academic Publishers, Amsterdam, The Netherlands, 1991.
2. Evans WH, Graham JM. *Membrane Structure and Function,* IRL Press, Oxford, UK, 1989.

Table 4
Summary of Significant Findings Regarding *Postia placenta*'s Hyphal Sheath

Investigator	Organism	Culture Condition	Observations
Green et al. (66)	*P. placenta* *Trametes versicolor* *Phanerochaete chrysosporium*	Wood blocks for 12 wk decayed by *P. placenta*, *Trametes versicolor*, and *Phanerochaete chrysosporium* using the ASTM soil–block procedure	Variety of complex sheath structures (sheets, filaments, and vesicles) often depended on the preparative method
Kim et al. (67)	*P. placenta*	Polyclonal antiserum was produced to *P. placenta* extracellular metabolites; red spruce and birch were degraded by *P. placenta* using the soil–block procedure; degraded wood–block samples were prepared for TEM and the immunoelectron localization of wood-degrading enzymes	Extracellular membrane structures (matrix) were observed surrounding hyphae, which degraded spruce and birch wood; the matrix labeled positively with antisera produced to *P. placenta* extracellular metabolites
Micales et al. (68)	*P. placenta*	Various isolates of *P. placenta* including a nondecay isolate, were used; decay was tested with soil block bottles incubated at 27°C and 70% RH for 12 wks; for collecting extracellular glucan, cultures were grown in abated liquid medium containing a basal salt solution and variable carbohydrates for 10 days at 25°C; extracellular glucan was collected from culture by alcohol precipitation and the chemical composition of the isolated glucan determined by HPLC after acid hydrolysis	Monokaryotic strain of *P. placenta* ME20, which is unable to degrade wood, failed to produce extracellular polysaccharides consisted primarily of glucose upon acid hydrolysis; isolate ME20 formed high levels of laminarinose and glucan degrading enzymes

(continued)

177

Table 4 (Continued)

Investigator	Organism	Culture Condition	Observations
Green et al. (69)	P. placenta Brown-rot fungus	P. placenta (Fr.) 7. Lars et Lomb (isolate MAD-698) was grown on sterile 12 mm glass slips placed on the surface of 2% (w/v) malt agar removed when about 50% covered	Extracellular wood degrading enzymes were localized using colloidal gold labeled monoclonal antibodies to the B-1, 4-xylanase fraction of P. placenta; enzymes were localized on the hyphal surface
Larsen and Green (70)	P. placenta	P. placenta inoculated onto 90 mm polycarbonate Petri plates containing 2% malt and 1.5% agar and inoculated at 24°C; four sterile cover slips placed 10 mm from the inoculation; microgram quantities of xylan placed on cover slips; fungus allowed to overgrow 2/3 of the coverslips	Existence of linear extracellular fibrillar elements; elements appeared as structural components of the hyphal sheath closely resemble mycofibrils than fungal fimbriae; extracellular structures have a diameter of 10–50 mm and are up to 25 µm in length
Nicole et al. (71)	Rigidoporus lignosus		Fungal sheaths had a dense or loose fibrillar appearance; close association between extracellular fibrils and wood cell walls was observed in both early and advanced stages of wood alteration; galactose residues and laccase present in the sheath

ASTM = American Society for Testing Materials; TEM = transmission electron microscope.

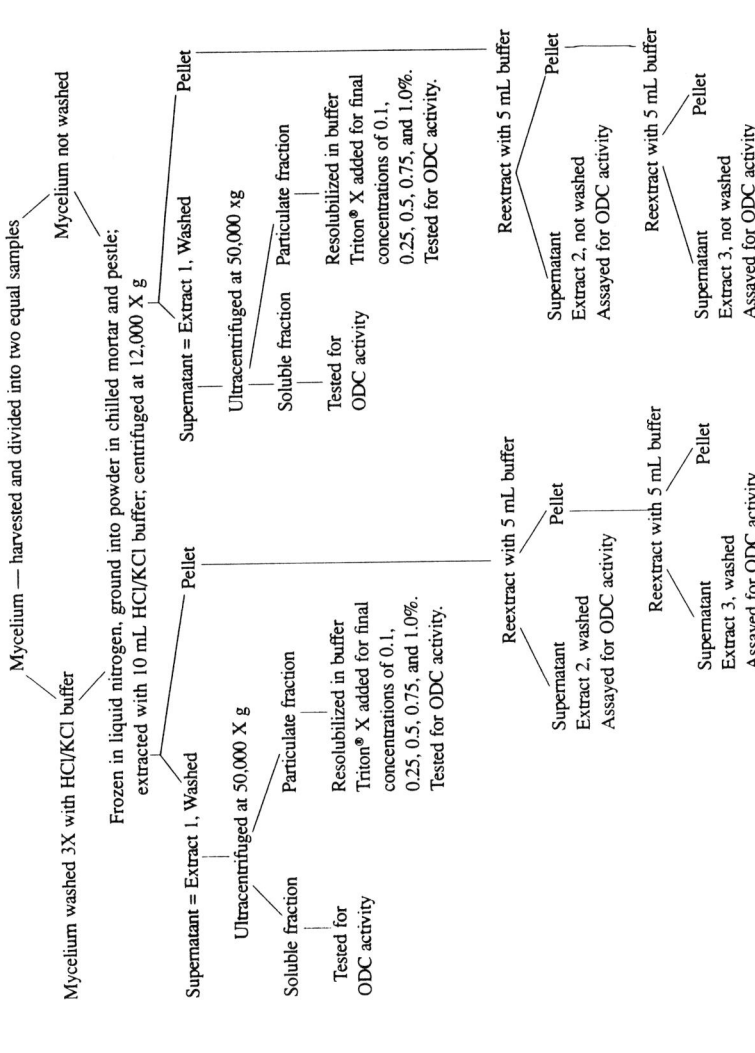

Fig. 3. Flowchart showing how to use Triton-X to solubilize hyphal ODC. From Micales (*65*), with permission.

3. Zubay G. *Principles of Biochemistry*, Wm. C. Brown Publishers. Dubuque, IA, 1993.
4. Martanosi A. *The Enzymes of Biological Membranes*, 4 Vols, Plenum Press, New York, 1985.
5. Azzi A. *Membrane Proteins: Isolation and Characterization*, Springer-Verlag, New York, 1986.
6. Gennis RB. *Biomembranes: Molecular Structure and Function*, Springer-Verlag, New York, 1989.
7. Turner AJ. *Molecular and Cell Biology of Membrane Proteins: Glycolipid Anchors of Cell Surface Proteins*, Ellis Horwood, New York, 1990.
8. Shinitzky M. *Biomembranes*, Verlagesellshaft, New York, 1993.
9. Lee A. *Biomembrane-bound Enzyme Systems*, Jai Press, London, UK, 1995.
10. Papa S, Tager JM. *Biochemistry of Cell Membranes: A Compendium of Selected Topics*, Birkhauser-Verlag, Basel, Switzerland, 1995.
11. Findlay JBC. *Membrane Protein Models*, Bios Scientific Publishers, Herndon, VA, 1996.
12. Van Heijne G. *Membrane Protein Assembly*, R. G. Landes, Austin, TX, 1997.
13. Higgins JA. Separation and analysis of membrane components, in *Biological Membranes: A Practical Approach* (Findlay JB, Evans WH, eds.), IRL Press, Oxford, UK, 1987, pp. 103–138.
14. Baker RW. *Membrane Separation Systems: Recent Developments and Future Directions*, Noyes, Westwood, NJ, 1991.
15. Graham J, Higgins JA. *Biomembrane Protocols. I. Isolation and Analysis*, Humana Press, Totowa, NJ, 1993.
16. Critser JR. *Membrane Separation Processes*, Lexington Data, Ashland, MA, 1995.
17. Nobel RD, Stern SA. *Membrane Separation. Technology Principles and Applications*, Elsevier, New York, 1995.
18. Schwuger MJ, Bartnik FG. Interaction of anionic surfactants with proteins, enzymes and membranes, in *Anionic Surfactants* (Gloxhuber C, ed.), Marcel Dekker, New York, 1980, pp. 1–49.
19. Bordier C. Phase separations of integral membrane proteins in Triton X-114 solution. *J Biol Chem* 1981; 256: 1604–1607.
20. Venter JC, Harrison LC. *Membranes, Detergents and Receptor Solubilization*, Alan R. Liss, New York, 1984.
21. Jones OT, Earnest JP, McNamee MG. Solubilization and reconstitution of membrane proteins, in *Biological Membranes: A Practical Approach* (Findlay JB, Evans WH, eds.), IRL Press, Oxford, UK, 1987, pp. 139–178.
22. Banerjee R, Jao JB, Bush JT, Dawson G. Differential solubilization of lipids along with membrane proteins by different classes of detergents. *Chem Phys Lipids* 1995; 77: 65–78.
23. Shepherd FM, Holzenburg A. The potential of fluorinated surfactants in membrane biochemistry. *Anal Biochem* 1995; 224: 21–27.
24. Dong M, Baggetto LG, Folson P, LeMaire M, Penin F. Complete removal and exchange of sodium dodecyl sulfate bound to soluble and membrane proteins and restoration of their activities, using ceramic hydroxyapatite chromatography. *Anal Biochem* 1997; 247: 333–341.
25. Chabaud E, Barthelemy P, Mora N, Popot JL, Pucci B. Stabilization of integral membrane proteins in aqueous solution using fluorinated surfactants. *Biochemie* 1998; 80: 515–530.
26. Chevallet M, Santoni V, Poinas A, Rouquie D, Fuchs A, Kieffer S, Rossignol M, Lunadi J, Gerin J, Rabilloud T. New zwitteronic detergents improve the analysis of membrane proteins by two-dimensional electrophoresis. *Electrophoresis* 1998; 19: 1901–1909.

27. Krogh-Hansen U, LeMarie M, Moller JV. The mechanism of detergent solubilization of liposomes and protein-containing membranes. *Biophys J* 1998; 75: 2932–2941.

28. Poste G, Nicolson G. *Membrane Reconstitution*, North Holland Publ. Co, Amsterdam, The Netherlands, 1982.

29. Sajo M Mar, Nunez-Delicado E, Garcia-Carmona F, Sanchez-Ferrer A. Partial purification of a bannana polyphenol oxidase using Triton X-114 and PEG 800G for removal of polyphenols. *J Agric Food Chem* 1998; 46: 4924–4930.

30. Nennstiel D, Scheel D, Nuernberger T. Characterization and partial purification of a oligopeptide elicitor receptor for parsley (*Petroselium crispum*). *FEBS Lett* 1998; 431: 405–410.

31. Galle AM, Demandre JM, Guercke C, Dibacq, JA. Ouised A, Mazliak P, Pelletier G, Koder JC. Biosynthesis of gamma-linolenic acid in developing solubilization of bio membrane enzymes of borage seeds (*Borago officinalis* L.) microsomes Δ^{12} and Δ^6 desaturases involved in the biosynthesis of polyunsaturated fatty acids. *Biochem Biophys Acta* 1993; 1158: 52–58.

32. Fath A, Boller T. Solubilization, partial purification and characterization of a binding site for a glycopeptide elicitor from microsomal membranes of tomato cells. *Plant Physiol* 1996; 112: 1659–1668.

33. Funk C, Schroder WP, Napiwotzki A, Tjus SE, Renger G, Andersson B. The PS 11-S protein of higher plants: A new type of pigment binding protein. *Biochemistry* 1995; 34: 11,133–11,141.

34. Jolivet P, Bergeron E, Meunier J-C. Evidence for sulphite oxidase activity in spinach leaves. *Phytochemistry* 1995; 40: 667–672.

35. Janson J-C, Ryder LG. *Protein Purification: Principles, High Resolution Methods and Applications*, Wiley, New York, 1977.

36. Righetti PG. *Isoelectic Focusing Theory, Methodology and Applications*, Elsevier, New York, 1983.

37. Henschev A. *High Performance Liquid Chromatography in Biochemistry*, Verlagesellshaft, New York, 1985.

38. Ragan CI, Cherry RJ. *Techniques for the Analysis of Membrane Proteins*, Chapman and Hall, London, UK, 1986.

39. Deutscher MP. *Guide to Protein Purification Methods in Enzymology*, Academic Press, New York, 1990.

40. Harris El L, Angal S. *Protein Purification: A Practical Approach*, IRL Press, Oxford, UK, 1990.

41. Doonan S. *Protein Purification Protocols*, Humana Press, Totowa, NJ, 1996.

42. Scopes RK. *Protein Purification: Principles and Practice*, Springer-Verlag, New York, 1996.

43. Walker JM. *Protein Purification Handbook*, Humana Press, Totowa, NJ, 1996.

44. Eisethat R, Danson MJ. *Enzyme Assays. A Practical Approach*, QUP, Oxford University Press, Oxford, UK, 1992.

45. Dashek WV, Micales J. Assay and purification of enzymes-oxalate decarboxylase, in *Methods in Plant Biochemistry and Molecular Biology* (Dashek WV, ed.), CRC Press, Boca Raton, FL, 1997, pp. 49–71.

46. Moore NL, Mariam DH, Williams AL, Dashek WV. Substrate specificity, *de novo* synthesis and partial purification of polyphenol oxidase derived from the wood-decay fungus, *Coriolus versicolor. J Indust Microbiol* 1989; 4: 349–364.

47. Moore NL, Brako LA, Claussen C, Jones BR, Dashek WV. Distribution of polyphenol oxidase in organelles of hyphae of the wood deteriorating fungus, *Coriolus versicolor. Biodeterioration Research 4.* Plenum Press, New York, 1994.

48. Martyn RD, Samuelson DA, Freeman TB. Ultrastructural localization of polyphenol oxidase activity in leaves of healthy and disease water hyacinth. *Phytopathology* 1979; 69: 1278–1287.
49. Vaughn KC, Duke SO. Tissue localization of polyphenoloxidase in *Sorghum. Protoplasma* 1981; 108: 319–327.
50. Ryan ID, Gregory R, Tingey WH. Phenolic oxidase activities in glandular trichomes of *Solanum berthauttii. Phytochemistry* 1983; 21: 1885–1887.
51. Flurkey WH. *In vitro* biosynthesis of *Vicia faba* polyphenoloxidse. *Plant Physiol* 1985; 79: 564– 567.
52. Flurkey WH. Polyphenoloxidase in higher plants immunological detection and analysis of *in vitro* translation products. *Plant Physiol* 1986; 86: 614–618.
53. Mayer AM, Harel E. Polyphenol oxidases and their significance in fruits and vegetables, in *Enzymes in Foods*, Elsevier, Amsterdam, The Netherlands, 1990.
54. Moore BM, Flurkey WH. SDS activation of a plant polyphenoloxidase effect of SDS on enzymatic and physical characteristics of purified broad bean polyphenoloxidase. *Biol Chem* 1990; 265: 4982–4988.
55. Kowalaksi SP. Bamberg J, Tringey WM, Steffens JC. Inheritance of polyphenol oxidase in type A glandular trichomes of *Solanum berthaultii. J Hered* 1990; 81: 475–478.
56. Menter JM, Moore CL, Willis I, Fisher MS. Melanin markedly accelerates the tyrosinase mediated oxidation of phenolic depigmenters. *Photochem Photobiol* 1986; 43: 255.
57. Hutcheson SW, Buchanan BB. Polyphenol oxidation by *Vicia faba* chloroplast membranes studies on the latent membrane-bound polyphenol oxidase and on the mechanism of photochemical polyphenol oxidation. *Plant Physiol* 1966; 66: 1150–1154.
58. Sánchez-Ferrer A, Bru R, Garcia-Carmora F. Novel procedure for extraction of a latent grape polyphenoloxidase using temperature-induced phase separation in Triton X-114. *Plant Physiol* 1989; 91: 1481–1487.
59. Nunez-Delicádo E, Bru R, Sánchez-Ferrer A, Garcia-Carmona F. Triton X-114-aided purification of latent tyrosinase. *J Chromatogr B Biomed Appl* 1996; 680,105–107.
60. Kieselbach T, Hagman A, Andersson B, Schroder WP. The thylakoid lumen chloroplasts. Isolation and characterization. *J Biol Chem* 1998; 273: 6710–6716.
61. Wilhelnova N. Thylakoid membranes fragmentation by means of different detergents during leaf antogeny. *Photosynthetica* 1994; 30: 415–424.
62. Highley TL, Dashek WV. Biotechnology in the study of white-rot and brown-rot decay, in *Forest Products Biotechnology* (Bruce A, Palfreyman JW, eds.), Taylor and Francis, London, UK, 1998, pp. 15–36.
63. Micales JA. Oxalate decarboxylase in the brown-rot wood decay fungus, *Postia placenta. Mat U Org* 1995; 29: 177.
64. Dashek W, Micales JA. Unpublished.
65. Micales JA. Localization and induction of oxalate decarboxylase in the brown-rot wood decay fungus, *Postia placenta. Int Biodeter Biodegrad* 1997; 39: 125–132.
66. Green F, Larsen M, Highley TL. Ultrastructural morphology of the hyphal sheath of wood-decay fungi modified by preparation for scanning electron microscopy, in *Biodeterioration Research*, Vol. III, Plenum Press, New York, 1990.
67. Kim YS, Goodell B, Jellison J. Immuno-electron microscopic localization of extracellular metabolites in spruce wood decayed by brown-rot fungus, *Postia placenta*. International Research Group on Wood Decay, Stockholm, Sweden, 1990; 1441.
68. Micales JA, Richter AL, and Highley TL. Extracellular glucan production by *Postia* (=*Poria*) placenta. *Mater Org* 1990; 24: 259–270.

69. Green F, Clausen CA, Larsen MJ, Highley TL. Immuno-scanning electron microscopic localization of extracellular wood-degrading enzymes within the fibullar sheath of the brown-rot fungus, *Postia placenta. Can J Microbiol* 1992; 38: 898–904.
70. Larsen M, Green F. Mycofibrillar cell wall extensions in the hyphal sheath of *Postia placenta. Can J Microbiol* 1992; 38: 905–911.
71. Nicole M, Chamberland H, Geiger JP, Lecours N, Valero J, Rio B, Quellette GB. Immunocytochemical localization of laccase in wood decayed by *Rigidoporus lignosus. Appl Environ Microbiol* 1992; 58: 1727–1739.
72. Labrou NE, Clonis YD. Biomimetic dye-ligand for oxalate requiring enzymes: Studies with oxalate oxidase and oxalate decarboxylase. *J Biochem* 1995; 40: 59.
73. Lambert O, Lev D, Ranck J-L, Leblanc G, Rigaud J-L. A new "gel-like" phase in dodecyl maltoside-lipid mixtures: implications in solubilization and reconstitution studies. *Biophys J* 1998; 74: 918–930.
74. Vermelho AB, Pereira MC, Meirelles M. Use of the detergent Triton in biochemical and biological research: a minireview. *Ciencia Cultura* 40: 45–51.
75. Vinogradova O, Sonnichsen F, Sanders CR II. On choosing a detergent for solution NMR studies of membrane proteins. *J Biomol NMR* 1998; 11: 381–386.

13 Scanning Electron Microscopy

Preparations for Diatoms

David W. Seaborn and Jennifer L. Wolny

1. DIATOMS AND THE SCANNING ELECTRON MICROSCOPE

In studying, describing, and classifying diatoms, the scanning electron microscope (SEM) has become an indispensable tool for diatom taxonomists *(1,2)*. The diatom surface, composed of organic materials and silica, provides excellent working material for use in scanning electron microscopy when properly preserved and fixed. The siliceous cell wall, also called a frustule, is often covered with organic material that can obscure key morphological details if not properly removed prior to viewing with the SEM *(3)*. There are several procedures that will allow diatoms to be stripped of their outer organic coverings while keeping the frustule intact. Here, we describe the most common techniques used for SEM analysis of diatoms. Whenever possible, we have included slight derivations suggested in the current literature to these classical techniques.

From: *Methods in Plant Electron Microscopy and Cytochemistry*
Edited by: W. V. Dashek © Humana Press Inc., Totowa, NJ

2. ACID TREATMENT METHOD

Hasle and Fryxell *(1)* developed the classical method of cleaning diatoms by acid treatment in the late 1960s. This technique is probably the most commonly used for both light and scanning electron microscopy analysis and can be found published in many current taxonomy and morphology papers such as Hasle *(4)* and Villac and Fryxell *(5)*. Examples of cells prepared using the acid treatment method are pictured in Fig. 1. The method published by Hasle and Fryxell *(1)* involves the following steps, which should be carried out under a chemical fume hood:

1. Pour equal amounts of concentrated sulfuric acid (H_2SO_4) and the diatom sample into a 150-mL beaker and agitate gently.
2. Add freshly made potassium permanganate ($KMnO_4$) dropwise until the sample turns from brown to purple. Agitate the solution gently between each addition of $KMnO_4$ to oxidize the organic matter. (Note: Round et al. *(3)* recommend a method of $KMnO_4$ oxidation that includes the addition of concentrated HCl until the solution becomes clear instead of using step 3).
3. Add freshly made oxalic acid $(COOH)_2$ dropwise until the solution becomes clear. Agitate the solution gently between each addition of $(COOH)_2$. The solution will bubble significantly.
4. Transfer the solution to a centrifuge tube and centrifuge down to a loose 5 mL pellet. We recommend a medium speed for 10 min. Decant off excess liquid. (Note: Miller and Scholin *[6]* developed a system using a filter tube and filtration manifold that reduces sample processing time and causes less cell damage than centrifuging.)
5. The sample is washed by adding distilled water, agitating the sample, centrifuging as described, and removing the supernatant several times.
6. Check the sample to see if any organic material remains on the cells by placing a drop under a light microscope. If organic material remains, repeat the procedure starting at step 2.

   If no organic material remains, the sample is ready to be analyzed under the SEM.

A final step that we have found useful, and recommended in Postek et al. *(7)*, involves rinsing the sample with distilled water at least 6 times to separate the cells before mounting the specimens.

3. HYDROGEN PEROXIDE METHOD

Removal of the outer organic layer can be achieved quickly and effectively through the introduction of hydrogen peroxide (H_2O_2) *(8)*. Preparation of diatom samples for SEM analysis using H_2O_2 is described in the following protocol, the results of which can be seen in Fig. 2.

Fig. 1. Scanning electron micrographs of diatoms prepared using the acid treatment method. Bars = 10 μm. (**A**) Unidentified centric diatom. (**B**) Interior of the frustule of *Actinoptychus* sp. (**C**) *Cyclotella meneghinii*. (**D**) *Actinoptychus* sp.

187

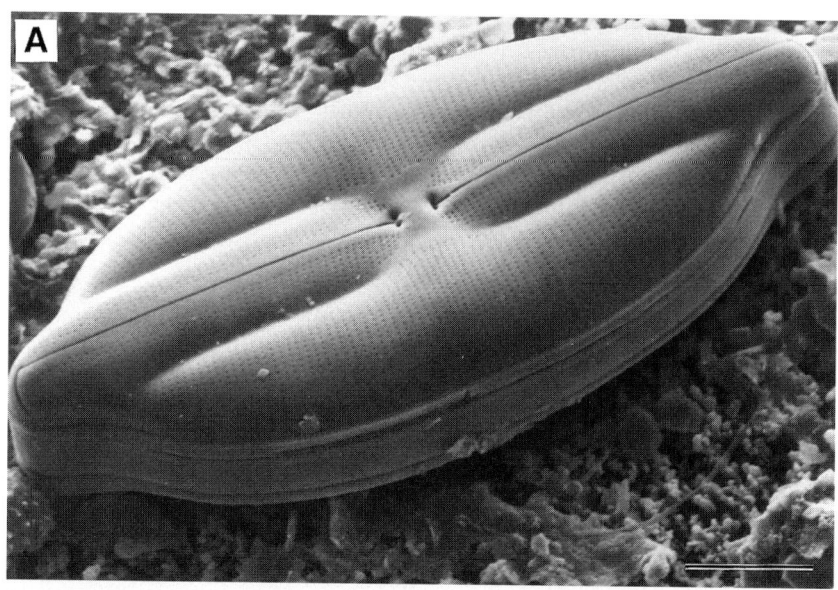

Fig. 2. Scanning electron micrographs of diatoms prepared using the hydrogen peroxide method. (**A**) A naviculoid diatom (bar = 10 μm). (**B**) *Fragilaria* sp. (bar = 13 μm).

1. Transfer 50 mL of cultured diatoms or 25 mL of a field sample to a 100-mL beaker.
2. Place the beaker on a hot plate, heat to 100°C, and let evaporate until a thin film of liquid remains on the bottom.
3. Adjust temperature to 60°C and let beaker cool for 15 min.
4. Add 20 mL of H_2O_2 to the beaker. (Note: We have found that a 30% H_2O_2 solution works best, as recommended by Edlund and Brant [9].)
5. Allow the solution to boil vigorously (1 h for culture samples or 2 h or more for field samples). Keep beaker covered with a watch glass while boiling to prevent evaporation. The oxidation is complete when the diatom fraction is white.
6. Wash the sample by dividing it into two centrifuge tubes, filling the tube with distilled water, and centrifuging at medium speed for 10 min.
7. Decant supernatant, add new distilled water, and centrifuge for an additional 10 min.
8. Repeat the procedure so that the sample is washed a total of five times. After all washes have been completed, the sample is ready for SEM analysis.

Edlund et al. *(10)* published a protocol similar to Vyverman et al. *(8)*, but recommend gently oxidizing diatoms overnight in cold 30% H_2O_2 at step 5 instead of boiling. At step 6, Edlund et al. *(10)* prefer to rinse the material with distilled water six times, each rinse followed by 6 h of settling to remove oxidation by-products.

4. OBTAINING DIATOM SPECIMENS FROM SEDIMENTS

We have found the method described by Sullivan *(11)* extremely useful for extracting diatoms out of marine sediment deposits (*see* Fig. 3). Using the following technique, Sullivan *(11)* was able to identify new genera and species of diatoms.

1. Clean the sample with a mixture of concentrated hydrochloric acid (HCl) and 20% H_2O_2. Let sit for 4 h at room temperature.
2. Dehydrate and carbonize sample in boiling concentrated H_2SO_4.
3. Oxidize sample by adding 1 mL of 80% nitric acid (HNO_3).
4. The sample is then leached with the addition of a 10% aqueous solution of sodium hydrogen phosphate/sodium bicarbonate at 55°C for 15 min.
5. Sieve the sample through a 27-μm mesh, wash with distilled water, purify with acetone, and dry at 100°C. The final result is a dry powder, which is used to make uncovered dry spreads.

5. PREPARATION OF ATTACHED DIATOMS

Many diatoms, especially pennate diatoms, are attached to substrates for much of their lives. This presents a unique problem for the electron

Fig. 3. Removal of diatoms from sediment using the method described by Sullivan *(11)*. (**A**) *Coscinodiscus* sp. (bar = 15 μm). (**B**) Unidentified centric diatom (bar = 10 μm).

microscopist. The success of the fixation procedure depends on the substrate type itself. We have found the method published by Postek et al. *(7)*, and modified in Holt et al. *(12)*, useful for preparing specimens both on living and nonliving substrates. The method works equally as well for diatoms and dinoflagellates. Diatoms fixed using this method can be seen in Fig. 4.

1. Harvested samples should be immediately prepared for the SEM by fixation in a 0.5% osmium tetroxide solution in the dark for 1 h. If the sample is collected in the field, store in buffered 3% glutaraldehyde solution (pH 7.2) until samples can be fixed.
2. Rinse sample in a phosphate buffer (pH 7.2) three times.
3. Dehydrate the sample using a standard ethanol series (30%, 50%, 70% (at which point the preparation can be stored for up to 24 h at 4°C), 90%, and 100% (wash sample in 100% EtOH three times).
4. Transfer samples from 100% EtOH to the critical point dryer to complete dehydration with liquid carbon dioxide.

Garduno et al. *(13)* developed a technique for preparing cultured diatoms attached to a solid agar. The method involves the following steps:

1. Diatom colonies formed in polystyrene culture dishes were fixed by adding a 2.5% glutaraldehyde solution to culture dish.
2. Specimens were dehydrated by the standard ethanol series as described.
3. Small discs (about 15 mm in diameter) were cut from the agar using a cork-boring machine.
4. Discs were critically point dried using liquid carbon dioxide.

Note: We recommend that after performing step 1 above that the specimens be rinsed three times in a phosphate buffer (pH 7.2), which will remove any residual glutaraldehyde before beginning the dehydration series. Garduno et al. *(13)* also use a second technique in which they cut small pieces of medium containing diatoms and immediately fix the pieces in liquid nitrogen. The pieces are freeze-dried and then gold sputter coated.

6. PREPARING CLEANED DIATOMS FOR VIEWING ON THE SCANNING ELECTRON MICROSCOPE

After using any of the foregoing techniques (except those for attached diatoms, which may or may not be attachable to standard SEM stubs), the sample must be readied for viewing on the SEM. A quick and effective method of doing this is as follows:

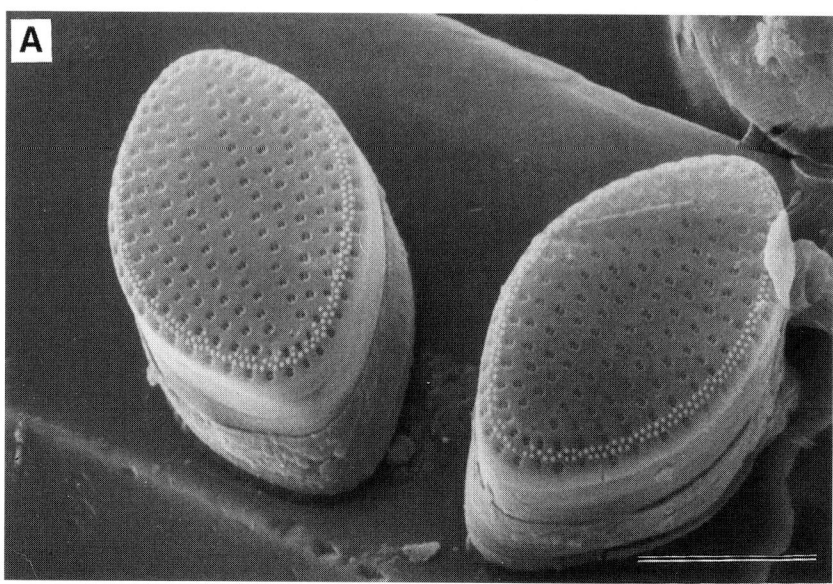

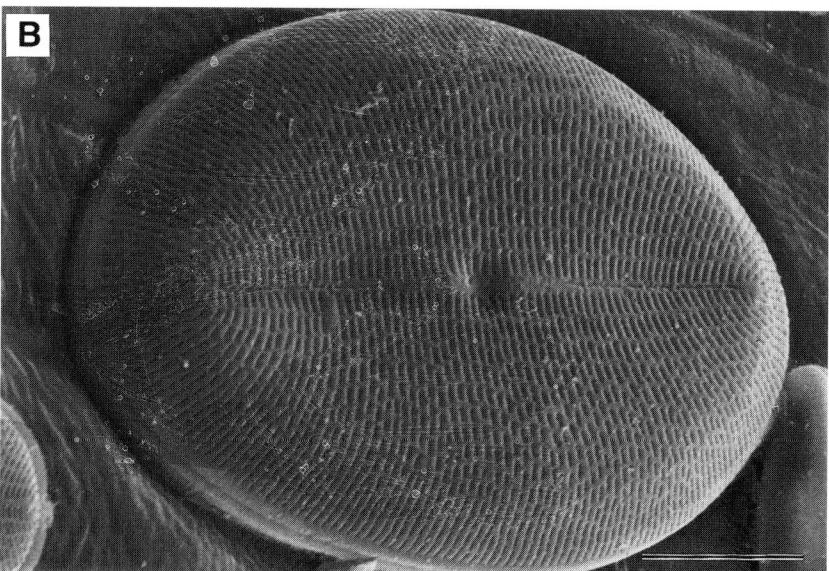

Fig. 4. Scanning electron micrographs of attached diatoms. (**A**) Attached pennate diatoms (bar = 6 μm). (**B**) *Cocconeis* sp. (bar = 5 μm).

1. An aliquot of the diatom sample is brought down onto a Nucleopore filter ($2.0\,\mu$m pore size will work for most diatoms) using a hand pump or light vacuum filter system.
2. Attach the filter to an aluminum stub using double stick tape.
3. Place stub in a dessicator over night to remove excess moisture.
4. Sputter coat the stub with a gold-palladium (Au/Pd) mixture to a thickness of 10 nm, as recommended by Lee et al. *(2)*.

When viewing the stubs, Lee et al. *(2)* recommend an accelerating voltage of 10–20 kV. We have found that using the foregoing procedures 15 kV gives the best resolution, as did Droop *(14)* and Garduno et al. *(13)*. When photographing images both Droop *(14)* and Round et al. *(3)* recommend tilting the specimens at a 40–45° angle for an added degree of dimensionality.

ACKNOWLEDGMENTS

The authors gratefully acknowledge Michael Adam in the Old Dominion University Electron Microscopy Laboratory for his patience and assistance with SEM preparations and the micrograph work in this text. We also would like to thank Dr. Harold Marshall for his assistance in identifying the diatoms pictured here.

REFERENCES

1. Hasle G, Fryxell, G. Diatoms: cleaning and mounting for light and electron microscopy. *Trans Am Microscopical Soc* 1970; 89: 435–468.
2. Lee J, Reimer C, Mahoney R. Some thoughts about the conservation of scanning electron microscopic preparations of diatoms in a museum repository. *Scanning Electron Microsc* 1986; 4: 1403–1406.
3. Round F, Crawford R, Mann D. *The Diatoms: Biology and Morphology of the Genera*, Cambridge University Press, Cambridge, UK, 1990.
4. Hasle G. *Pseudo-nitzschia pungens* and *P. multiseries* (Bacillariophyceae): nomenclatural history, morphology, and distribution. *J Phycol* 1995; 31: 428–435.
5. Villac M, Fryxell G. *Pseudo-nitzschia pungens* var. *cingulata* var. nov. (Bacillariophyceae) based on field and culture observations. *Phycologia* 1998; 37: 269–274.
6. Miller P, Scholin C. Identification and enumeration of cultured and wild *Pseudo-nitzschia* (Bacillariophyceae) using species-specific LSU rRNA-targeted fluorescent probes and filter-based whole cell hybridization. *J Phycol* 1998; 34: 371–382.
7. Postek M, Howard K, Johnson A, McMichael K. *Scanning Electron Microscopy: A Student's Handbook*, Ladd Research Industries, Inc., Burlington, VT, 1980.
8. Vyverman W, Sabbe K, Vyverman R. Five new freshwater species of *Biremis* (Bacillariophyta) from Tasmania. *Phycologia* 1997; 36: 91–102.
9. Edlund M, Brant L. *Frustulia bahlshii* sp. nov, a freshwater diatom from the eastern USA. *Diatom Res* 1997; 12: 207–216.
10. Edlund M, Stoermer E, Taylor C. *Aulacoseira skvortzowii* sp. nov. (Bacillariophyta), a poorly understood diatom from Lake Baikal, Russia. *J Phycol* 1996; 32: 165–175.

11. Sullivan M. *Porguenia perviana* gen. et. sp. nov., a marine centric diatom with an unusual ocellus. *J Phycol* 1997; 33: 881–887.

12. Holt J, Merrell J, Seaborn D, Hartranft J. Population dynamics and substrate selection by three *Peridinium* species. *J Freshwater Ecol* 1994; 9: 117–128.

13. Garduno R, Hall B, Brown L, Robinson M. Two distinct morphotypes of *Amphora coffeaeformis* (Bacillariophyceae) cultured on solid media. *J Phycol* 1996; 32: 469–478.

14. Droop S. *Diploneis sejuncta* (Bacillariophyta) and some new species from an ancient lineage. *Phycologia* 1998; 37: 340–356.

14 Methods for the Ultrastructural Analysis of Plant Cells and Tissues

William V. Dashek and John E. Mayfield

1. INTRODUCTION

Transmission electron microscopy (TEM) has yielded a copious amount of detailed information regarding the subcellular organization of plant cells. The morphological details of certain organelles just visible to the light microscopist were rendered apparent by TEM. In other instances, the enhanced magnification and resolving power revealed organelles not observed with conventional light microscopy. The subsequent coupling of TEM with biochemistry has resulted in a profound understanding of structure–function relationships providing for a rather thorough knowledge of how cellular organelles function.

Currently, transmission and scanning electron (*see* Chapter 13) microscopies have become the two workhorses for day in and day out electron microscopy in both research and clinical laboratories. Many of the other techniques discussed in this volume are more recent offshoots of these two widely employed electron microscopies. They have further enriched our appreciation of plant cell structure-function.

Although there is an abundant supply of general em textbooks and advanced monographs *(1–14)*, only a few have been dedicated to the

From: *Methods in Plant Electron Microscopy and Cytochemistry*
Edited by: W. V. Dashek © Humana Press Inc., Totowa, NJ

Table 1
List of Fixation Conditions for Preparation
of Plant Cells and Tissues for Transmission Electron Microscopy

---

Fixative/Application and Conditions for Use

Prefixation[a]

   1% Paraformaldehyde/2% glutaraldehyde in 0.1M cacodylate buffer, pH
   6.9, or 0.1 M PIPES, pH 6.9. To optimize quality of fixation, one can add
   1–3 mM CaCl$_2$ and 1% sucrose. Fix in 1 h, but optimum fixation time can
   vary, i.e, as long as overnight.

Postfixation

   2% Buffered OsO$_4$ for 2 h.

Prefixation

   2–6.25% Glutaraldehyde with the possible addition of 0.5–4% paraformaldehyde in 0.2–0.02 M phosphate or cacodylate buffer, pH 6.8–7.2; can add
   1–3 mM CaCl$_2$ or 1–3% mM sucrose; prefix 1–2 h or, in some cases,
   overnight.

Postfixation

   1% OsO$_4$ in 0.2–0.02 M veronal, phosphate, or cacodylate buffers, pH
   6.8–7.2, 30–90 min.

In some instances, e.g., fungi, potassium permanganate can be useful as a
   fixative (*see* ref. *[62]*) or as a postfixative after aldehyde prefixation.

---

   [a]Tannic acid can be added to the prefixative (1–4% w/v) in cacodylate buffer, or
tannic acid in 0.1 M buffer 1–2 h can be used after osmication between fixation and
dehydration.
   Adapted from ref. *(21)*.

plant sciences *(15–20)*. The preparation of plant cells and tissues requires
special precautions because of the unique properties of plant cell walls
and vacuoles. Because this preparation has been published *(15–20)*, the
current discussion focuses on recent improvements in fixation, dehydration, embedding, sectioning, positive staining, viewing, photography,
and darkroom procedures. The reader is referred to Chapter 16 for a pres-
entation of attempts to develop specific stains at the ultrastructural level.

2. TISSUE PREPARATION

2.1. Fixation, Dehydration, and Embedding

As mentioned, chemical fixation of plant cells has been reviewed many
times *(15–20)* and the reader is referred to these citations for a variety
of fixation procedures for preserving plant cells and tissues. One of the
most recent references regarding the topic is that of Hopwood and Milne
(21). Table 1 presents their recommendations regarding fixation of plant
cells and tissues for electron microscopy.

Table 2
Cryogens Used for Freezing in Order of Efficiency

Cryogen	Temperature	Notes
Liquid nitrogen (subcooled)	−210°C	Only useful for freezing well-cryopro-protected specimens, e.g., sucrose-infused tissue
Liquid freon or Arcton 12 or 22	−150°C	Freon is environmentally friendly
Liquid propane	−185°C	Highly hazardous, as it can be explosive when mixed with oxygen
Liquid ethane		Can be explosive
Liquid helium		Useful in coating a metal freezing surface

Because chemical fixation can result in cellular components being displaced from their in vivo state or completely extracted and because the chemical nature of molecules may be altered, rapid freezing methods (Tables 2 and 3) for preserving the in vivo structure of plant cells and tissues, e.g., cytoplasmic streaming or vesicle movement, have been developed. This topic, which warrants a chapter in itself, has been reviewed thoroughly *(21–29)*. In addition, a monograph exists regarding the cryopreservation of plant cells and organs *(30)*.

Recently, microwave fixation *(21)* has been utilized as an alternative to chemical fixation as another effort to circumvent artefacts induced by chemical fixation. During microwave fixation, heat is delivered in a controlled manner uniformly throughout the tissue, i.e., thin slices. The penetration of microwaves into the tissue is limited to approx 2 cm from the surface.

Dehydration occurs subsequent to prefixation, postfixation, and buffer washes. The dehydrating agents which are most frequently employed are ethanol or acetone depending upon the embedding resin. It is imperative that em-grade ethanol or acetone be used and that the ethanol is 200 proof. A typical alcohol dehydration scheme is presented in Fig. 1, which compares tissue preparation for TEM and SEM. Following dehydration, specimens are usually passed through a transitional solvent such as propylene oxide (Fig. 1) and then progressively embedded in a graded transitional solvent/embedding media series with rotation to ensure infiltration. It should be noted that Spurr's embedding medium *(32)* and Polybed are miscible with acetone and, thus, the tissue can be dehydrated through a graded acetone series and then progressively embedded in acetone/resin with final embedment in pure resin. Table 4

Table 3
Summary of Procedures for Rapid Freezing of Tissues

Methods of rapid freezing	Notes
Plunge freezing	Plunge specimen into a container of cryogen with forceps; cell suspensions can be adhered to coverslips coated with poly-L-lysine; design of cryogen baths is important (27); baths are cheaper than other freezing methods
Jet freezing	Jet freezers emit two fine, synchronized jets of propane on either side of the specimen; applicable to small specimens sandwiched between two thin supporting films
Spray freezing	Specimens in the form of an emulsion, suspension or solution; formation of 10–50 μm microdroplets that can be sprayed into liquid propane
Cold block slamming	Rapid slamming of a specimen against a copper block cooled with liquid helium or liquid nitrogen; complicated and expensive method
High-pressure freezing	Specimens are subjected to a pressure of 2100 bar, which depresses the melting point of water, reduces the formation of ice nuclei, and retards the growth rate of crystals
Freeze-substitution	Replacement of ice within the specimen with another solvent while keeping the specimen at low temperature
Freeze-fracture and replication	Fixation of specimens and cryoprotected prior to freezing

The reader is referred to Hawes (31) and Robards (27) for in-depth discussions and protocols for rapid freezing.

Transmission electron microscopy	*Scanning electron microscopy*
Prefix in buffered glutaraldehyde (*see* Table 1)	Prefix in buffered glutaraldehyde (*see* Table 1)
Wash in buffer	Wash in buffer
Postfix in buffered OsO_4 (*see* Table 1)	Postfix in buffered OsO_4 (*see* Table 1)
Wash in buffer	Wash in buffer
Dehydrate in a graded alcohol series (30%, 50%, 70%, 90%, 100%), for 100% use in 200 proof alcohol	Dehydrate in a graded alcohol series series (30%, 50%, 70%, 90%, 100%), for 100% use in 200 proof alcohol
Propylene oxide (2X, 100%)	Freon or amyl acetate (use a
Propylene oxide, Polybed 812 or Spurr's (1:1, 1:2, 1:3)	graded series, e.g., 1:3, 1:2, 1:1 Freon/alcohol and then pure Freon)
Infiltrate with pure embedding medium using a rotating device	Critically point dry using liquid CO_2 soaks and flushes
Embed pure Polybed 812 or Spurr's	Mount on aluminum stubs using silver paint or double-stick tape[a]
Polymerize according to instructions for Polybed 812 or Spurr's (*see* company technical bulletins)	Sputter coat with gold/palladium in a sputter coater flushed with argon
Trim block to yield a trapezoidal face	View directly or store specimens in a desiccator until viewing
Section with glass or diamond knives 1 mm thick section	Photograph
Stain with Richardson's stain or toluidine blue	
Reface block so that trapezoid contains desired tissue	
Thin section using glass or diamond knives	
Collect sections on acetone-washed copper grids (*see* Subheading 2.2.)	
View directly or positive stain with uranyl acetate and lead citrate (sections can be stored in a grid box)	

[a]Note that in some instances, spores and pollen grains can be air-dried and mounted onto stubs.

Fig. 1. Comparison of transmission and scanning electron microscopy protocols.

offers information beneficial in the selection of an appropriate embedding medium. Dehydration and embedding are discussed in most standard em books. In addition, Smith and Croft *(33)* and Robards and Wilson *(16)* discuss these topics in depth. Embedding media are often marketed as kits that contain instructions for their uses.

Table 4
Resins Commonly Used for Embedding
Plant Tissues for Transmission Electron Microscopy

Resin	Application and conditions for use

Acrylic
  n-Butyl and methyl methacrylate
  Glycol methacrylate
	Water-miscible, add *n*-butyl-methacrylate or styrene to improve sectioning

LR White	Low viscosity as well as low toxicity suited for immunocytochemistry; available as hard, medium, or soft. Hard is best for biological samples; resin is polymerized by heat, UV light, or chemicals
LR Gold Lowicryl	Developed for use with unfixed material can be cured at room temperature by the addition of 1% dry bezoin peroxide

Epoxy
Araldite
  Araldite CY212 10 mL,
  DDSA 10 mL, DBPth 0.25 mL;
  mix for 20 min, add 0.15 mL
  BDMA and mix for 20 min.	

Spurr's formulation	ERL/VCD 4206 10.0 g, DER 736 6.0 g, NSA 26.0 g, dimethylamino-ethanol 0.4 g. DER component can be varied from 4–8 g; use the same day or store at −20°C; polymerize 9 h at 70°C
Epon formulation	Epon 812 15 mL, DDSA 8 mL, MNA 8 mL, mix well and add the BDMA 0.6 mL; alternatively, mix A Epon 812 6.2 mL, DDSA 10.0 mL; Mix B Epon 812 10.0 mL, MNA 8.9 mL; to obtain different hardnesses mix Epon A and B as per kit instructions

Melamine
  FB 101
  AME 01
  MUV 116
Polyester
  Vestopal	

Table 4 (Continued)

Resin	Application and conditions for use
Others	
Urea-aldehyde	
(50% glutaraldehyde mixed with	
4% urea using oxalic acid)	
Gelatin	
Bacteriological-grade gelatin	
mixed with 2% glycerine	
Glutaraldehyde/carbohydrazide	
Lipid-retaining embedding medium	

Adapted from *(33)*.

Referred to Roos *(34)*.

The directions for thorough mixing of the reagents in disposable plastic beakers are included with commercially available resin kits. Most embedding media are toxic, requiring the use of chemical hoods. In addition, certain reagents can cause dermatitis and thus require the use of gloves.

2.2. Sectioning and Staining

2.2.1. ULTRAMICROTOMY AND STAINING

Subsequent to curing (polymerization of the resin), the embedded specimens are ready for sectioning. The process of trimming the em block into a trapezoid to encompass the tissue followed by thick sectioning, reshaping the trapezoid, and thin sectioning (Fig. 2) have received repeated attention (*see* any conventional TEM book). Similarly, methods for section "pickup" onto grids (Fig. 2) and subsequent positive staining (Table 5) have also been detailed.

Although 200- or 300-mesh copper grids are most popular, grids composed of nickel, gold, palladium, stainless steel, beryllium, and carbon-coated nylon can be used. The choice of a grid other than copper is dictated by the em procedure being employed and, thus, the reader should consult the appropriate chapters of this volume for grid choice. Reid *(35)*, Reid and Beesley *(36)*, Robards and Wilson *(16)*, and Smith and Croft *(33)* describe ultramicrotomy. Cryosectioning has been treated by Reid *(35)*, Richter *(37)*, and Erk et al. *(38)*.

Seasoned electron microscopists are aware of artefacts such as those resulting from improper fixation, dehydration, embedding, and sectioning. In addition, contamination of sections during positive staining (Table 5) is quite common. Crang and Klumparens *(39)* address artefacts in depth and the beginning electron microscopist should consult these authors.

2.2.2.1. Negative Staining. This technique differs from positive staining in that heavy metals surround the specimen leaving the latter unstained

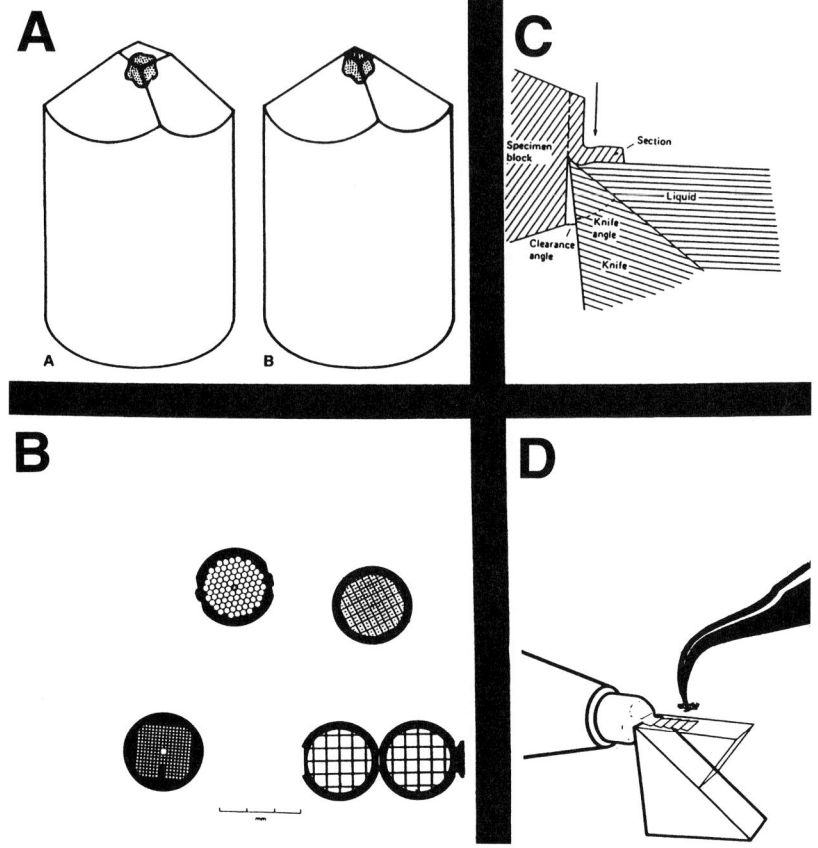

Fig. 2. Composite figure illustrating (**A**) block trimming to a trapezoid, (**B**) TEM grids, (**C**) ultramicrotomy, and (**D**) section pickup. (**B**) and (**C**) from Watt *(54)* with permission.

and visible against its stained background. The technique is employed for revealing surface structure on particles, such as viruses, macromolecules, or isolated cellular organelles. Table 6 presents some of the commonly employed negative stains together with their applications. Prior to negative staining, em grids must be coated with support films such as carbon films, formvar plastic films or carbon-coated plastic films. The preparation of support films has been extensively reviewed. The methods for negative staining are summarized in Table 7. The reader is referred to Benedetti and Emmelot *(42)*, Spiess et al. *(41)*, and Harris and Horne *(43)* for the many intricacies regarding negative staining. Chapter 11 presents negatively stained profiles of isolated Golgi bodies.

Table 5

Summary of Positive Stains Routinely Employed in TEM[a]

Stain	Conditions	Specificity
Lead citrate/uranyl acetate[b]	Step 1: Float or immerse sections for 10–30 min on filtered 1–2% aqueous uranyl acetate (or in EtOH); wash with ultrapure H_2O (three beakers of 50 mL each) by dipping grids held with a forceps; dry for 5 min Step 2: Place drops of lead citrate (lead carbonate free) onto a wax surface (parafilm or dental wax) in a Petri dish; line edges of dish with pellets of KOH; float grid with sections (sections face down) for 4–5 min (if overstained 2–3 min and dilute stain); wash grids with sections in ultrapure H_2O	Nonselective enhancement of membrane contrast, ribosomes, and nuclear material; proteins and lipid droplets
Periodic acid–thiocarbozide–Ag proteinate (PATAg test)	Use glutaraldehyde-osmium-fixed, epoxy-embedded sections; collect sections on gold grids; float for 20–30 min on 1% periodic acid and then wash with H_2O; float sections successively at room temperature on (1) 0.2% TCH or TSC in 20% acetic acid for 3–4 h (or overnight) (e.g., 2 mL acetic acid + 0.02 g TCH (or TSC) adjusted to 10 mL with water); (2) an aqueous solution of acetic acid with decreasing concentration for 30 min; (3) water for 30 min; (4) 1% aqueous silver proteinate (in the dark; the stain must be made at least 12 h before use) for 20–30 min; and (5) water, two washes of 10 min. Sections are mounted on grids and allowed to dry	Polysaccharides

Table 5 (Continued)

Stain	Conditions	Specificity
Phosphotungstic acid (PTA) or silicotungstic acid (STA)	1 g STA or PTA in 100 mL of 10% aqueous solution of chromic acid or 100 mL 10% HCl; store in refrigerator. Float sections for 2–15 min at room temperature on the solution and rinse with dH$_2$O; if specimen OsO$_4$ —fixed, float sections on 10% H$_2$O$_2$ (or 1% periodic acid) for bleaching	Plasma membrane; Golgi complex-derived vesicles
Tannic acid	Tannic acid can be added at 1–4% w/v in either cacodylate-buffered glutaraldehyde or glutaraldehyde-formaldehyde or between pre-fixation and postfixation (immerse specimens 1–2 h in tannic acid in 0.1M buffer)	Tonoplast
Zinc/Iodide—OsO$_4$	See Chapter 16	Golgi apparatus

[a] Summarized from Roland and Vian (19). The reader is referred to Chapter 16 and ref. (40) for discussions of ultrastructural cytochemistry.
[b] There is a number of lead citrate formulations. A modification of the Reynold's procedure (60a) is that of Sato (61). Meticulous care must be taken to keep the lead citrate solution free of lead carbonate, which originates via reaction of lead citrate with atmospheric CO$_2$. Once prepared, the solution can be stored in capped syringes at 4°C.

Table 6
Summary of Negative Stains

Negative stains[a]	Concentration and useful pH range	Application(s)	Reference(s)
Phosphotungstic acid	2% aqueous, pH 6.5–7.5 (note some investigators employ 4%)	Virus particles, proteins molecules, liposomes	Harris and Horne (43)
Uranyl acetate[b]	1% in 75% ethanol	Chromatin	Zentgraft et al. (44)
	1–2%, pH 4.2–4.5	Protein–nucleic acid complexes	Sogo et al. (45)
		Ribosomes	Khulbrandt and Unwin (47)
		Tubulin	Mandelkow and Mandelkow (46)
Ammonium molybdate	2%, pH 5.7	Osmotically sensitive organelles	Harris and Horne (43)
Sodium silicotungstate	2%, pH 5–8	Macromolecules, membranes, and viruses	Roland and Vian (19)
Methylamine tungstate			Harris and Horne (43)

[a] Maintain all stains at room temperature; filtration of negative staining solutions may be required before use. (Note: do not invert solution prior to use.)

[b] Some investigators have employed ammonium uranyl oxalate, uranyl formate, or uranyl nitrate—the choice of which depends on the graining image the investigator is willing to accept. (Note: uranyl salts are radioactive—consult your radiation safety officer for your institution's use.)

Table 7
Methods for Negative Staining

Negative staining procedure[a]	Application	Reference
On-grid droplet procedure	Rapid preparation of a single sample rather than multiple samples	Huxley and Zubay (48), Roland and Vian (19)
Single-droplet procedure	Multiple samples	Benedetti and Emmelot (42); Harris and Aguther (49)
Negative staining–carbon film procedure	Useful for purified proteins	Horne and Pasquali-Rochetti (50); Harris (51)
Immobilized cell negative staining–carbon film procedure	Cultured cells	Harris (52)
Spraying methods onto support films	Viruses, macromolecules, and liposomes	Spiess et al. (41)

Referred to Harris and Horne (43), Roland and Vian (19), Sogo et al. (45), and Speiss et al. (41) for protocols.

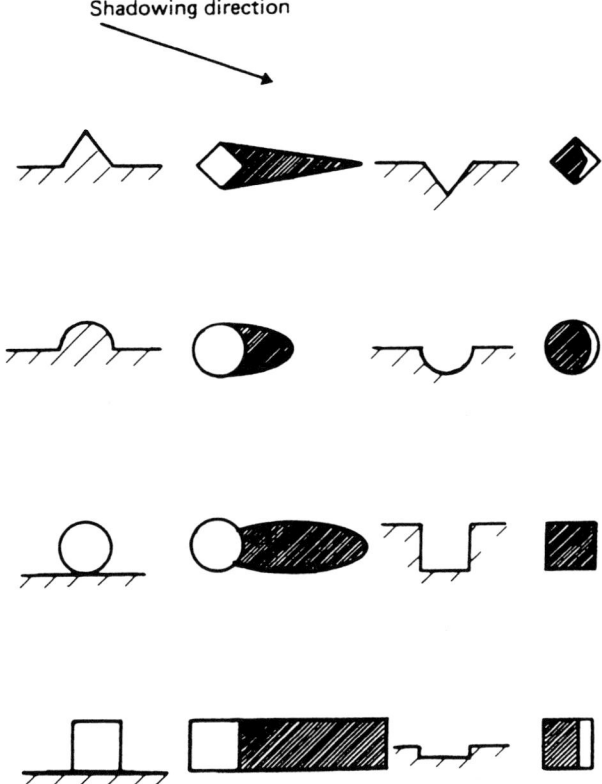

Fig. 3. Shadows cast by simple solid profiles. It can be seen that the shadows cast in positive and negative replicas are not interchangeable in either shape or tone. (From ref. *[54]* with permission.)

2.3. Shadowcasting

Shadowcasting is a technique that yields a three-dimensional look to TEM images. This is achieved by obliquely coating a specimen with a very thin deposit of heavy metal evaporated from a point source (*see* Fig. 3). Although platinum, tungsten and carbon have been widely used as the coating materials, a variety of other heavy metals can be employed. These coating materials have been arranged in an order of decreasing refractiveness *(53)*. Metal deposit is maximal on portions of the specimen's surface normal to the direction of deposition and minimal on areas sloping away from the source *(54)*. Although the angle of incidence is normally between 10° and 45°, other angles have been employed to vary the mass of metal evaporated. Slayter *(53)* has discussed this critical aspect of shadowcasting in depth.

The net effect of shadowcasting is to create highlights and shadows in an electron image enabling the investigator to determine both the size and form of a surface feature as well as the thickness of a particle. Thus, shadowcasting can be employed in either a quantitative or qualitative manner yielding considerable information regarding biological structure of macromolecules. Perhaps, this has been most apparent in the revelation of the orientation of cellulosic microfibrils in plant cell walls (*see* Figs. 1–10 in ref. *[19]*). In addition, the current author employed this technique to determine the orientation of cellulosic microfibils in elongating pollen tubes (*55*).

Shadowcasting protocols for electron microscopy can be found in Glenny (*56*), Coggins (*57*), and Slayter (*53*). The two former references concern proteins and nucleic acids, and the latter reference is dedicated to a comprehensive application of the technique.

3. PROTOCOL

Transmission Electron Microscopy of *Lilium longiflorum*[a] Pollen Tubes (58)

Chemicals	Supplies	Equipment
Acetone (EM grade)	50 mL beakers	Darkroom
Dektol	Beem capsules	Enlarger and negative
D-19	Copper grids—	holder
Ethanol(95% for	200–300 mesh	Hot plate
dilutions and	Dental wax	Knife maker
200 proof)	EM forceps	Magnetic stirrer
Epon or Spurr's	Glass for glass knives	Print trays
embedding resins	Glassine envelopes	Transmission electron
Glutaraldehyde	Grid boxes	microscope
Lead citrate	65° incubator	TEM film, e.g., Estar
Kodak Rapid Fixer	Plastic beakers	
Osmium tetroxide	Plastic boats for	
Phosphate	glass knives	
Dibasic	Plastic Petri dishes	
Monobasic	Q-tips	
Photoflow	Razor blades—	
Propylene oxide	single edge	
Stop bath solution	Rotation device for	
Tongs for print	embedding	
development	Snap cap vials	
Uranyl acetate	Syringes—1 cc	
	Waste containers	
	Ethanol	

[a] *See* Chapter 3.

1. Procure, grow, and harvest lily pollen tubes as in Chapter 3.
2. Fix 2 h germinated pollen in 2.5% EM-grade glutaraldehyde in 0.1 M phosphate buffer, pH 7.2–7.4 for 2 h at 4°C. (*See* Table 1 for alternative

fixations for other than pollen tubes.) Wash fixed pollen with phosphate buffer 2X, 10 min each time.

3. Dehydrate through a graded ethanol series and cap vials each time.
   30%, 2X, 10 min each time
   50%, 2X, 10 min each time
   70%, 2X, 10 min each time
   95%, 2X, 10 min each time
   100% (200 proof), 3X, 30 min each time

4. Pass through a transitional solvent, e.g., EM-grade propylene oxide 2X, 30 min each time.

5. Infiltrate with a resin (*see* Table 4 for alternative embedding media for other than pollen tubes).
   Progressive Infiltration with Propylene Oxide: Polybed
   > 2:1 Propylene oxide: Polybed, 1 hr
   > 1:1 Propylene oxide: Polybed, 1 hr
   > 1:2 Propylene oxide: Polybed, 1 hr

   (Polybed formulation procedures accompany commercially available polybed kits.)

6. Infiltrate with pure polybed, overnight use rotating device to ensure maximum resin infiltration.

7. Embed pure polybed in beem capsules or flat-bed molds. (Spurr's embedding medium is especially suitable for plant material. If Spurr's medium is used, material can be dehydrated through EM-grade acetone, as acetone and Spurr's are compatible.)

8. Polymerize 65°C for 48 h.

9. Trim block face to a trapezoid with greaseless razor blade.

10. Section with glass knives or diamond knife and ultramicrotome. *Note*: Usually thick sections (1 μm) are cut, affixed to glass microscope slides, and stained 5–15 min at 40–50°C with toluidine blue solution (1 g toluidine blue and 1 g of sodium borate in 100 mL H_2O), The block is then refaced so that the trapezoid encompasses the desired tissue. Thin sections of gray, silver, or gold interference colors are cut.

11. Pickup sections onto 200–300 mesh, acetone-washed copper grids. (Alternatively, one can use formvar-coated slot grids to enhance the viewing area. Recently, Santos et al. *(59)* published a simpler method for fixing, dehydrating, and embedding pollen tubes.) (See any of the EM volumes for methods of sectioning and section pickup.)

12. Stain with uranyl acetate and lead citrate.

13. Place drops of 2% aqueous uranyl acetate onto parafilm in a Petri dish; use a 1-mL syringe equipped with a 0.2-μm filter.

14. Float grids with attached thin sections on stain droplets for 20 min (float section side down; cover Petri dish with box to protect from light).

15. Wash with ultrapure H_2O using three separate 50 mL beakers.

16. Dry sections at least 5 min.

17. Examine in TEM or store in grid box.

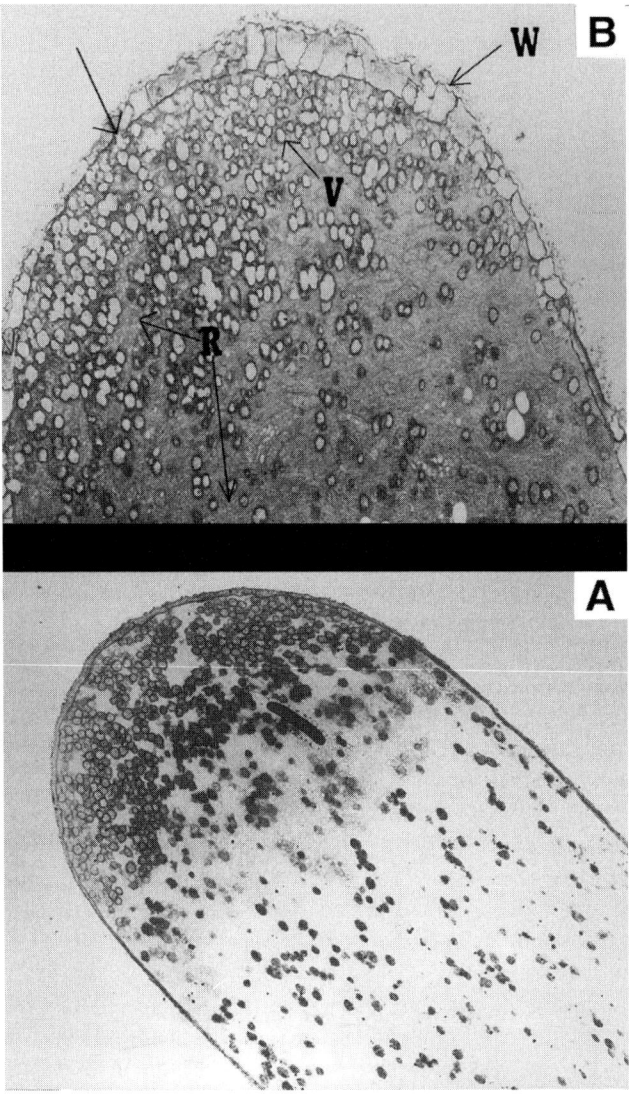

Fig. 4. Comparison of osmium-fixed (**A**) and glutaraldehyde-prefixed and osmium-postfixed (**B**) pollen tubes. (From Dashek and Rosen *[58]*, with permission.) (**A**) = 5800× (**B**) = 9300×.

18. Place drops of Sato's *(61)* lead citrate onto parafilm in a Petri dish lined with NaOH pellets; use a 1-cc syringe equipped with a 0.2-μm filter.

19. Float grids with attached thin sections on stain droplets for 3 min (float section side down).

20. Wash with ultrapure H_2O using three separate 50 mL beakers.

21. Dry sections.

22. Store in grid box or view directly (Fig. 4).

REFERENCES

1. Bozzola JJ, Russell LD. *Electron Microscopy Techniques for Biologists*, Jones and Bartlett Publishers, Boston, MA, 1992.
2. Cheng PC, Lin TH, Wu WL, Wu JL. *Multidimensional Microscopy*, Springer Verlag, New York, 1993.
3. Dykstra MJ. *Biological Electron Microscopy*, Plenum Press, New York, 1992.
4. Dykstra MJ. *A Manual of Applied Techniques of Biological Electron Microscopy*, Plenum Press, New York, 1993.
5. Gabriel BL. *Biological Electron Microscopy*, Van Nostrand Reinhold, New York, 1982.
6. Griffin RL. *Using the Transmission Electron Microscope in the Biological Sciences*, Ellis Horwood, New York, 1990.
7. Harris R. *Electron Microscopy in Biology. A Practical Approach*, IRL Press, Oxford, UK, 1991.
8. Hawkes P, Valdre V. *Biophysical Electron Microscopy, Basic Concepts and Modern Techniques*, Academic Press, New York, 1990.
9. Hayat MA. *Basic Techniques for Transmission Electron Microscopy*, Academic Press, New York, 1986.
10. Hayat MA. *Principles and Techniques of Electron Microscopy*, CRC Press, Boca Raton, FL, 1989.
11. Joy DC, Romig AD, Goldstein JI. *Principles of Analytical Electron Microscopy*, Plenum Press, New York, 1986.
12. Mohanty SB. *Electron Microscopy for Biologists*, Charles C. Thomas, Springfield, IL, 1982.
13. Slayter EM. *Light and Electron Microscopy*, Cambridge University Press, Cambridge, UK, 1993.
14. Tribe MA, Evant MR, Snook RK. *Electron Microscopy and Cell Structure*, Cambridge University Press, Cambridge, UK, 1975.
15. Juniper BC, Cox CC, Gilchrist AJ, Williams PR. *Techniques for Plant Electron Microscopy*, Blackwell Scientific Publications, Oxford, UK, 1970.
16. Robards AW, Wilson AJ. *Procedures in Electron Microscopy*, Wiley, New York, 1993.
17. Hall JL. *Electron Microscopy and Cytochemistry of Plant Cells*, Elsevier-North Holland, Amsterdam, The Netherlands, 1978.
18. Hall JL, Hawes C. *Electron Microscopy of Plant Cells*, Academic Press, New York, 1991.
19. Roland JC, Vian B. General preparation and staining of thin sections, in *Electron Microscopy of Plant Cells* (Hall JL, Hawes C, eds.), Academic Press, San Diego, CA, 1991, pp. 2–66.
20. Harris N, Oparka KJ. *Plant Cell Biology*, IRL Press, Oxford, UK, 1994, pp. 1–15.
21. Hopwood D, Milne G. Fixation, in *Electron Microscopy in Biology. A Practical Approach* (Harris JR, ed.), IRL Press, Oxford, UK, 1991, pp. 1–15.
22. Day JG, McLellan MR. *Cryopreservation and Freeze-Drying Protocols*, Humana Press, Totowa, NJ, 1995.
23. Finkle BJ, Zavola ME, Ulrich JM. Cryoprotective compounds in the viable freezing of plant tissues, in *Cryopreservation of Plant Cells and Organs* (Kartha KK, ed.), CRC Press, Boca Raton, FL, 1985, pp. 75–113.
24. Steponkus P. Cryobiology of isolated protoplasts, in *Cryopreservation of Plant Cells and Organs* (Kartha KK, ed.), CRC Press, Boca Raton, FL, 1985, pp. 49–60.

25. Withers LA. Cryopreservation of cultured plant cells and protoplasts, in *Cryopreservaton of Plant Cells and Organs* (Kartha KK, ed.), CRC Press, Boca Raton, FL, 1985, pp. 243–267.

26. Roos N, Morgan JR. *Cryopreparation of Thin Biological Specimens for Electron Microscopy: Methods and Applications*, Oxford University Press, New York, 1990.

27. Robards AW. Rapid-freezing methods and their application, in *Electron Microscopy of Plant Cells* (Hall JL, Hawes C, eds.), Academic Press, London, UK, 1991, pp. 257–312.

28. Echlin P. *Low Temperature Microscopy and Analysis*, Plenum Press, New York, 1992.

29. Sitte H. Advandled instrumentation and methodology related to cryoultramicrotomy: a review. *Scanning Microsc Suppl* 1996; 10: 387–466.

30. Kartha KK. *Cryopreservation of Plant Cells and Organs*, CRC Press, Boca Raton, FL, 1985.

31. Hawes C. Electron microscopy, in *Plant Cell Biology* (Harris N, Oparka KJ, eds.) IRL Press, Oxford, UK, 1994, pp. 69–96.

32. Spurr AR. A low viscosity epoxy resin embedding medium for electron microscopy. *J Ultrastruc Res* 1969; 25: 31.

33. Smith M, Croft S. Embedding and thin section preparation, in *Electron Microscopy in Biology. A Practical Approach* (Harris J, ed.), IRL Press, Oxford, UK, 1991, pp. 17–37.

34. Roos N. Freeze-substitution and other low temperature embedding methods, in *Electron Microscopy in Biology. A Practical Approach* (Harris JR, ed.). IRL Press, Oxford, UK, 1991, pp. 39–58.

35. Reid N. *Ultramicrotomy*, American Elsevier, New York, 1975.

36. Reid N, Beesley JE. *Sectioning and Cryosectioning for Electron Microscopy*, Elsevier, New York, 1991.

37. Richter K. Aspects of cryofixation and cryosectioning for the observation of bulk biological samples in the hydrated state by cryoelectron. *Scanning Microsc Suppl* 1996; 10: 375–386.

38. Erk I, Nicolas G, Caroff A, Leparet J. Electron microscopy of frozen biological objects: a study using cryosectioning and cryosubstitution. *J Microsc* 1998; 189: 236–248.

39. Crang RFE, Klomparens KL. *Artifacts in Biological Electron Microscopy*, Plenum Press, New York, 1988, pp. 39–58.

40. Lewis PR. *Cytochemical Staining Methods for Electron Microscopy*, Elsevier, Amsterdam, The Netherlands, 1992.

41. Spiess E, Zimmerman H-P, Lunsdrof H. Negative staining of protein molecules and filaments, in *Electron Microscopy in Molecular Biology: A Practical Approach* (Sommerville J, Scheer U, eds.), IRL Press, Oxford, UK, 1987, pp. 147–166.

42. Benedetti E, Emmelot P. Electron microscopic observations on negatively- stained plasma membranes isolated from rat liver. *J Cell Biol* 1965; 26: 299–304.

43. Harris R, Horne R. Negative staining, in *Electron Microscopy in Biology. A Practical Approach* (Harris R, ed.), IRL Press, Oxford, UK, 1991, pp. 203–228.

44. Zentgraft H, Bock C-T, Schenk M. Chromatin spreading, in *Electron Microscopy in Molecular Biology. A Practical Approach* (Sommerville J, Scheer U, eds.). IRL Press, Oxford, UK, 1991, pp. 81–100.

45. Sogo J, Stasiak A, DeBernardin W, Losa R, Koller T. Binding of protein to nucleic acids, in *Electron Microscopy in Molecular Biology. A Practical Approach* (Sommerville J, Scheer U, eds.). IRL Press, Oxford, UK, 1991, pp. 61–79.

46. Khulbrandt W, Unwin PNT. Structural analysis of stained and unstained two-dimensional ribosome crystals, in *Electron Microscopy at Molecular Dimensions, State of the Art and Strategies for the Future* (Baumeister W, Vogell W, eds.), Springer-Verlag, Berlin, Germany, 1980, pp. 108–116.

47. Mandelkow EM, Mandelkow E. (1980) Subunit structure and conformations of tubulin protofilaments, in *Electron Microscopy at Molecular Dimensions, State of the Art and Strategies for the Future* (Baumeister W, Vogell W, eds.), Springer-Verlag, Berlin, Germany, 1980, pp. 117–125.

48. Huxley HE, Zubay G. Electron microscope observations on the structure of microsomal particles from *Escherichia coli*. *J Mol Biol* 1960; 2: 10–18.

50. Horne RW, Pasquali-Ronchetti I. A negative staining carbon film technique for studying viruses in the electron microscope. I. Preparative procedure for examining icosahedral and filamentous viruses. *J Ultrastructure Res* 1974; 47: 361–383.

51. Harris JR. The production of paracrystalline two-dimensional monolayers of purified protein molecules. *Micron* 1982; 13: 147–168.

52. Harris JR. *J Electron Microsc Tech* 1991; 18: 269–276, as cited in Harris R, Horne R. Negative staining, in *Electron Microscopy in Biology* (Harris JR, ed.), IRL Press, Oxford, England, pp. 203–228.

53. Slayter HS. High resolution shadowing, in *Electron Microscopy in Biology. A Practical Approach* (Sommerville J, Scheer U, eds.), IRL Press, Oxford, UK, 1991, pp. 151–172.

54. Watt I. *The Principles and Practice of Electron Microscopy*, 2nd ed,, Cambridge University Press, Cambridge, UK, 1997.

55. Dashek WV. The lily pollen tube: aspects of chemistry and nutrition in relation to fine structure. Ph.D. Dissertation, Marquette University, Milwaukee, WI, 1966.

56. Glenny JR. Rotary metal shadowing for visualizing rod-shaped proteins, in *Electron Microscopy in Molecular Biology. A Practical Approach* (Sommerville J, Scheer U, eds.), IRL Press, Oxford, UK, 1987, pp. 167–178.

57. Coggins LW. Preparation of nucleic acids for electron microscopy, in *Electron Microscopy in Molecular Biology* (Sommerville J, Scheer U, eds.), IRL Press, Oxford, UK, 1987, pp. 1–29.

58. Dashek WV, Rosen WG. Electron microscopical localization of chemicals components in the growth zone of *Lilium* pollen tubes. *Protoplasma* 1966; 61: 192–204.

59. Santos RPD, Mariath DeAraujo JE. A simple method for fixing dehydrating pollen tubes cultivated in vitro for transmission electron microscopy. *Histochemistry* 1997; 76: 315–319.

60. Hayat MA. *Stains and Cytochemical Methods*, Plenum Press, New York, 1993.

60a. Reynolds ES. The use of lead citrate at high pH as an electron opaque stain in electron microscopy. *J Cell Biol* 1963; 17: 208–219.

61. Sato T. A modified method for lead staining of thin sections. *J Electron Microsc* 1967; 16: 133.

62. Luft JH. Permanganate–a new fixative for electron microscopy. *J Biophys Biochem Cytol* 1956; 2: 799–801.

15 Methods for Atomic Force and Scanning Tunneling Microscopies

William V. Dashek

CONTENTS

1. INTRODUCTION

1.1. Principles

Scanning tunneling microscopy (STM) is a relatively new tool for examining surface topography on a subatomic scale. Watt *(1)* summarized the main principles of STM as follows (*see* Fig. 1): The scanning tunneling microscope utilizes a very fine stylus with an atomically sharp tip and does not require an ultrahigh vacuum. The tip is positioned 1 nm or thereabouts from a conducting specimen's surface *(2)*. When a small voltage is applied between the tip and the specimen, a limited but measurable tunneling current passes between the two. Watt *(1)* further explained that the tip and specimen are traversed relative to one another. Either the tip is kept at a constant height and the variation in tunneling current plotted or the tip is moved up and down to maintain a constant tunneling current. An impression of an area of the surface is compiled through oscillations of the tip tracing the surface topography. Welland and Taylor *(2)* pointed out that the interpretation of STM images is not straightforward, requiring comparison with other surface image technologies. A summary of the advantages and disadvantages of STM are

From: *Methods in Plant Electron Microscopy and Cytochemistry*
Edited by: W. V. Dashek © Humana Press Inc., Totowa, NJ

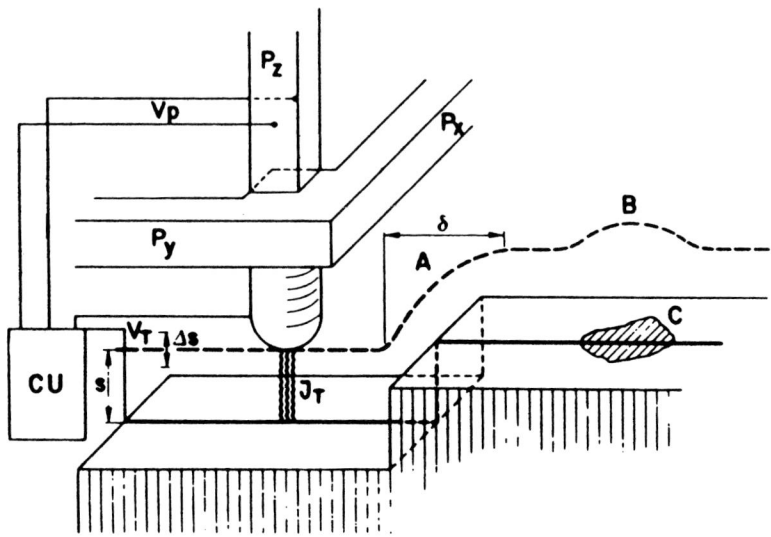

Fig. 1. Principle of scanning tunneling microscopy. A sharp needlelike tip probes the surface atomic structure of a specimen by closely scanning the surface, utilizing extreme sensitivity of the vacuum tunneling current to the tunneling gap. From Sakurai et al. *(10)* with permission.

Table 1
**Summary of the Advantages and Disadvantages
of Scanning Tunneling Microscopy**

Advantages	*Disadvantages*
1. Yields atomic resolution images of surface in real space surfaces	1. Interpretation of STM images is not straightforward
2. Specimen is not damaged, as electrons that tunnel forming the STM image possess energies of a few electron volts	2. Comparison of STM data with those derived from other surface techniques is essential
3. STM can operate in air or in a liquid, making images of molecules in their natural environment possible	3. STM is directed to conducting samples on thin insulating films, bulk insulators have to be coated with metal

presented in Table 1. Should the reader require additional information regarding principles, the reviews and monographs of Neddermeyer *(3)*, Stroscio and Kaiser *(4)*, Chiang and Wilson *(5)*, Hansma and Tersoff *(6)*, Park and Quate *(7)*, Welland and Taylor *(2)*, Wiesendanger and Gountherodt *(8)* and Cohen et al. *(9)* can be consulted. Sakurai et al. *(10)* consider STM instrumentation and resolution.

$$\Delta A(\omega) \approx \frac{1}{2} A(\omega_0) \frac{Q^2}{k^2} \Delta F^2$$

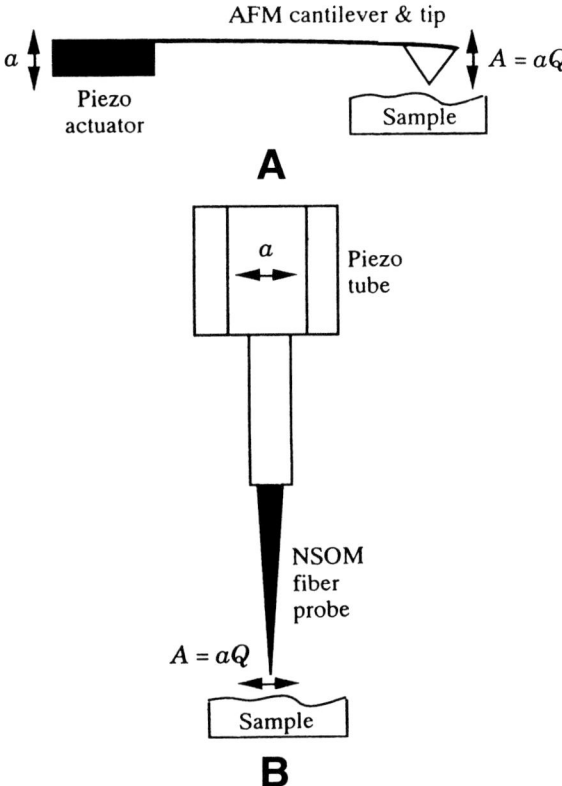

Fig. 2. Schematic diagrams of alternative force feedback modes. (**A**) Illustrates noncontact atomic force microscope feedback. (**B**) Shear-force feedback frequently used in NSOM imaging applications. From Paesler and Moyer *(41)*, with permission.

Atomic force microscopy (AFM) is a variant of STM and was introduced in 1986 by Binnig et al. *(11)*. AFM belongs to a family of near-field microscopies and is capable of imaging a wide variety of specimens' surface down to an atomic scale. The technique employs a probe (pyramidal tip) mounted at the end of a sensitive but rigid cantilever (*see* Fig. 2). The probe is drawn across the specimen under very light mechanical loading *(1)*. Measurements of the probe's interaction with the sample's surface are accomplished with a laser beam reflected from the cantilever.

Table 2
Summary of Main Features of Atomic Force Microscopy[a]

AFM images the surface with the signal originating from the outermost surface. Thus, AFM does not provide subsurface structure in contrast to TEM.

AFM can employ a contact-imaging mode, i.e., the tip's end comes in close contact with the sample's surface. The tip may touch the sample.

AFM can be performed in air but also *in vacuo* or with the sample immersed in a liquid.

AFM can image surfaces of a wide range of biological material.

[a]Adapted from Canet et al. *(21)*.

The main features of AFM are displayed in Table 2. In-depth discussions of scanning force microscopies including AFM can be found in Orr and Morris *(12)*, Bustamante et al. *(13)*, Cohen et al. *(9)*, Maddox *(14)*, Yang and Shao *(15)*, Shao and Zhang *(16)*, Ushiki et al. *(17)*, Butt *(18)*, and Vesenka et al. *(19)*. The reader is referred to Allen et al. *(20)* for the applications of AFM to biotechnology. To enhance resolution of AFM, Han et al. *(22)* employed cryo-AFM. Therefore, these authors operated a cryo-AFM in liquid nitrogen for biological applications. In this connection, reviews concerning cryo-AFM have been published by Shao and Yang *(23)* and Sheng et al. *(24)*.

2. PROTOCOLS

Because AFM and STM are more recently developed microscopic technologies, they have not been employed as extensively by plant scientists as other microscopies. In fact, STM has not witnessed much application to biology. Thus, standardized protocols applicable to diverse plant systems have not been reported. In this regard, Table 3 provides a summary of some recent AFM and STM applications to plant systems. Through critical comparison of published Materials and Methods for these papers, the reader should be able to originate AFM and STM protocols suitable for his/her plant system.

3. CONCLUSIONS

Currently, both AFM and STM are in their infancies regarding applications to plant systems. Thus, the future most likely will involve adapting these methods to plant investigations. Furthermore, it is anticipated that cryo-AFM will be more widely employed as investigators continue to improve AFM resolution. In this regard, the future of probe-based microscopies is discussed by Paesler and Moyer *(41)*.

Table 3
**Summary of Some Recent Applications of Atomic Force
and Scanning Tunneling Microscopies to Plant Systems**

Technique	Application	Reference
Atomic force	Imaging of starch granule surfaces	Baldwin et al. *(25)*
	Surface analysis of photosystem 1 complex	Fotiadis et al. *(26)*
	Surface topography and roughness of onion and garlic skins	Hershko et al. *(27)*
	Analysis of two-dimensional crystal of photosystem II core complex	Nakazato et al. *(28)*
	Molecular imaging of phytochrome	Sato et al. *(29)*
	High resolution of *Valonia* cellulose	Baker et al. *(30)*
	Analyses of plant cell walls	Morris et al. *(31)*
	Investigations of pectin branching	Round et al. *(32)*
	Study of isolated ivy leaf cuticles	Canet et al. *(21)*
	Imaging pollen grains	Demanet and Sankar *(33)*
	Visualization of plant cell walls	Kirby et al. *(34)*
	Mapping leaf surface landscapes	Mechaber et al. *(35)*
	Analyses of pollen grains, cellulose microfibrils and protoplasts	Van Der Wel et al. *(36)*
	Pollen exine substructure	Rowley et al. *(37)*
Scanning tunneling	Structure and electron conduction of photosystem II	Luskins and Oates *(38)*
	Microscopic analysis of DNA and RNA	Mueller-Richert and Gross *(39)*
	Molecular electronics of a single photosystem I reaction	Lee et al. *(40)*

REFERENCES

1. Watt I. *The Principles and Practice of Electron Microscopy*, Cambridge University Press, Cambridge, UK, 1977.
2. Welland ME, Taylor ME. Scanning tunneling microscopy, in *Modern Microscopies Techniques and Applications* (Duke PJ, Michette AG, eds.), Plenum Press, New York, 1990, pp. 231–254.
3. Neddermeyer H. *Scanning Tunneling Microscopy*, Kluwer Academic Publishers, Boston, MA, 1993.
4. Stroscio JA, Kaiser W. *Scanning Tunneling Microscopy*, Academic Press, New York, 1994.
5. Chiang S, Wilson RJ. Scanning tunneling microscopy, in *Examining the Submicron World* (Feder R, McGowan JW, Shinozaki DM, eds.), Plenum Press, New York, 1986.
6. Hansma PK, Tersoff JJ. Scanning tunneling microscopy. *App Phys* 1987; 61: R1–R23.

7. Park S-I, Quate CF. Scanning tunneling microscopy. *Rev Sci Instrum* 1987; 58: 2010–2017.

8. Wiesendanger R, Gountherodt HJ. *Scanning Tunneling Microscopy III. Theory of STM and Related Scanning Probe Methods*, Springer Verlag, New York, 1996.

9. Cohen SH, Bray MT, Lightbody ML. *Atomic Force Microscopy-Scanning Tunneling Microscopy*. Proceedings of the first U.S. Army Natik Research Development and Engineering Center. Plenum Press, New York, 1995.

10. Sakurai T, Saki A, Pickering HW. *Atom-Probe Field for Microscopy and the Applications*. Advances in Electronics and Electron Physics Supplement. Academic Press, Boston, MA, 1989.

11. Binnig G, Quate C, Geber C. Atomic force microscope. *Phys Rev Lett* 1986; 56: 930–933.

12. Orr BG, Morris MD. Fundamentals of scanning probe microscopies. *Spectroscopy Sciences* 1993; 16: 131-158.

13. Bustamante JO, Liepers A, Prendergast RA, Hanover JA, Oberliethner H. Patch clamp and atomic force microscopy demonstrate TATA-binding protein (TBP) interactions with the nuclear pore complex. *J Membrane Biol* 1995; 146: X263–X272.

14. Maddox J. Atom microscopy comes of age. *Nature* 1995; 373: 653.

15. Yang J, Shao Z. Recent advances in biological atomic force microscopy. *Micron* 1995; 26: 35–49.

16. Shao Z, Zhang X. Biological cryo atomic force microscopy: A brief review. *Ultramicroscopy* 1996; 66: 141–152.

17. Ushiki T, Hitomi J, Ogura S, Umemoto T, Shigero M. Atomic force microscopy in histology and cytology. *Arch Histol Cytol* 1996; 59: 421–431.

18. Butt H-J. Atomic force microscopy: Non-imaging applications in biology. *FASEB J* 1997; 11: A1130.

19. Vesenka J. Biological applications of scanning probe microscopies. *Scanning* 1997; 19: 135–136.

20. Allen S, Davies MC, Roberts CJ, Tendler SJB, Williams PM. Atomic force microscopy. *Trends Biotechnol* 1997; 15: 101–105.

21. Canet D, Rohr R, Chanel A, Guilliam F. Atomic force microscopy study of isolated ivy leaf cuticles observed directly and other embedding in Epon. *New Phytologist* 1996; 134: 571–577.

22. Han W, Mou J, Sheng J, Yang J, Shao Z. Cryo atomic force microscopy: A new approach for biological imaging at high resolution. *Biochemistry* 1995; 34: 8215–8220.

23. Shao Z, Yang J. Progress in high resolution atomic force microscopy in biology. *Quarterly Rev Biophysics* 1995; 28: 195–251.

24. Sheng S, Zhang Y, Shao Z. Cryo atomic force microscopy: A new tool for cell biology. *Mol Biol Cell* 1997; (Suppl) 8: 346A.

25. Baldwin PM, Adler J, Davies MC, Melia CD. High resolution imaging of starch granule surfaces by atomic force microscopy. *J Cereal Sci* 1998; 27: 255–265.

26. Fotiadis D, Mueller DJ, Tsiotis G, Hasler L, Tittmann P, Mini T, Jeno P, Gross H, Engel A. Surface analysis of the photosystem I complex by electron and atomic force microscopy. *J Mol Biol* 1998; 283: 83–94.

27. Hershko V, Weisman D, Nussinovitch A. Method for studying surface topography and roughness of onion and garlic skins for coating purposes. *J Food Sci* 1998; 63: 317–321.

28. Nakazato K, Ichimwa T, Mayanagi K, Ishikawa T, Inoue Y. Atomic force microscopy of two-dimensional crystal of photosystem II core complex. *Plant Cell Physiol* 1998; (Suppl) 39: S12.

29. Sato M, Tokumasu F, Takeyasu K, Manabe K. Molecular imaging of phytochrome a using atomic force microscopy. *Plant Cell Physiol* 1998; (Suppl) 39: S150.

30. Baker AA, Helbert W. Sugiyama J, Miles MJ. High-resolution atomic force microscopy of native *Valonia* cellulose I. Microcrystals. *J Struct Biol* 1997; 119: 129–138.

31. Morris VJ, Gunnig AP, Kirby AR, Waldron K, Ng A. Atomic force microscopy of plant cell walls, plant cell wall polysaccharides and gels. *J Biol Macromol* 1997; 21: 61–66.

32. Round A, MacDougall AJ, Ring SG, Morris VJ. Unexpected branching in pectin observed by atomic force microscopy. *Carbohydr Res* 1997; 303: 251–253.

33. Demanet CM, Sankar KV. Atomic force microscopy images of a pollen grain: A preliminary study. *S Afr J Bot* 1996; 62: 221–-223.

34. Kirby AR, Gunnig AP, Waldron KW, Morris VJ, Ng A. Visualization of plant cell walls by atomic force microscopy. *Biophys J* 1996; 70: 1138–1143.

35. Mechaber WL, Marshall CB, Mechaber RA, Jobe RT, Chew FS. Mapping leaf surface landscapes. *Proc Natl Acad Sci USA* 1996; 93: 4600–4603.

36. Van Der Wel NN, Putman CAJ, Van Noort SJT, Degroth BG, Emons AMC. Atomic force microscopy of pollen grains, cellulose microfibrils and protoplasts. *Protoplasma* 1996; 194: 29–39.

37. Rowley JR, Flynn JJ, Takahashi M. Atomic force microscope information on pollen exine substructure in *Nuphar Bot Acta* 1995; 108: 300–308.

38. Luskins PB, Oates T. Single molecule high-resolution structure and electron conduction of photosystem II from scanning tunneling microscopy and spectroscopy. *Biochem Biophys Acta* 1998; 1409: 1–11.

39. Mueller-Reichert T, Gross H. Microscopic analysis of DNA and DNA protein assembly by transmission electron microscopy, scanning tunneling microscopy and scanning force microscopy. *Scanning Microscopy* 1996; (Suppl) 10: 111–121.

40. Lee I, Lee JW, Wamack RJ, Allison DP, Greenbaum E. Molecular electronics of a single photosystem I reaction center: Studies with scanning tunneling microscopy and spectroscopy. *Proc Natl Acad Sci USA* 1995; 97: 1965–1969.

41. Paesler MA, Moyer PJ. *Near Field Optics Theory, Instrumentation and Applications*, Wiley, New York, 1996.

16 Visualization of Golgi Apparatus by Zinc Iodide–Osmium Tetroxide (ZIO) Staining

Takako S. Kaneko, Mamiko Sato, and Masako Osumi

CONTENTS

1. INTRODUCTION

In 1959, Maillet *(1)* developed the zinc iodide–osmium tetroxide (ZIO) staining method for nerve tissues from the original staining method of Champy *(2)*, which basically consisted of mixing osmium tetroxide with alkaline iodides such as potassium iodide, sodium iodide, or ammonium iodide. The immersion technique that was modified by Jabonero *(3)*, involves the preparation of zinc iodide solution not by using commercial preparations of zinc iodide but by mixing zinc powder, iodine crystals, and water prior to use. Akert and Sandri *(4)* adapted the ZIO immersion technique for electron microscopy of nerve tissues. The affinity of the ZIO immersion technique for synaptic vesicles in nerve terminals of vertebrate and invertebrate nervous systems has been confirmed *(4–11)*. Since Akert and Sandri *(4)* applied the ZIO immersion technique

From: *Methods in Plant Electron Microscopy and Cytochemistry*
Edited by: W. V. Dashek © Humana Press Inc., Totowa, NJ

to electron microscopy, a number of modified ZIO techniques have been introduced to study the Golgi apparatus in various tissues or cells in animals, plants, and bacteria, as shown in the Subheading 5 *(12–20)*.

In the nerve tissues of vertebrate and invertebrates, the Golgi complexes have been visualized by treatment with a ZIO mixture *(4,8,16, 21–26)*. Cytochemical studies of the Golgi apparatus in animal tissues and cells have been carried out using ZIO techniques *(13–15, 27)*. It has been reported that the intensity of ZIO staining appeared to be selectively strong in the *cis* cisternae of the Golgi apparatus and rather weak in the *trans* side and condensing vacuoles in the Golgi stack of mouse pancreatic exocrine cell *(18)*. The proposed three-dimensional structure of the Golgi apparatus in the cells was based on the microscopic observation of the Golgi apparatus. Using ZIO staining, the mechanism involved in the selective transport of different molecules in the retrograde and antegrade pathways between the Golgi apparatus and the rough endoplasmic reticulum (ER) was suggested in brefeldin-A-treated mouse cells *(28)*. The staining was used as a marker of the *cis*/intermediate compartment of the Golgi apparatus that actively transports proteins through the Golgi complex *(29)*.

In microorganisms, visualization of the lysosomal system of different species of malaria parasites, *(19)*, the endoplasmic reticulum–Golgi complex system of *Tritrichomonas foetus (20)* and the Golgi complex and primary lysosomes of *Pneumocystis carinii (30)* has been attempted using ZIO staining.

The process of infection of lupine nodule cells by Rhizobia was examined by the thin-section electron microscopic technique, as well as the freeze–fracture technique. Different membranes such as infection thread membranes, peribacterioid membranes, plasma membranes, membranes of cytoplasmic vesicles, and membranes of the Golgi bodies and ER were stained with uranium-lead, silver, phosphotungstic acid, and ZIO *(31)*. ZIO stained the membranes of the proximal face of the Golgi bodies and endoplasmic reticulum. ZIO staining has given good contrast in thick sections such as a cotyledon cell, a root cell, and an aleurone layer for ER, dictyosomes cisternae, mitochondria, and nuclear envelopes *(17,32–37)*.

The ER, Golgi bodies, and nuclear envelopes in cotyledon cells and cells of *Chara corallina* showed a marked increase in density by ZIO staining *(38,39)*. We investigated the continuity of the Golgi apparatus into the ER of cultured tobacco protoplasts by ZIO staining *(40)*.

The study first attempted to elucidate whether the membranes of the Golgi cisternae of tobacco cells could be differentially stained by the ZIO method, and whether they were more clearly visible than the sections stained by uranyl acetate. For this purpose, we analyzed the density of the membranes using a computer. We also confirmed by the ZIO method that the Golgi cisternae treated with brefeldin A extended into the ER *(40)*.

2. NATURE OF THE ZIO REACTION

It is commonly known that lipids, carbohydrates, and glycolipids are present in the Golgi apparatus *(27)*. The determination of the components that react with the ZIO mixture was carried out by removing each component from tissues before incubation in the ZIO mixture. After lipid extraction by acetone *(14)*, chloroform-methanol *(15)*, or propylene oxide *(27)*, no osmium–zinc precipitates could be detected in structures that normally reacted with ZIO. Blümcke et al. *(15)* summarized the nature of the lipids that react with the ZIO mixture as follows: lipids and lipoproteins of cell membranes, neutral fat droplets *(41)*, and lipid globules of type II pneumocytes and alveolar macrophages were, however, not as electron dense as the normally reactive lamellae containing highly unsaturated fatty acids.

Digestion of cells or tissues with hyaluronidase prevented the staining of Golgi elements of *Hydra* cells *(14)*, although no apparent change was observed in the ZIO immersion pattern observed in the oocytes tissues *(27)*. The difference in the digestion test between the *Hydra* cells and the oocytes might be because the oocytes were different from somatic tissues *(27)*. The digestion of *Hydra* cells with neuraminidase had no effect on the staining of the Golgi cisternae *(14)*.

The reaction of OsO_4 and OsKI (1% osmium tetroxide-1% potassium iodide) with or without cysteine (-SH) and cystine (-S.S-) was observed on each spot to determine participation of -SH groups in proteins in the reaction *(29)*. Both reagents were reduced instantly by the -SH groups of cysteine, but only OsKI was reduced by longer incubation with cystine. OsKI reactions with tissues and test molecules suggested that the stain detected labile -S.S- bridges in the ER and Golgi complexes before they formed the tertiary configurations of proteins *(29)*.

3. EFFECT OF pH IN THE ZIO MIXTURE

The pH of the ZIO mixture is the most important factor for the successful ZIO staining of the Golgi apparatus and other structures. The effect of pH on the ZIO reactions in rat nerve tissues was studied in detail by Rodríguez and Giménez *(23)* at five different pHs: ZI in distilled water (pH 5.5), ZI in 0.2 *M* monosodium–disodium phosphate buffer (pH 7.4), ZI in 0.2 *M* veronal sodium–HCl buffer (pH 7.4), ZI in 0.2 *M* veronal sodium–HCl buffer (pH 9.0), and ZI in 0.2 M citric acid–disodium phosphate buffer (pH 4.4). After ZIO immersion for 18 h, the nuclear envelopes, the external cisternae of the Golgi complex, the internal cisternae of the Golgi complex, the vesicles of the Golgi complex, and the smooth ER of all types of cells appeared to be densely stained in 0.2 M veronal sodium–HCl buffer (pH 7.4). According to Maillet (42), the ionized iodine promotes the precipitation of osmium tetroxide (OsO4). When the relative iodine concentrations were determined, the ZIO mixture with the lowest pH (4.4) was found to have the highest concentrations of free iodine. A possible explanation for the behavior of ZIO with varying pH is that the essential reagent involved in the reaction is Zn(OH)(IO), and the chemical equilibrium relation is

$$2 \text{ Zn(OH)}^+ + \text{I}_2 + 2 \text{ H}_2\text{O} \leftrightarrow 2 \text{ Zn(OH)(IO)} + 2 \text{ H}^+$$

The increase in the density of H$^+$ results in the decrease in the objective Zn(OH)(IO), so the ZIO reaction rate is decreased.

4. EFFECT OF TEMPERATURE OF THE ZIO MIXTURE

It has been demonstrated previously that temperature for the ZIO reaction plays an important role in promoting the reactivity of certain structures to the ZIO mixture (43). Most investigations in which the ZIO method has been applied have been performed either at room temperature or at 4°C.

5. SUMMARY OF THE ZIO TECHNIQUE IN VISUALIZING THE GOLGI APPARATUS

1. Akert and Sandri (4)
   a. Material: nerve tissue (subfornical organs of cats, muscle end plates from the diaphragm of albino mice, ventral gray matter of the spinal cord in cat and rat) (4).
      The ZIO mixture was prepared as follows: 15 g of zinc (powder) and 5 g of iodide (crystal) were dissolved in 200 mL of distilled water. Six to eight milliliters of the filtered solution was added to 2 mL of 2% OsO$_4$ solution shortly before use.
   b. Material: pineal gland of adult golden hamster (*Mesocricetus auratus*) (22).
      The ZIO treatment was carried out for 18 h at 4°C without prefixation.
   c. Material: adult Mongolian gerbils, pineal gland (25).
      Following perfusion of 2% glutaraldehyde and 2% paraformaldehyde in 0.1 M cacodylate buffer, ZIO treatment was carried out for 18 h at 4°C.

     d. Material: adult Long–Evans strain rat, small granule containing cell, SGC cell in the superior cervical ganglion *(26)*.
ZIO treatment was carried out for 18 h at 4°C without prefixation.
     e. Material: tobacco (*Nicotiana tabacum*) cell *(40)*.
ZIO treatment was carried out for 16 h at 4°C.
The ZIO mixture was prepared by the method of Akert and Sandri *(44)* with modifications.

2. Kawana et al. *(12)*
     a. Material: adult rat brain specimens, cervical spinal cord, ventral gray matter *(12)*.
The specimens were immersed in 0.1 *M* phosphate buffer containing 6.25% glutaraldehyde (pH 7.4) for 2 h at room temperature. The tissue blocks were impregnated with the ZIO mixture for 16 h at 4°C. The ZIO mixture was prepared as follows: 6 g zinc (powder) and 2 g resublimed iodine crystals were dissolved in 40 mL distilled water. Four milliliters of the filtered solution was added to 4 mL of Tris–HCl buffer (pH 7.4) containing 3.3 g NaCl, 0.06 g $CaCl_2$, 0.31 g $MgCl_2.6H_2O$, and 0.605 g trisaminomethane in 50 mL distilled water. This solution was then mixed with 2 mL of 2% OsO_4 solution prior to use. The pH of the final solution was about 6.25.
     b. Material: the lumbosacral enlargements of albino rats *(8)*.
The lumbosacral enlargements were fixed with 0.1 *M* cacodylate buffer containing 2% paraformaldehyde and 0.2% $CaCl_2$ (pH 7.5). One milliliter sections of the lumbosacral enlargements were refrigerated overnight in ZIO solution. Two series of newborn to 2 d, 7 d, 14 d, 21 d and adult rats were treated similarly. Transverse slices from three other 7-d-old rats were treated for different periods in ZIO solution (4, 6, 8, 12, 16, or 24 h).
     c. Material: nerve tissues; cerebral or visceral ganglia from both the ter-restrial stylommatophoran pulmonate, *Helix aspersa* and the freshwater basommatophoran snail, *Limmaea stagnailis*; thoracic and abdominal ganglia from the ventral nerve cord of the locust, *Schistocerca gregaria (21)*.
     d. The nerve tissues were immersed in 0.1 *M* phosphate buffer containing 6.25% glutaraldehyde (pH 7.4) for 2 h at room temperature. The ZIO mixture was prepared according to the method of Akert and Sandri *(4)* and Kawana et al. *(12)* as follows: 6 g of zinc (powder) and 2 g of iodine crystals were dissolved in 40 mL of water; 4 mL of the solution was filtered and added to 4 mL of 0.2 *M* phosphate buffer (pH 7.4). To this was added 2 mL of 2% aqueous osmium tetroxide. The treatment with this ZIO mixture was carried out at 4°C for 16 h.

3. Niebauer et al. *(13)*
     a. Material: human skin *(13)*.

Biopsy specimens were immersed in a solution of OsO_4-ZnI_2. The ZIO mixture was prepared as follows: 5 g of metallic iodide was mixed with 10–15 g of zinc metal. The combined powders were dissolved in 200 mL of distilled water. Eight milliliters of the filtered solution was combined with 2 mL of unbuffered 2% OsO_4. The biopsy specimens were immersed in the ZIO solution for 24 h at room temperature in the dark.

   b. Material: *Hydra, H. pseudoligactis,* and *H. littoralis (14).*

Treatment with this ZIO mixture was carried out for 1, 2, 4, 8, and 18 h at room temperature in the dark. Most animals were treated for 4 h. Treatment with the ZIO mixture neutralized to pH 7.0 with 0.1 N NaOH was also carried out for 4 h at room temperature in the dark.

   c. Material: wheat (*Triticum sativum*), root (*Euphorbia characians*) *(33–35).*

Treatment with the ZIO mixture was carried out for 12, 24, 36, and 48 h at room temperature in the dark without prefixation.

   d. Material: lupine seed (*Lupinus angustifolius* L. cv. Uniwhite or Britter blue) *(31).*

Blocks were fixed in ZIO mixture for 22 h at room temperature in the dark.

   e. Material: juvenile hamsters (8,12, and 21 d after birth) and young adults (2–3 mo after birth) *(27).*

Fixation was carried out at room temperature or at 4°C for 1–2 h in 0.1 M cacodylate buffer (pH 7.4) containing 1% glutaraldehyde and 1% paraformaldehyde. The pH of the ZIO mixture was approximately 5.7. For comparison, some tissues were impregnated in solution, the pH of which had been raised to 7.0 with NaOH.

4. Blümcke et al. *(15)*

   a. Material: lung tissue of rat and dog *(15).*

Two-thirds of the blocks of 1 mm diameter were immersed for 12 h in a ZIO mixture prepared according to the method of Maillet *(1)*, and was first utilized in electron microscopy by Stach *(15a)*. The remaining blocks were prefixed at 4°C for 2 h in 0.1 M cacodylate buffer containing 5% glutaraldehyde for 15 min, and immersed in the ZIO mixture at 25°C for 12 h. The ZIO mixture was prepared as follows: twelve to 15 g of zinc (powder) and 5 g of iodine (crystal) were dissolved in 200 mL of distilled water. Eight milliliters of the filtered solution was added to 2 mL of 2% OsO_4 solution prior to use.

5. Joó et al. *(16)*

   a. Material: rat nerve cells, olfactory bulb, cerebellar vermis parietal cortex *(16).*

The ZIO mixture consisted of one part of 2% osmium tetroxide and three parts of 3% sodium iodide.

Treatment with the ZIO mixture was carried out for 24 h at room temperature or for 2 h at 4°C.

6. Harris *(17)*
    a. Material: *Vicia faba*, cotyledon *(17)*.
       The tissues were fixed in 0.05 *M* cacodylate buffer containing 2.5% glutaraldehyde and 1.5% formaldehyde (pH 7.0) for 16 h. The ZIO mixture was prepared as follows: 3 g zinc (powder) and 1 g resublimed iodine crystals were dissolved in 20 mL distilled water. After stirring for 5 min, the zinc was filtered off. The filtered solution was mixed with an equal volume of 2% OsO_4 solution and the solution used immediately. Treatment with the ZIO mixture was carried out for 4 h at room temperature.
    b. Material: wheat (*Triticum aestivum L.*), endosperm *(37)*.
       The tissues were fixed in 0.05 *M* cacodylate buffer containing 3% glutaraldehyde (pH 7.2) for 2 h at 20°C. Treatment with the ZIO mixture was carried out for 16 h at 40°C in the dark.
    c. Material: *Chara corallina* Klein ex Wild *(39)*.
       The cells were fixed by immersion in 0.05 *M* cacodylate buffer containing 3% formaldehyde, 0.5% glutaraldehyde and 10 mM $CaCl_2$ (pH 6.8) at 21–23°C. Treatment with the ZIO mixture was carried out at 30°C for 18 h.
7. Noda and Ogawa *(18)*
    a. Material: mouse, pancreatic exocrine cell *(18)*.
       The tissue blocks were immersed in 0.1 *M* phosphate buffer containing 2% glutaraldehyde and 5% sucrose (pH 7.4). The tissue blocks were immersed in ZIO solution. The ZIO solution was prepared as follows: 3 g of zinc (powder) was dissolved ultrasonically in 20 mL distilled water and 1 g of iodine crystals was slowly added to the zinc suspension with stirring. After cooling, 4 mL of the filtered solution was mixed with 2 mL of Tris-HCl buffer (pH 7.5) and 2 mL of 2% OsO_4 solution. Treatment with this ZIO mixture was carried out for 16–20 h at 4°C in the dark.
    b. Material: adult male Mongolian gerbil *(24)*.
       Animals were fixed by perfusion with 0.1 *M* phosphate buffer containing 2% glutaraldehyde and 4% saccharose (pH 7.4). Treatment with the ZIO solution was carried out for 20 h at 4°C.
    c. Material: Mouse monoclonal antibody producing cell line H35 *(28)*.
       The cells were prefixed in 0.1 *M* cacodylate buffer containing 2% glutaraldehyde and 2% paraformaldehyde (pH 7.4) for 60 min. Treatment with the ZIO solution was carried out for 12–15 h at 4°C.
8. Benchimol and De Souza *(20)*
    a. Material: *Tritrichomonas foetus (20)*.
       Cells were fixed in 0.1 *M* phosphate buffer containing 2.5% glutaraldehyde (pH 7.4) for 1 h at room temperature. The ZIO mixture was prepared as follows: 3 g of zinc powder and 1 g of resublimed iodine crystals were separately suspended in 10 mL of distilled water.

The two solutions were then mixed and stirred for 5 min. The filtered solution was mixed with 0.01 M Tris buffer (pH 4.5) in a 1:1 ratio. Four parts of this solution were mixed with one part 2% OsO_4 solution in distilled water 5–10 min before use. They were incubated in the dark for 17–22 h at 4°C in ZIO reagent.

b. Material: *P. carinii* parasitized rabbit lung *(30)*.

The tissues were fixed by immersion in 0.1 M cacodylate buffer containing 2.5% glutaraldehyde (pH 7.3) for 2 h at 4°C.

9. Slomianny and Prensier *(19)*

a. Material: *Plasmodium berghei, P. yoelii, P. falciparum (19)*.

The cells were fixed in 0.1 M phosphate buffer containing 2.5% glutaraldehyde (pH 7.3) for 1 h at room temperature. The ZIO mixture was prepared as follows: 3 g of zinc (powder) and 1 g of resublimed iodine crystals were suspended separately in 10 mL water. The filtered solution was mixed with an equal volume of Tris-HCl buffer (pH 7.4). Four parts of this solution were then mixed with one part of 2% (w/v) OsO_4 solution. Treatment with the ZIO mixture was carried out for 14–24 h at 4°C in the dark.

6. TOBACCO PROTOPLAST–ZIO STAINING

6.1. Procedures

6.1.1. ISOLATION OF PROTOPLASTS

Protoplasts of tobacco cells (cell line XD-6, *N. tabacum* L. var. Xanthi) were isolated by the method described previously *(40)*.

Cells were then resuspended in a solution that contained 2% (w/v) cellulase Onozuka R-10 (Yakult Honsha Co., Ltd., Tokyo, Japan), 0.05% (w/v) pectolyase Y-23 (Seishin Pharmaceutical Co., Ltd., Tokyo, Japan), and 0.1% carboxymethyl cellulose dissolved in 0.48 M mannitol (pH 5.2). After incubation for 40 min at 30°C, protoplasts were separated from aggregates of undigested cells by successive passages through 1.70 mm and 510 μm mesh stainless steel and a 60 μm mesh nylon sieve. The population of protoplasts was then purified by sedimentation through 13.7% sucrose at 173g for 1 min. The sedimented protoplasts were then immobilized in beads of calcium alginate by the method of Dragct et al. *(45)* with modifications.

6.1.2. ZIO STAINING METHOD

6.1.2.1. Steps in the Procedure

1. Transfer the alginate beads into a solution of 4% (v/v) glutaraldehyde in 0.1 M 2-(N-morpholino) ethanesulphonic acid-NaOH (pH 5.8) that contained 0.4 M glucose.

2. After being immersed for 2 h at 4°C, the alginate beads are washed three times with the buffer and kept in the buffer at 4°C for 18 h.

3. Cut the alginate beads into small cubes of 1.0×1.0 cm^2 square using a razor blade.

4. Immerse the alginate beads according to a modified method of Akert and Sandri *(44)* for 3, 16, or 20 h at 4°C in the ZIO mixture, which consisted of 3.75% zinc (powder) and 1.25% resublimed iodine crystals in 50 mM 2-(N-morpholino) ethanesulphonic acid-NaOH buffer (pH 5.8) containing 0.2 M glucose and 1% OsO$_4$ (MG buffer).

5. The ZIO mixture was prepared as follows:
   a. Ultrasonicate 3 g of zinc (powder, Wako) in 20 mL of distilled water at 30S (Bransonic 220) for 1 min five times.
   b. Add 1 g of iodine crystals gradually to 40 mL of distilled water, as the reaction is exothermic.
   c. Agitate this suspension rapidly and repeatedly during a cooling period of 20 min, then filter.
   d. Immediately before use, add 4 mL of this solution to 4 mL of MG buffer that contains 2% OsO$_4$.

6. After incubation, wash the beads three times with MG buffer at 4°C and postfix the samples with 2% OsO$_4$ in MG buffer for 90 min.

7. Wash the beads and then dehydrate them with increasing concentrations of ethanol, followed by acetone.

8. Embed the dehydrated samples in epoxy resin (Quetol 653), cut into thin sections, stain with 4% uranyl acetate and 0.4% lead citrate, and examine with a Jeol 1200EXS electron microscope.

9. Feed electron micrographs into an image scanner (Sharp Jx-330M, Sharp Co., Japan, 200 pixel/2.54 cm) and obtain a line profile of the gray level (Golgi, ER of Golgi apparatus-ER) in each image using software: NIH Image 1.60 on Power Macintosh 8500/120 (Apple Computer Co.).

6.2. ZIO Immersion of Membranes of Golgi vesicles and ER

We first determined the appropriate periods of ZIO–mixture immersion for staining differentially the membranes of Golgi cisternae. Throughout the immersions (3, 16, and 20 h) the cisternae of the Golgi apparatus were kept parallel and the shape of the ER was unchanged. Time dependence of the stain intensities of the membranes of Golgi cisternae and of the ER are shown in Figs. 2 and 3. Three hours of the ZIO immersion appeared insufficient to stain the Golgi cisternae(data not shown), while the density of the membranes of the Golgi cisternae appeared higher than that of the ER after 16 h of immersion (Fig. 2). The membranes stained by the ZIO method were denser than those stained with 4% uranyl acetate and 0.4% lead citrate (compare Fig. 1 with Fig. 2).

To determine the difference in density between the membranes of the Golgi cisternae and the ER, the electron micrograph was fed to an image scanner, and a line profile of gray level in the image was taken using software: NIH Image 1.60 (Figs. 1–3, line profile of each gray level).

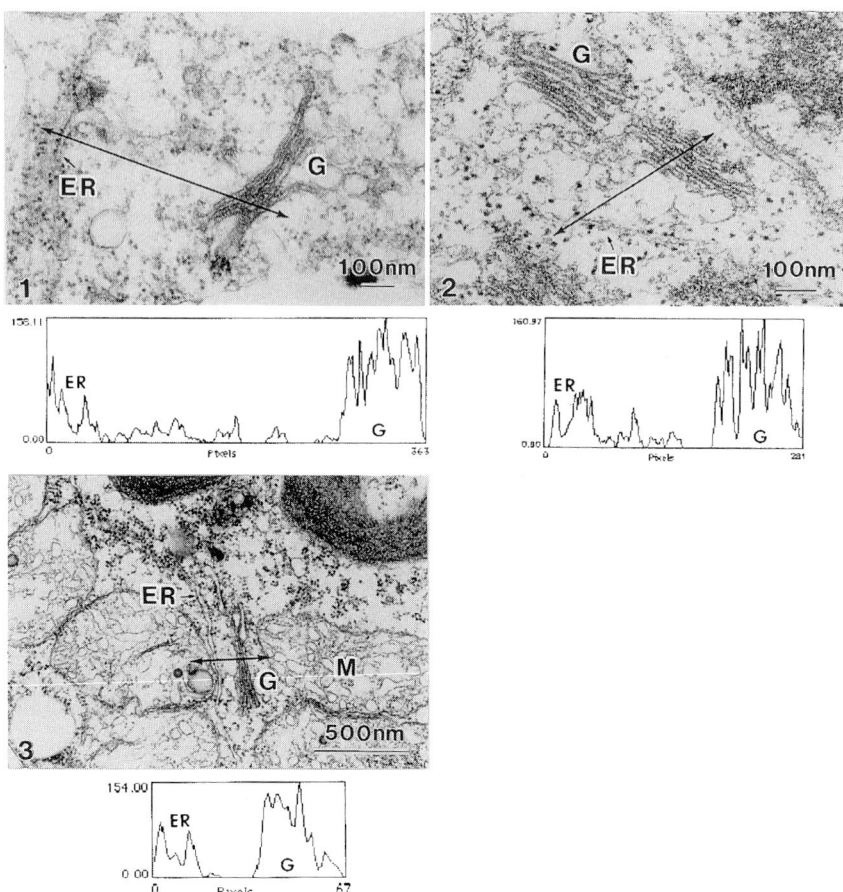

Fig. 1. An ultrathin section of an untreated fresh protoplast with 4% uranyl acetate and 0.4% lead citrate. A line profile of gray level (below) was taken along the double-headed arrow through the Golgi cisternae (G) and the endoplasmic reticulum (ER). (From ref. *40.*)

Fig. 2. A section of a fresh protoplast stained for 16 h by the ZIO method. A line profile shows that membranes of the Golgi cisternae (G) and ER are differentially stained. (From ref. *40.*)

Fig. 3. A section of a fresh protoplast stained for 20 h by the ZIO method. A line profile is through Golgi cisternae (G) and ER. M = mitochondrion. (From ref. *40.*)

Mean value of peak contrast (MP) can be expressed as: total variation/ number of peaks, where total variation is the sum of gray level difference between the peaks and the valleys in a line profile. The difference of the MP between the membranes of the Golgi cisternae and of the ER is expressed by the ratio VG/VER, where VG is the MP of the Golgi apparatus, and VER the MP of the ER. As shown in Table 1, the ratio VG/VER was about 1.2 times higher than the control, confirming that the differ-

Table 1
Time Dependence of ZIO
Impregnation of Membranes of Golgi Vesicles and ER

Z10 impregnation	Mean value of peak contrasts		
(h)	VER	VG	VG/VER
0	101.55	179.13	1.76
16	93.17	193.43	2.07
20	147.65	220.70	1.49

The difference of the mean value of peak contrasts (MP) between the membranes of the Golgi cisternae and of the ER is expressed by the ratio *VG/VER*, where *VG* is the MP of the Golgi apparatus, and *VER* is the MP of the ER. MP can be expressed as (total variation) / (number of peaks), where total variation is the sum of the gray level differences between the neighboring peaks and valleys (From ref. *40.*).

ential ZIO staining was effective in discriminating Golgi cisternae from the ER. However, after 20 h the Golgi cisternae could not be identified by this staining because the VG/VER ratio was not as high as the control (*see* Fig. 3 and Table 1). We thus concluded that membranes of the Golgi cisternae were differentially stained and distinguished from those of the ER when the tobacco protoplasts were immersed in the ZIO mixture for 16 h.

Finally, the reader is referred to Van Nooden and Hulstaert *(46)* and Lewis and Knight *(47)* for reviews of cytochemical staining methods for electron microscopy. Continual improvements in ultrastructural cytochemistry are anticipated.

REFERENCES

1. Maillet M. Modification de la technique de Champy au tétraoxyde d'osmium-iodure de K. Résultats de son application à l'étude des fibres nerveuses. *C R Soc Biol* 1959; 153: 939–941.
2. Champy C. Granules et substances reduisant l'iodure d'osmium. *J Anat Physiol Norm Pathol* 1913; 49: 323–343.
3. Jabonero V. Über die Brauchbarkeit der Osmiumtetroxyd-Zinkjodid-Methode zur Analyse der vegetativen Peripherie. *Acta Neuroveg* 1964; 26: 184–210.
4. Akert K, Sandri C. An electron-microscopic study of zinc iodide-osmium impregnation of neurons. I. Staining of synaptic vesicles at cholinergic junctions. *Brain Res* 1968; 7: 286–295.
5. Pellegrio de Iralde A, Gueudet R. Action of reserpine on the osmium tetroxide zinc iodide reactive site of synaptic vesicles in the pineal nerves of the rats. *Z Zellforsch* 1968; 91: 178–185.
6. Matus AI. Ultrastructure of the superior cervical ganglion fixed with zinc iodide and osmium tetroxide. *Brain Res* 1970; 17: 195–203.
7. Dennison ME. Electron stereoscopy as a means of classifying synaptic vesicles. *J Cell Sci* 1971; 8: 525–539.

8. Stelzner DJ. The relationship between synaptic vesicles, Golgi apparatus, and smooth endoplasmic reticulum: A developmental study using the zinc iodide-osmium technique. *Z Zellforsch* 1971; 120: 332–345.

9. Lamparter HE, Steiger U, Sandri C, Akert K. Zum Feinbau der synapsen im Zentralnervensystem der Insekten. *Z Zellforsch* 1969; 99: 435–442.

10. Martin R, Barlow J, Miralto A. Application of the zinc-iodide-osmium tetroxide impregnation of synaptic vesicles in cephalopod nerves. *Brain Res* 1969; 15: 1–16.

11. Barlow J, Martin R. Structural identification and distribution of synaptic profiles in the octopus brain using the zinc iodide-osmium method. *Brain Res* 1971; 25: 241–253.

12. Kawana E, Akert K, Sandri C. Zinc iodide-osmium tetroxide impregnation of nerve terminals in the spinal cord. *Brain Res* 1969; 16: 325–331.

13. Niebauer G, Krawczyk WS, Kidd RL, Wilgram GF. Osmium zinc iodide reactive sites in the epidermal langerhans cell. *J Cell Biol* 1969; 43: 80–89.

14. Elias PM, Park HD, Patterson AE, Lutzner MA, Wetzel BK. Osmium tetroxide-zinc iodide staining of Golgi elements and surface coats of hydras. *J Ultrastruct Res* 1972; 40: 87–102.

15. Blümcke WD, Kessler HR, Niedorf NH, Veith FJ. Ultrastructure of lamellar bodies of type II Pneumocytes after osmium-zinc impregnation. *J Ultrastruct Res* 1973; 42: 417–433.

15a. Von Stach W. Die Malletsche Osmiumsöure—Zinkjodatum—Technik als Routinemethode zur Darstellung spezieller strunkturen. *Z Med Labortechnik* 1963; 4: 7–20.

16. Joó, F Halász N, Párducz Á. Studies on the fine structural localization of zinc-iodide-osmium reaction in the brain I. Some characteristics of localization in the perikarya of identified neurons. *J Neurocytol* 1973; 2: 393–405.

17. Harris N. Nuclear pore distribution and relation to adjacent cytoplasmic organelles in cotyledon cells of developing *Vicia faba*. *Planta* 1978; 141: 121–128.

18. Noda T, Ogawa K. Golgi apparatus is one continuous organelle in pancreatic exocrine cell of mouse. *Acta Histochem Cytochem* 1984; 17: 435–451.

19. Slomianny C, Prensier G. A cytochemical ultrastructural study of the lysosomal system of different species of malaria parasites. *J Protozool* 1990; 37: 465–470.

20. Benchimol M, De Souza W. *Tritrichomonas foetus*: cytochemical visualization of the endoplasmic reticulum-Golgi complex and lipids. *Exper Parasitol* 1985; 59: 51–58.

21. Lane NJ, Swales LS. Interrelationships between Golgi, GERL and synaptic vesicles in the nerve cells of insect and gastropod ganglia. *J Cell Sci* 1976; 22: 435–453.

22. Lu K-S, Lin H-S. Cytochemical studies on cytoplasmic granular elements in the hamster pineal gland. *Histochemistry* 1979; 61: 177–187.

23. Rodríguez EM, Giménez AR. Zinc-iodide-osmium procedures as markers of subcellular structures. I. Standardization of staining of transmitter containing vesicles. *Z mikrosk-anat Forsch* 1981; 95: 257–275.

24. Krstic R. Ultracytochemical evidence for the presence of GERL in pinealocytes of the Mongolian gerbil (*Meriones unguiculatus*). *Cell Tissue Res* 1986; 246: 583–588.

25. Chau Y-P, Liao K-K, Kao M-H, Huang B-N, Kao Y-S, Lu K-S. Ultrastructure, ZIO-staining and chromaffinity of gerbil pinelocytes. *Kaohsiung J Med Sci* 1994; 10: 613–623.

26. Chau Y-P, Lu K-S. ZIO impregnation and cytochemical localization of thiamine pyrophosphatase and acid phosphatase activities in small granule-containing (SGC) cells of rat superior cervical ganglia. *Histol Histopathol* 1994; 9: 649–656.

27. Weakley BS, Webb P, James JL. Cytochemistry of the Golgi apparatus in developing ovarian germ cells of the Syrian hamster. *Cell Tissue Res* 1981; 220: 349–372.

28. Hashimoto R, Tanaka O, Otani H. Selective translocation of different makers in the ante- and retrograde pathways between the Golgi apparatus and the rough endoplasmic reticulum in a hybridoma cell line. *Ann Anat* 1997; 179: 105–116.

29. Locke M, Huie P. The mystery of the unstained Golgi complex cisternae. *J Histochem Cytochem* 1983; 31: 1019–1032.

30. Palluault F, Dei-Cas E, Slomianny C, Soulez B, Camus D. Golgi complex and lysosomes in rabbit derived *Pneumocystis carinii*. *Biol Cell* 1990; 70: 73–82.

31. Robertson JG, Lyttleton P, Bullivant S, Grayston GF. Membranes in lupine root nodules. I. The role of Golgi bodies in the biogenesis of infection threads and peribacteroid membranes. *J Cell Sci* 1978; 30: 129–149.

32. Harris N. Endoplasmic reticulum in developing seeds of *Vicia faba*. *Planta* 1979; 146: 63–69.

33. Marty F. Observation au microscope électronique à haute tension (3 Me V) de cellules végétales en coupes épaisses de 1 à 5μ. *C R Hebd L Seances Acad Sci* 1973; 277: 2681–2684.

34. Marty F. Cytochemical studies on GERL, provacuoles, and vacuoles in root meristematic cells of *Euphorbia*. *Proc Natl Acad Sci USA* 1978; 75: 852–856.

35. Marty F. High voltage electron microscopy of membrane interactions in wheat. *J Histochem Cytochem* 1980; 28: 1129–1132.

36. Hawes CR. Applications of high voltage electron microscopy to botanical ultrastructure. *Micron* 1981; 12: 227–257.

37. Parker ML, Hawes CR. The Golgi apparatus in developing endosperm of wheat (*Triticum aestivum* L.) *Planta* 1982; 154: 277–283.

38. Harris N, Oparka KJ. Connections between dictyosomes, ER, and GERL in cotyledons of mung bean (*Vigna radiata* L.). *Protoplasma* 1983; 114: 93–102.

39. Pesacreta TC, Lucas WJ. Plasma membrane coat and a coated vesicle-associated reticulum of membranes: Their structure and possible interrelationship in *Chara corallina*. *J Cell Biol* 1984; 98: 1537–1545.

40. Kaneko TS, Sato M, Watanabe R, Konomi M, Osumi M, Takatsuki A. Observation of the continuity of Golgi apparatus with endoplasmic reticulum induced by brefeldin A studied by zinc iodide-osmium staining in tobacco protoplasts. *J Electron Microsc* 1998; 47: 93–97.

41. Ostendorf ML, Niedorf HR, Blümcke S. Elektronenmikroskopische untersuchungen an menschlichen leukozyten nach osmium-zinc-imprägnation. *Z Zellforsch* 1971; 121: 358–376.

42. Maillet M. La technique de Champy à l'osmium iodure de potassium et la modification de Maillet à l'osmium-iodure de zinc. *Trab Inst Cajal Invest Biol* 1962; 54: 1–36.

43. Pellegrino De Iraldi A. ZIO staining of monoaminergic granulated vesicles. *Brain Res* 1974; 66: 227–233.

44. Akert K, Sandri C. Significance of the Maillet method for cytochemical studies of synapses, in *Golgi Centennial Symposium. Proceeding* (Santini M, ed.), Raven Press, New York, 1975, pp. 387–399.

45. Draget KI, Myhre S, Ostgaard K. Plant protoplast immobilized in calcium alginate. A simple method of preparing fragile cells for transmission electron microscopy. *Stain Technol* 1988; 63: 159–163.

46. Van Nooden CFF, Hulstaert CE. Electron microscopical enzyme histochemistry, in *Electron Microscopy in Biology* (Harris JR, ed.), IRL Press, Oxford, UK, 1991, pp. 125–149.

47. Knight DR, Lewis PR. General cytochemical methods, in *Cytochemical Staining Methods for Electron Microscopy. Practical Methods in Electron Microscopy* (Glauert A, ed.), Elsevier, Amsterdam, 1992, pp. 79–145.

17 Methods in Electron Microscope Autoradiography

William V. Dashek

1. INTRODUCTION

Electron microscope autoradiography (radioautography) is a technique that involves administration of [³H]-labeled compounds to cells in culture, excised tissues, or whole plants followed by processing of plant material for conventional transmission electron microscopy (TEM) and subsequent application of fine-grain photographic emulsions to thin sections mounted on copper grids (1–9). Tritiated-labeled compounds are employed because the emitted β-particle resulting from [³H]-decay ($t^{1/2} = 12.3$ yr) is weak and possesses a short pathlength. This is in contrast to the energies and β-particles resulting from decay of other radionuclides commonly used in life science research (*see* Chapter 3).

2. ADMINISTRATION OF RADIOISOTOPES, TISSUE PREPARATION, AND EMULSIONS

Cells, tissues or plants, are exposed to an appropriate quantity of [³H]-labeled compound for defined periods of time (varies from system to

From: *Methods in Plant Electron Microscopy and Cytochemistry*
Edited by: W. V. Dashek © Humana Press Inc., Totowa, NJ

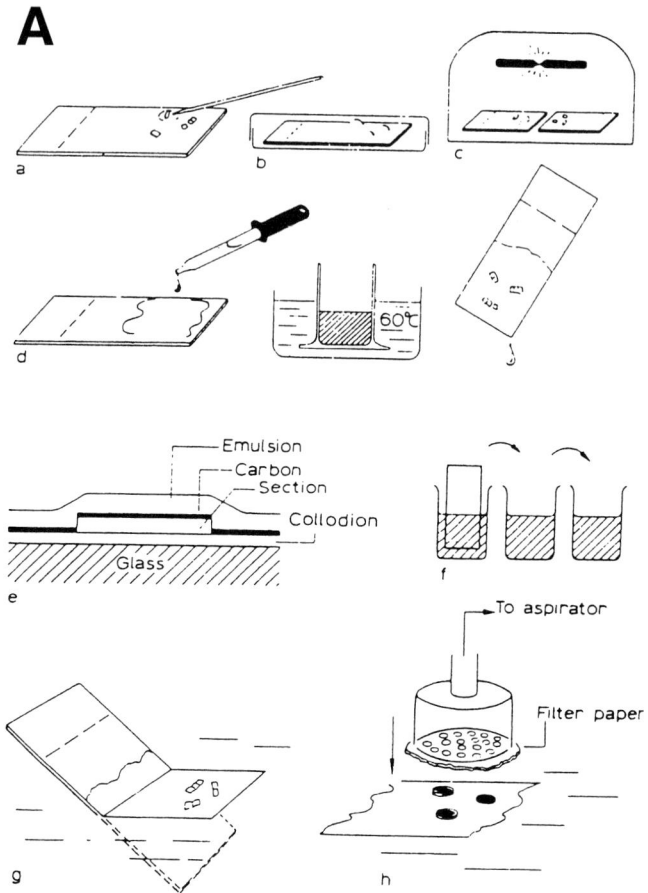

Fig. 1. (**A**) Emulsion application by the flat substrate method. (From Rogers [20], with permission.)

system). After washing the plant material with unlabeled compound to remove nonspecifically bound compound, the material is processed for conventional TEM (3,7). Thin sections (approx 100 nm), mounted on uncoated or coated copper grids, are coated with fine-grain photographic emulsions (*see* Chapter 3). Figure 1 depicts the flat substrate and loop coating methods which are performed in a light-tight darkroom equipped with a safelight appropriate to the employed emulsion. Subsequent to the application of the emulsion, the grids are transferred to a light-tight box, sealed, and maintained in total darkness. At regular intervals (to be determined for each system), grids are withdrawn and the emulsion developed with a suitable developer (*see* Table 1). As with coating, emulsion development must be performed in the darkroom with the appropriate safelight. Following drying, the grids containing the emulsion-coated

B

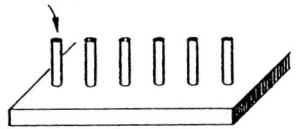

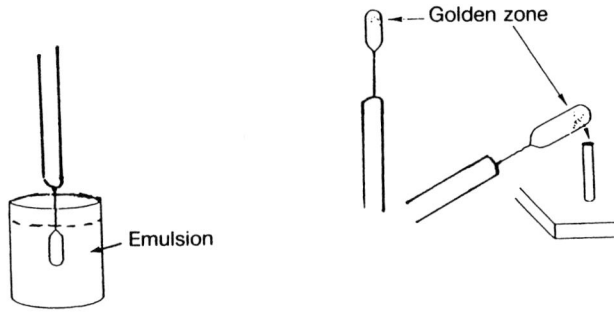

Golden zone

Emulsion

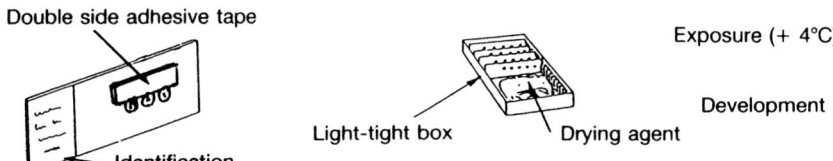

Double side adhesive tape

Identification

Light-tight box

Drying agent

Exposure (+ 4°C)

Development

Fig. 1. (B) Emulsion application by the loop method. (From Morel *[7]*, with permission.)

Table 1
Developers for Electron Microscope Emulsions [a]

Emulsions	Developer
Ilford L-4	Microdol X (3 min, 24°C), p-phenylenediamine, D19
Kodak NTE	Dektol (2 min, 24°C), gold latensification and elon-ascorbic acid (8 min, 24°C)

[a] The reader is referred to Evans and Callow *(3)* for a discussion of efficiency, i.e., the number of silver grains in the developed emulsion relative to the number of disintegrations occurring in the specimen during exposure. These authors present data representing estimates of efficiency values for tritium in EM autoradiographs.

sections can be inserted into the electron microscope (EM) and viewed. A "typical" EM autoradiograph is illustrated in Fig. 2. It is apparent that the appearance of the silver grain is quite different from that for light

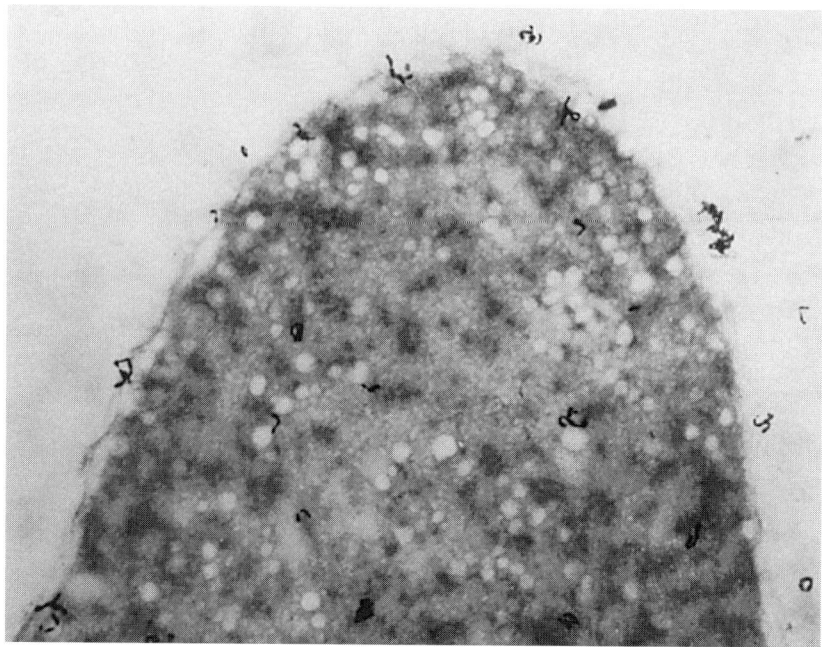

Fig. 2. EM autoradiograph of a thin section of a pollen tube labeled with [³H]-*myo*-inositol. (From Dashek and Rosen *[2]*. Numerous other examples of EM autoradiographs can be found in the listed literature.) Magnification ×12,100.

microscope autoradiographs (*see* Chapter 3). Figure 3 depicts a diagrammatic model of an EM autoradiograph.

3. RESOLUTION AND QUANTITATIVE AUTORADIOGRAPHY

Both resolution and statistical analyses of EM radioautographs have been discussed in depth by Evans and Callow *(3)* and Morel *(7)*. Table 2 reveals that reducing section thickness as well as diameters of halide crystals and developed grains improve resolution of EM radioautographs. With regard to statistical analysis, Morel *(7)* considers the probability–circle method and cross-fire method for the quantitative analysis of EM autoradiographs.

4. LIMITATIONS OF EM AUTORADIOGRAPHS

EM radioautography possesses certain limitations. For example, the observation that silver grains representing decay of a [³H]-labeled compound over an organelle may or may not mean that the compound was incorporated intact. Concomitant biochemistry is required to determine

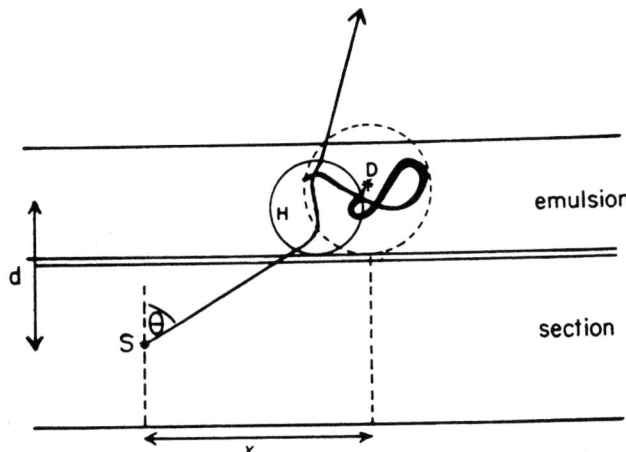

Fig. 3. Diagram of an EM radioautograph. (From Rogers [20]. with permission.)

Table 2
Effect of Section Thickness Emulsion Type and Development on Resolution of EM Radioautographs

Isotope	Emulsion	Section thickness	Development	HD
^3H	L4 monolayer	1200	Microdol X	1650
		500	Paraphenylenediamine	1450
	NTE double layer	1200	Microdol X	1450
		500	Paraphenylenediamine	1300
	NTE monolayer	1200	Dektol	1250
		500	Dektol	1000
^{14}C	L4 double layer	1000	Microdol X	1000
		500		800
	L4 monolayer	1000		2850
		500		2350
	NTE douhle layer	1000	Dektol	2300
	NTE monolayer	1000		1800
				2500
^{55}Fe	L4 monolayer	1000	Microdol X	2000
				1300

From Rogers *(20)*, with permission.

HD by definition, refers to the 50% distance for a line source only. The distribution of developed grains differs for different sources and so does the distance within which 50% of the grains lie. In each case that distance can be expressed as a multiple of HD. By limiting the term HD to refer only to the line source, it retains a unique value for each specimen, and can thus serve as a normalizing unit. From Salpeter and McHenry *(9)*.

whether the compound was metabolized. Thus, cell/tissue fractionation followed by re-isolation of the administered compound is a beneficial adjunct to EM autoradiography. Nevertheless, EM radioautography has

Table 3
Some Contemporary EM Autoradiographic Investigations

System investigated	Reference
Circadian rhythm of total protein synthesis—cytoplasm and chloroplasts of *Gonyaulax polyhedra*	Donner et al. *(13)*
Sulfate incorporation in *Udotea petiolata* thalli	Mariani and Favoli *(14)*
Photosynthetic activity of phytoplankton	Bourdier and Bohatier *(15)*
Thymidine incorporation into *Nostoc*	Favoli and Grilli *(16)*
Chloroplast nucleoids of *Zea mays*	Lindbeck et al. *(17)*
Localization of chloroplast binding protein in cells of wheat leaves	
Distributon of [^{14}C]-dichlorophenoxyacetic acid	Bronsema et al. *(18)*

yielded a wealth of information regarding the role of intracellular organelles in the transfer of radiolabeled compounds to the plant cell walls. For example, Dashek and Rosen *(2)* employed the technique (protocol 5.1) to follow the fate of [^{3}H]-myo-inositol, a pectin precursor, in elongating pollen tubes (*see* Fig. 2). Protocol 5.2 lists the concomitant preparation of light microscope autoradiographs from plastic-embedded plant material. In this connection, the reader is referred to the papers of Pickett-Heaps *(10–12)* for the application of autoradiography to investigating the role of the Golgi apparatus in cell wall deposition. Evans and Callow *(3)* catalog the pioneering EM autoradiographic investigations (1964–1976) of polysaccharide, protein, nucleic acid, and lignin synthesis. Table 3 summarizes more contemporary applications of EM auto-radiographs.

5. PROTOCOLS

5.1. Incorporation of [^{3}H]-Myo-Inositol into Germinating Lilium longiflorum *Pollen*

Supplies and chemicals	Plant material	Equipment

Chemicals
  [^{3}H]-*Myo*-inositol
    (specific activity 14.3
    millicuries/millimole)
Sucrose
Boron
Yeast
Acetone—EM grade
Alcohol—EM grade
Developing solutions—
  Microdol X,
  Kodak D-19,
	Source of *Lilium longiflorum*, cv. 'Ace' pollen—greenhouse, botanical gardens, or supermarket at Easter	Electron microscope
		Knife breaker
		Light-tight darkroom
		Tabletop centrifuge
		Ultramicrotome
		Water bath, 45°C
		Wire loop of tungsten
		  or platinum

Supplies and chemicals	Plant material	Equipment

paraphenylenediamine, or
   phenidon
Epon 812 (Polybed)
Formvar
Glutaraldehyde
Ilford L-4 emulsion with
  appropriate safelight
Osmium tetroxide
Phosphate
  Dibasic
  Monobasic
Propylene oxide
Sodium thiosulfate fixing
  solution
Spurr's embedding medium
Supplies
  Boats for glass knives
  Conical centrifuge tubes
  Copper grids
  Dental wax
  Glass or diamond knives
  Graduated cylinders
  Grid box
  Light microscope slides
  Small box with drierite
  Sterile Petri dishes

1. Procure, store, grow, and harvest *Lilium longiflorum* pollen, as in Dashek and Rosen *(2)*.
2. Augment medium with 10–40 µCi/mL of *myo*-inositol -2-[H^3] (14.3 millicuries/millimole).
3. Fix pollen tubes 30 min in 6.25% glutaraldehyde in phosphate buffer or veronal-acetate buffer, pH 7.4.
4. Postfix 15 min in 2% OsO_4 buffered as above.
5. Dehydrate through a graded acetone series.
6. Embed in Vestopal W. More contemporary embedding procedures would include Spurr's or Epon embedding media. The protocols for both Spurr's and Polybed (a recent em formulation) use can be found in a variety of EM books (*see* Chapter 1). In addition, these protocols accompany the commercially available kits.
7. Polymerize 2–3 d at 65°C.
8. Section with ultramicrotome and glass knives.
9. Collect sections on 200 mesh copper grids.
10. Coat with Ilford L-4 nuclear emulsion.

11. Store in dark.
12. Develop and process as in Dashek and Rosen *(2)*.

If Epon as Polybed 812 is employed, a graded alcohol series followed by a transitional solvent such as propylene oxide and progressive propylene oxide-Polybed is desired. (Updated from: Dashek and Rosen *[2]*.)

5.2. Preparation of Thick Sections from EM-Embedded Blocks for Concomitant Light Microscope Radioautography

1. Prepare 1-μm thick sections by ultramicrotomy (thin sections from ultramicrotomy of same blocks used for obtaining thick sections can be employed for EM radioautography).
2. Expand sections with solvent vapor while sections are in water bath.
3. Transfer expanded sections to a glass slide with a drop of H_2O.
4. Warm slides on a hot plate at 40–50°C; remove epoxy resin by inserting slide into methanolic sodium methoxide for 3 min (2 g sodium metal to 110 mL methanol; then add 100 mL benzene after dissolution—use immediately).
5. Wash 5 min in methanol-benzene and then 5 min in acetone and air dry.

5.3. Protocol for Preparing Light Microscopic Radioautographs from Resin-Embedded Tissues (Adapted from Morel [7])

1. Apply emulsion of choice, e.g., Amersham LM-1, Ilford KS, Kodak NTB2 diluted 1:1v/v in distilled H_2O or apply Kodak AR10 Stripping Film.
2. Dry at room temperature.
3. Place slide at 4°C in light-tight container containing drierite for 2–30 d.
4. Develop the emulsion at 18°C in either D-19 or Dektol for 3 min.
5. Wash briefly in water and then in 30% sodium thiosulphate.
6. Rinse carefully in water.
7. View directly or stain sections.
8. For staining, apply a drop of stain and warm on a hot plate 60°C for 30 s (1% toluidine blue in 1% sodium borax).

6. CONCLUSIONS

The future of EM autoradiography most likely will center about the utilization of frozen sections for the localization of soluble compounds within tissues. Of special interest is the expanding application of EM autoradiography to molecular biology. Fakan and Fakan *(4)* have employed the technique for investigating spread molecular complexes. The reader is referred to Sigre *(19)* for additional applications of EM autoradiography.

REFERENCES

1. Bouteille M, Dupuy-Coin AM, Moyne G. Techniques of localization of proteins and nucleoproteins in the cell nucleus by high resolution autoradiography and cytochemistry, in *Methods in Enzymology* (O'Malley BW, Hardman JG, eds.), Academic Press, New York, 1975, pp. 3–41.
2. Dashek WV, Rosen WG. Electron microscopical localization of chemical components in the growth zone of *Lilium pollen* tubes. *Am J Bot* 1966; 51: 61–67.
3. Evans LV, Callow ME. Autoradiography, in *Electron Microscopy and Cytochemistry of Plant Cells* (Hall JL, ed.), Elsevier, North-Holland, Amsterdam, The Netherlands, 1978, pp. 235–277.
4. Fakan S, Fakan J. Autoradiography of spread molecular complexes, in *Electron Microscopy in Molecular Biology. A Practical Approach* (Sommerville J, Scheer U, eds.), IRL Press, Oxford, UK, 1987, pp. 201–214.
5. Heaps JD. Autoradiography with the electron microscope experimental techniques and considerations using plant tissues, in *Microautoradiography and Electron Probe Analysis, Their Application to Plant Physiology* (Luttage U, ed.), Springer-Verlag, New York, 1972, pp. 167–190.
6. Moran DT, Rowley JC. Biological specimen preparation for correlative light and electron microscopy. in *Correlative Microscopy in Biology, Instrumentation and Methods* (Hayat MA, ed.), Academic Press, New York, 1987, pp. 2–22.
7. Morel G. Electron microscopic autoradiographic techniques, in *Electron Microscopy in Biology. A Practical Approach* (Harris JR, ed.), IRL Press, Oxford, UK, 1991, pp. 83–123.
8. Negata T. Electron microscope radioautography with cryo-fixation and dry mounting procedure. *Acta Histochim Cytochem* 1994; 27: 471–489.
9. Salpeter MM, McHenry FA. Electron microscope radioautography, in *Advanced Techniques in Biological Electron Microscopy* (Koeller JK, ed.), Springer-Verlag, New York, 1973, pp. 113–152.
10. Northcote DH, Pickett-Heaps JD. A function of the Golgi apparatus in polysaccharide synthesis and transport in the root-cap cells of wheat. *Biochem J* 1966; 98: 159.
11. Pickett-Heaps JD. Further observations on the Golgi apparatus and its functions in cells of the wheat seedling. *J Ultrastruct Res* 1967; 18: 287.
12. Pickett-Heaps JD. Further ultrastructural observations on polysaccharide localization in plant cells. *J Cell Sci* 1968; 3: 55.
13. Donner B, Helmboldt-Caesar U, Resing L. Ciradian rhythm of total protein synthesis in the cytoplasm and chloroplasts of *Gonyaulax polyedra*. *Chronobio Int* 1985; 2: 1–10.
14. Mariani P, Favoli MA. Sulfur-35 sulfate incorporation in *Udotea petiolata* thalli and in the endophytic bacteria on electron microscope autoradiographic study. *Cytobios* 1985; 42: 243–250.
15. Bourdier G, Bohatier J. Photosynthetic activity of phytoplankton visualized by autoradiography. *Am Sci Mob Zoo Biol Anim* 1986–1987; 8: 75–80.
16. Favoli MA, Grilli CM. Autoradiography and ultrastructural study of *Nostoc* species labeled with tritiated thymidine. *Microbios* 1986; 45: 33–40.
17. Lindbeck AGC, Rose RJ, Lawrence ME, Possingham JV. The chloroplast nucleoids of the bundle sheath and meosophyll cells of *Zea mays*. *Physiol Plant* 1989; 75: 7–12.
18. Bronsema FBF, Vanuvustveen WJF, Pinsen E, Van Lammeren AAM. Distribution of [14C] dichlorophenoxyacetic acid in cultured zygotic embryos of *Zea mays* L. *J Plant Growth Regul* 1998; 17: 81–88.
19. Sigre DC. Electron microscope autoradiography. *Pure Appl Chem* 1991; 63: 1277–1283.
20. Rogers AW. *Techniques of Autoradiography*, Elsevier Scientific Publishing, Amsterdam, The Netherlands.

18 Electron Microscopic Immunogold Localization

Eliot M. Herman

CONTENTS

1. INTRODUCTION

Even simple eukaryotic organisms such as yeast or nematodes have been shown to possess 4000–19,000 distinct genes, most of which encode proteins. Although all of these genes are not expressed at any particular time, any individual cell may synthesize several thousand different proteins. Determining the intracellular localization of proteins is an essential part of elucidating function. Localization can present further complications because proteins may be synthesized at one intracellular site but are subsequently translocated to other cellular sites for utilization or function. During the translocation process, proteins may be further modified by attachment or removal of sugars, phosphate, fatty acids, enzyme cofactors, and metals as well as proteolytically processed to mediate protein maturation. This has important implications for protein function—the localization of a single protein may be simultaneously

From: *Methods in Plant Electron Microscopy and Cytochemistry*
Edited by: W. V. Dashek © Humana Press Inc., Totowa, NJ

in multiple cellular sites depending on the protein's trafficking, processing, and function. The movement and alteration of proteins further complicates the process of localizing a protein because critical antigenic sites can be either modified or removed. The designs of experiments to localize proteins appear to be superficially simple; however, technical considerations in the conduct of the assay as well as interpretation of the results can greatly complicate the final results.

There are two primary approaches by which a protein's intracellular localization may be determined. These are loosely characterized as either biochemical approaches where cellular compartments are isolated and identified, or structural approaches where cells are examined by microscopy with a protein's localization determined either by its activity or with a specific constituent/sequence labeled by a visualizing agent. Localization is best defined by the parameters of resolution and specificity, which are defined as how well the protein's position can be determined in spatial terms. The resolution of an assay is limited by the technique used and is roughly equivalent to spatial accuracy. For instance, a biochemical assay has a crude resolution because it is necessary to fragment cells before separating its components into a few dozen fractions. At its very best, a biochemical assay can colocalize a specific protein within the same fractions(s) as known markers. Microscopic approaches provide increased resolution of many orders of magnitude. Light microscopic assays can easily resolve structures of 1 μm or less, and electron microscopes offer the prospect of resolution at least 100-fold better. Only electron microscopic assays (see Figs. 1 and 2A) can localize a protein to specific membranes and subdomains of organelles (1).

2. ANTIBODIES

Antibodies are uniquely suitable probes for localization studies. Antibodies recognize the cross-reacting antigen and do not require any specific activity of the target as is necessary for many histochemical assays. Because antibodies can be developed with a very narrow specificity, it is possible to localize individual members of gene families or proteins with a particular conformation or processing status. Antibodies used for localization may be derived from a wide variety of animals, with the most common being chickens, rabbits, mice, and rats, although hamster, goat, sheep, and donkey are other possibilities. Antibodies obtained from human sera can also be used and are useful for a variety of pathology and allergy studies (see Fig. 2B). Antibodies are either polyclonal or monoclonal that are, respectively, either the antibodies derived from crude serum directed at many epitopes, or the antibodies expressed from a

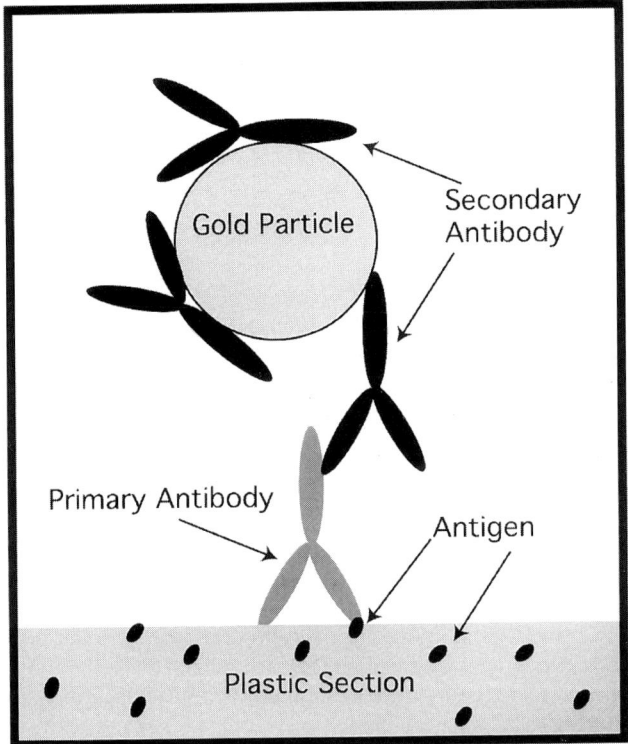

Fig. 1. Diagram of an EM immunogold assay localizing a protein on plastic sections. The primary antibody binds to an exposed surface epitope of the embedded cells. The antibody is then visualized by binding a second antibody coupled to a colloidal gold particle. The electron-dense gold particle visibly marks the position of the bound antibodies when visualized with the electron microscope.

cloned cell line directed at a single antigenic epitope. Polyclonal sera can either be used as crude sera or the antibodies can be purified as a class, such as IgGs or antigen-immunoaffinity purified. Class purification enriches for a single type of antibodies by the antibodies directed against the immunizing antigen are still only a small fraction of the total antibodies directed solely at the antigen of interest eliminating other potential cross-reacting antibodies that might yield spurious results.

The requirement for high-specificity antibodies cannot be understated. Testing the antibodies for specificity is an important aspect of any immunocytochemical study. Antibodies used for immunocytochemical assays may be developed especially for the localization project, or more likely developed for other aspects of the project or obtained from other investigators. Sharing probes such as antibodies is now widespread and indeed an expectation for publication in most journals. Antibodies obtained

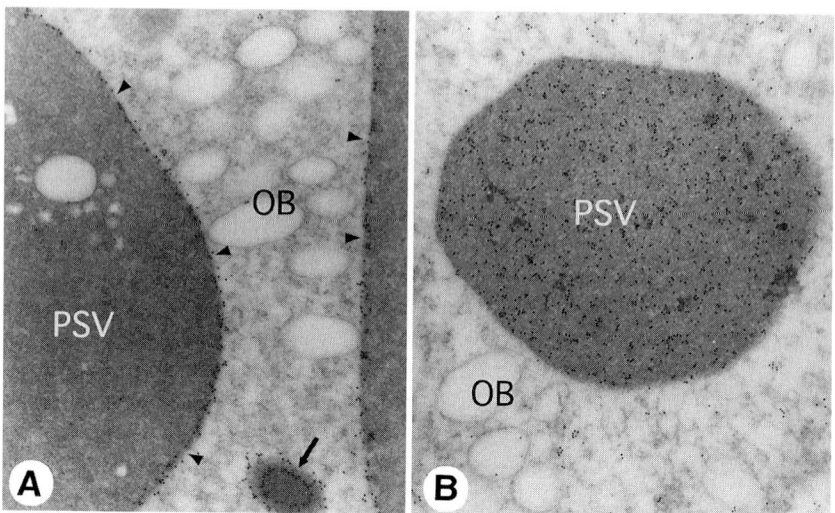

Fig. 2. Two examples of immunogold localization protein antigens in soybean seed cells is shown (see ref. *[19]* for background on seed storage organelle ontogeny). (**A**) Localization of α-TIP (tonoplast intrinsic protein), a water channel protein on the limiting membrane of the protein storage vacuole (PSV) of a soybean seed *(20)*. Note that the gold label is localized on the membrane (arrowheads). A grazing section of the membrane reveals labeling in two concentric circles (arrow), which indicates that the polyclonal sera against α-TIP recognizes epitopes on both sides of the intrinsic membrane protein. TIPs have six membrane-spanning domains with linking segments on the surface. The labeling pattern indicates that two or more of these linking segments are cross-reactive with the anti-α-TIP serum. (**B**) Localization of allergenic proteins in a soybean seed cell using human IgEs derived from a pool of soybean-sensitive persons *(21)*. The primary soybean allergen, P34, is an outlying member of the papain superfamily of cysteine proteases that is localized in the PSV, which is labeled with most of the gold label.

from other investigators may be useful but should always be viewed with some caution for applicability. It should never be assumed that an antibody that recognized a protein in one species also specifically recognized the same protein in another species. The same caution can also apply to localizing proteins in other organs or even other developmental stages of the same species. Because a large percentage of proteins are members of gene families, differences in immunological cross-reactivity is to be expected and should be tested by appropriate immunochemical assay. This is critical in cross-species assays where an antibody was developed against an antigen of one species and will be used to assay the protein localization in another species. It is absolutely essential to run an sodium dodecyl sulfate-polyacrylamide gel electrophoresis (SDS/PAGE)-

immunoblot to confirm cross-reactivity of the antibody and to examine the specificity. Antibodies that label numerous bands unrelated to the target antigen are not useful and the antisera must either be purified or not used.

An immunoblot is also useful to assess the titer of the antisera. Immunocytochemical assays are direct assays where the antibody labels a site that in turn is labeled by an indicator reagent. This means that there is essentially no amplification as is present in enzyme-linked immunoassays including immunoblots. Low-titer antisera in an immunoblot assay is rarely useful in immunocytochemical assays. It is the writer's experience that antibodies that are not suitable for immunoblots at least 1:5000–1:10000 will not be useful for immunocytochemical assay. Immunoblot assays do not confirm that a particular antibody will be useful for immunocytochemistry, but rather that the immunoblot assay is useful to disqualify some antibody samples. SDS/PAGE-based assays have a flaw in that the proteins assayed are fully denatured. This exposes antigenic epitopes that would otherwise be hidden within the correctly folded protein and eliminates structural conformation dependent epitopes which can easily be accomplished by dot blots of SDS-denatured and native proteins. Differences in labeling, such as elimination by SDS, indicate that there is likely a conformation dependent epitope(s). This in itself does not disqualify the antisera for immunocytochemical assays because the process of fixation and preparing tissue preserves much of the normal protein conformation so that antibodies that recognize native and not denatured proteins are still quite useful. The problem is primarily one of controls where it is much more difficult to substantiate that antibody directed at a native conformation is specific for a particular protein. Because of this, it is very useful to immunopurify an antibody directed at a native conformation using an antigen known to be completely purified. This infers, but does not absolutely prove that the antibodies directed at a native conformation are monospecific for a single antigen.

A further complication for immunolocalization studies is that the fixatives used to preserve the tissue chemically modify the antigens that are the target of the antibodies. The aldehyde fixatives attack the amino side chains of proteins that eliminate many epitopes. Osmium postfixation is far more destructive of antigenicity and, as a general rule, osmium most often permanently destroys the possibility of immunocytochemical assay. There are many examples of immunocytochemical assay being conducted on osmicated tissue *(2)* and many others in which the osmium is chemically removed from the sections by oxidation with periodate restoring antigenicity that had previously been masked by the bound osmium *(3)*. In general, immunocytochemistry is best attempted without osmium postfixation initially, and only after the assay appears to be working smoothly should further assays of osmium postfixed material be attempted.

3. PREPARATION AND ANALYSIS
OF MATERIAL FOR IMMUNOGOLD ASSAYS

3.1. Tissue Preparation

3.1.1. CHEMICAL FIXATION AND EMBEDDING

The initial steps in immunogold assays are essentially identical to producing material for conventional EM observation *(4,5)*. The general protocol for obtaining material for immunocytochemistry is to fix tissue in glutaraldehyde or glutaraldehyde and formaldehyde, followed by dehydration with a solvent such as alcohol and then infiltration and embedding in a plastic resin. A few caveats are useful in understanding the fixation requirements for immunocytochemistry. A general principle is that the least possible modification of the target antigen would be best; however, without chemical cross-linking, structure cannot be preserved. At best, all fixation is a compromise between the need to preserve the chemical characteristics of the antigen and the need to chemically modify the target antigen to preserve structure that may eliminate antigenicity. Formaldehyde is the least destructive of antigens but a primary formaldehyde fixation does not preserve structures sufficiently for EM analysis. Bifunctional glutaraldehyde is necessary to cross-link and preserve structures. There are two different approaches to preserving structure and antigenicity. The first concept is to "lightly" fix using low concentrations of fixative and sometimes short fixation times as well. The concept is to limit cross-linking elimination of antigenic sites by limiting the fixation. However, this is very difficult to quantify and there is often more of a faith in this approach than there is experimental evidence.

The second approach is to accept that some antigens will be lost due to fixation, and to accept this as a fact that needs to be incorporated into the experimental protocol. In this case, fixation is optimized by use of more concentrated fixation reagents to obtain the best possible structural preservation. Antibodies are then selected that will work on this tissue and, if necessary, new antibodies can be selected that will recognize antigens in well-fixed tissues. The advantage of optimizing fixation first is that structures are easier to identify and the data have more validity because structures and antigens are much more likely to be firmly in place as in the living cell. In conventional electron microscopy (EM), postfixation with osmium tetroxide is frequently employed. This increases structural preservation, especially of membranes, as well as increasing the visual contrast of the cells due to the electron-dense characteristic of the osmium. However, osmium eliminates many to most peptide antigens and osmium postfixed tissue has limited use in immunocytochemical studies.

Conventional EM is often accomplished with epoxy plastic resins that have trade names such as Araldite®, Epon®, and Spurr's®. These resins can yield remarkable structural preservation due to their extensive cross-linking, and because these resins can be quite hard very thin sections obtained with diamond knives show remarkable detail. The epoxy resins have not proven best for immunocytochemical studies. The extensively cross-linked structure of the plastic appears to inhibit accessibility of the antigens in the tissue to the antibodies, which results in low sensitivity and labeling density. A number of acrylic-based resins have been developed that offer a number of advantages over epoxy resins. Acrylic resins are much more hydrophilic, which means the section is better wetted in the immunocytochemical assay improving the accessibility of the antigen. Acrylic resins are not as tightly cross-linked as epoxy resins, which improves accessibility to the antigen. Acrylic resins can be polymerized by longwave UV light, which can eliminate thermo-denaturation of proteins by hardening the plastic at room temperature or even subzero °C. Commonly used acrylic resins go by the trade names of L.R. White® *(6,7)*, Lowicryl® *(8–10)*, and Unicryl®. L.R. White and Unicryl can be polymerized by either UV or heat, whereas Lowicryl is polymerized only with UV. Using acrylic resins can be quite similar to using epoxy resins in conventional EM. Fixed tissue is dehydrated and infiltrated with resin that is then set up for polymerization in molds. Acrylic resins can tolerate some residual water in the tissue sample and many investigators dehydrate to a level of 95% solvent, leaving some water to retain a hydration shell around the proteins. This might prevent solvent denaturation of the protein and retain additional epitopes that would otherwise be lost. However, in side-by-side tests we have not shown that this enhances immunolabeling with the tissue and antibodies that we have employed and the effective enhancement may be atypical.

Sectioning material for immunocytochemistry with glass and diamond knives is identical to the process used for conventional EM. One difficulty frequently encountered is that unosmicated tissue embedded in plastic is sufficiently cleared by the solvent and embedding process as to be almost invisible. This makes orientation of the tissue in the trimming of the blocks difficult and increases the chances of mistakes of cutting too much. A small amount of Sudan Black dye can be added to

L.R. White that is thermopolymerized to color the plastic dark brown. This will result in the cleared sample becoming visible as a white/clear area within the dark plastic, increasing its visibility. This technique cannot be used in UV-polymerized plastic preparations, as the dye will impede the catalysis. Another technique to enhance visibility is to alter the illumination using back illumination and intense side illumination with fiber-optic lamps. The question of whether or not to mount sections on a supporting membrane has resulted in differing opinions. In the writer's experience, the supporting membranes can often be a source of nonspecific binding and further complicates developing a specific assay. Other investigators have the contrary opinion. It is usually the writer's choice to use finer mesh grids up to 600, rather than a supporting membrane. It is useful to have many grids available, as adjustments to the assay will likely be necessary and having several grids to use for refining the assay is essential. Before undertaking an immunogold assay, it is very important to look at a grid before labeling to ascertain that the material contains the structures to be assayed and that it presents material at the necessary developmental stage or physiological state. The most important item is to look at the tissue with a critical eye for its structural preservation. Poorly preserved material, aside from not being pleasing to the eye, may also contain redistributed constituents invalidating the assay. Structural preservation is an essential criteria for reviewers and editors, and with the large investment of time required to accomplish an immunocytochemical assay it is best that the time is well spent by starting with the best possible material. The added time investment to produce well-fixed and preserved material will be rewarded in the end product presented for publication.

3.1.2. CRYOFIXATION ALTERNATIVE

Cryofixation offers many advantages in rapidly fixing cells within milliseconds and permits embedding with either mild or no fixation *(11–13)*. Cryofixation in principle is very simple, where the tissue to be frozen is placed in contact with a supercooled surface or fluid. Here, the physical chemistry of water intervenes to make the simple process of freezing extremely complicated. Rapid freezing is essential to produce amorphous, essentially noncrystalline, water. Slower freezing allows crystal formation with the rate of freezing controlling the crystal size. The heat content of the tissue to be frozen and the speed by which the cold temperature propagates through the tissue act to impede rapid freezing permitting the crystallization of water and destruction of the tissue. Direct contact of the tissue with a cold surface or fluid generally allows no more than 10–20 μm depth of preserved cells. Beyond this depth, the heat content of the tissue slows the freezing to the point where destructive crystals form. Devices to freeze tissue by contact freezing can be constructed in the lab or purchased for several thousand dollars. More effective freezing results from the use of high-pressure freezing devices. These devices take advantage of the thermodynamics of water freezing so that crystallization is reduced for a longer period of time by increasing the pressure of the freezing. These devices have proven very useful to cryofix cells but have a disadvantage of being expensive to acquire and operate. If cryofixation options are available, it should be explored, offering the possibility of enhanced antigen preservation and structural preservation that may be free of fixation artifacts.

3.2. Labeling

3.2.1. STEP ONE: BLOCKING

Antibodies are proteins that have a tendency to nonspecifically stick to the plastic surfaces of the section. If left unimpeded, this nonspecific binding will eventually be labeled with the visualizing regents producing an unacceptable background for the assay. Blocking these nonspecific sites is usually accomplished with a solution containing a high concentration of proteins. Various protein solutions that are often used include 1–5% bovine serum albumin (BSA), 5% nonfat dry milk, and 10–20% fetal or blood serum made up in either phosphate or Tris-buffered saline (PBS and TBS). One effective blocking solution is to use 10% whole serum of the same species as the origin of the second antibody. To all of these solutions a low concentration (0.05%) of nonionic detergent such as Tween-20 is usually added to improve wetting the plastic sections. These various solutions have different effects on the assay that are related to the "stringency" of the assay. A highly stringent blocking solution is very effective at blocking nonspecific binding at the price of also reducing the specific labeling density. Conversely, a low-stringency blocking solution allows a higher density of specific labeling, but usually at the price of a significantly higher background. Among blocking solutions, the nonfat milk solutions are highly stringent, whereas BSA and low concentrations of serum are much less stringent. It is more useful to begin to define an assay with a lower-stringency blocking solution even if the background is higher in order to determine if the antibody is a useful immunological probe.

3.2.2. STEP TWO: PRIMARY ANTIBODY LABELING

After blocking, the initial step for immunolabeling is to incubate grids in blocking solution in which is diluted the primary antibody. Grids can either be floated on the solution or immersed within in it. We use immersion in order to be certain that the sections are well exposed to the labeling solution. It is useful to use a clean surface that assists in picking up grids with tweezers. A layer of Parafilm in a dish provides a clean new surface for each reaction at very little cost. Because each antibody and antisera is a unique preparation, there is no single guideline that can be used for the initial dilution. This must be determined experimentally. However, an initial dilution of 1:50 for 1 H at room temperature will usually yield information on how to proceed with developing an assay. This is usually a sufficient dilution to minimize some of the nonspecific artifacts while being concentrated enough to see if there is specific labeling. An evaluation of the sample will then reveal whether it is desirable to rerun the assay with more or less concentrated antibody. An initial assay in a low or moderate stringency blocking solution will also reveal whether there is a serious problem with background and/or pseudo specific artifacts. We normally treat all sera as potentially biologically hazardous whether derived from animals or humans and although we often add sodium azide for preservation, this will not necessarily preclude possible viral hazards. Unless working with known biological hazards, normal laboratory precautions including latex gloves are sufficient.

3.2.3. Step Three: Wash

The primary antibody is then removed and excess antibodies washed by immersion in buffer. PBS or TBS with added Tween-20 will effectively remove unbound antibodies. A brief wash of a few minutes will be sufficient.

3.2.4. Step Four: Indirect Labeling

The next step is to indirectly label the bound primary antibody with a second antibody (or protein A/G) bound to colloidal gold (14,15). Colloidal gold is an ideal reagent for immunocytochemistry. It is highly electron dense and can be produced in sizes ranging from 2–20 nm with little variation in size (16,17). Gold particles of 5 or 10 nm are the most useful for EM as a good compromise between visibility and labeling density. As a general rule, smaller gold particles more densely label a structure. Colloidal gold conjugates can be produced from either second antibodies directed at the same species as the primary antibody or with protein A or G that are bacterial proteins that specifically bind immunoglobulins. There are many published protocols for producing colloidal gold particles and conjugates. However, high-quality colloidal gold reagents are now widely available from commercial sources. Although there is some economy in producing reagents in the laboratory, this is balanced against the convenience of the purchased reagents. Indirect labeling with the gold particles is usually brief with a 1:1–1:4 dilution of the commercial reagents in the same blocking solution for 5–10 min at room temperature. This is, of course, a starting point with longer times and more or less concentrated reagents used to further optimize the assay depending on the observed results.

3.2.5. STEP FIVE: WASH

The grids are then washed with buffer as before to remove the excess gold particles. The grids should then be dipped in distilled water for a few seconds to remove excess buffer before staining. Uranium stains can form precipitates with a variety of substances and the removal of the buffer helps ensure a good staining of the section.

3.2.6. STEP SIX: STAIN

After the water wash, the grids are stained with aqueous uranyl acetate (5%). We always either centrifuge or filter the uranium solution immediately before use. A brief spin in a microfuge is sufficient. Alternatively, the solution can be stored in a syringe and passed through a syringe filter immediately before use. This will remove aggregates of the stain and help ensure that the sections are free of particle contamination. The grids are immersed in a large drop of the stain for 30 min at room temperature. This can be varied depending on the observed results. In general, we have found that longer staining times are more useful in increasing observed section contrast. We rarely use lead citrate as a counterstain. The lead often forms aggregates on the sections possibly by interacting with the blocking reagents. After staining, the grids are immersed in distilled water to remove the excess stain and the grids are then dried on filter paper with the section side up. At this point, the protocol is finished and the grids (highly stable experimental material) can be immediately visualized or they can be stored for months to years for future use.

3.2.7. STEP SEVEN: VISUALIZE, PHOTOGRAPH, CRITICALLY EXAMINE

It is at this point that the real fun begins, to visualize the material and determine whether the obtained results are scientifically useful. Instrument accessibility is an important aspect of EM immunocytochemical studies. EM facilities are expensive to acquire and maintain, which induces many universities and other organizations to centralize the EMS and require advance scheduling. This can impede developing an immunochemical assay where multiple samples must be examined in order to refine an assay protocol. It may be necessary to repeat the assays many times before obtaining a satisfactory result. It is very useful to have access to an EM that allows an assay to be conducted, followed by a quick examination of the grid, and then to conduct additional rounds of assays and examination. In most cases, it will take several assays before the labeling is near optimum. With good access to the EM, this development period can be limited to one or at most a few days.

The unosmicated sections used for immunogold studies have much lower contrast than conventionally prepared EM material. This makes looking at the material more difficult and much more exhausting with increased eye strain. If the contrast of the section is poor, then resetting the objective, which is usually set for conventional material, may improve visualization. Similarly, when developing the film, altering the processing time for increased contrast can help improve the photographed images.

A more dramatic improvement can be obtained with digital imaging. Photography and microscopy are undergoing a revolution in imaging methods. Digital charge-coupled device (CCD) cameras are now widely available for installation on electron microscopes, although the current high cost has limited wide-scale acquisition. Although space does not permit a complete discussion of digital imagery, its application and growing importance warrants a brief discussion. Digital images are particularly useful in immunocytochemical EM where the structural preservation is a compromise with preservation of antigenicity. The lack of some structural details in material fixed only with aldehydes is highly compatible with the lower resolution of $1K \times 1K$ megapixel CCD camera now commonly available. $1 K \times 1 K$ digital images occupy about 1.5 MB of storage with a 8-bit depth in grayscale (256 gray tones) that is sufficient for most use. Compared with film and processing costs, the acquisition and storage of digital images can reduce running costs by about 100-fold compared to conventional photography. For instance, a $2 CD-ROM disk can store over 300 images and associated notes. Digital darkroom programs are used to enhance, group, and label photographs with all the complex manipulations formally done in the darkroom and assembling finished photographs are instead accomplished on a computer screen. We have found digital imagery particularly effective with EM immunocytochemical projects because the digital enhancement allows contrast adjustment of the material, making it much easier to see gold particle labeling and underlying structures. The enhancement provides for immediate correction for the low contrast of unosmicated sections. Multiple windows can remain open on the computer at the same time. One application is to capture a low-magnification image of the section being examined and then to use this image as a reference for further examination of the section at higher magnification. With 100 MB or more of RAM available it is possible to have a dozen or more images open simultaneously on the computer.

Digital images are particularly effective in quantitative immunogold projects where it is necessary to sample large amounts of material. It is quite simple to acquire large numbers of images, overlay them with grids, and to count the gold particles per unit area. For manual particle counting, the images can be exported from the computer to a laser printer even while using the EM. Large contiguous areas can be photographed, saved in digital form, and printed all at the same time. A laser print has more than sufficient resolution for particle counting and the cost per print is then insignificant. We have not yet attempted automated gold particle counting, but this would appear to be quite feasible to develop.

4. CONTROLS AND ARTIFACTS

Perhaps the single most important control occurs in the mind of the investigator. This is simply to ask if the results obtained appear reasonable. Is the localization discrete to a single cellular site? Or if it is on multiple sites are those sites related by ontogeny or membrane flow (*see* Fig. 2)? Very disparate localization patterns where organelles are labeled that should not be related call for a critical examination with the view of a possible artifact. If there are background data on the protein that suggest a possible localization, and this is then confirmed by immunogold assay, this increases the confidence that the assay may be accurate. Further controls and other critical evaluations should further substantiate the quality of the assay. Controls and possible artifacts must be carefully considered in evaluating the results from each localization assay. Immunological controls including preimmune serum from the same animal are most common. These are run at the same time and under the same conditions as the specific antisera. Good specificity of the labeling with immune serum and a lack of labeling with the preimmune serum is usually considered to be appropriate control. Localization of a presumably intracellular protein on cell surface or in the cell wall can be a red flag indication that further caution and analysis is needed. Glycan polymers are frequently labeled by crude serum or purified classes of antibodies. Glycans present in food and bedding may be a source of immunization. Intense labeling of the cell surface may be an indication of this type of artifact. SDS/PAGE immunoblots usually do not control for this possible artifact, as glycan polymers are not fractionated by SDS/PAGE. A simple control for antiglycan antibodies is to incubate a grid in periodate/HCl as would be used to oxidize osmium and the wash with water before conducting the assay as previously discussed. This treatment oxidizes carbohydrates, and if this eliminates the cell surface immunolabeling then the labeling is the consequence of recognition of glycans.

This treatment does not usually harm polypeptide epitopes so the protein labeling should remain at about the same level as the sample not treated with periodate. The other type of common artifact is the pseudo-specific artifact in which a specific labeling is obtained, but not of the protein targeted. A pseudo-specific artifact is most often the result of either a polyspecific primary antibody or antiglycan antibody. The polyspecific antibody results from a protein that was impure for the immunization and then elicited antibodies against multiple antigens. The only way to salvage such antisera is by immunoaffinity purifying the antibodies using a column from the pure antigen.

5. ANTIBODIES AGAINST OTHER TYPES OF MOLECULES

There is no reason why any immobilized substance that is capable of eliciting a specific antibody cannot be the subject of an immunocyto-chemical study. For instance, many natural substances, lipids, pigments, and other abundant biochemicals can be effective immunogens. The primary limitation is the ability to effectively immobilize the antigen in the cell with fixatives that do not abolish immunoreactivity with the antibody. Sugar residues in the form of polymers or glycan side chains of proteins are often highly immunogenic and are widely used for immu-nocytochemical assays. These glycan polymers constitute a large frac-tion of the artifacts that can plague immunogold assays. Many proteins are naturally glycoproteins and when used to elicit antibodies in their glycosylated form. For studies where it is appropriate, glycans offer the best prospects for immunolabeling molecules other than proteins.

If the specificity of the antiglycan antibody is known, then suitable controls may consist of either coincubating with a hapten or preincubat-ing the reaction mixture with an excess of the glycan or glycoprotein. However, the specificity of glycan antibodies may not be known and further the use of whole serum for labeling reactions presents the poten-tial that other antiglycan antibodies may be present that are not part of the targeted antigen. For this reason, the single best general control for glycan cross-reactivity is to preincubate sections in periodate/HCl that will oxidize the glycans altering its structure so that it will no longer be recognized by a specific antibody.

6. FUTURE USES OF IMMUNOCYTOCHEMISTRY

Immunogold assays are now widely used to localize proteins within cells and, in particular, to achieve the detail and resolution that cannot be obtained with alternative assays such as immunofluoresence or biochemical fractionation *(1,19)*. There is great potential to elaborate on immunogold assays to obtain dynamic information of cellular processes. For instance, physiological studies using immunogold methods should become much more prominent. Immunogold assays can be easily adapted to study protein trafficking, posttranslational modification, and function in a wide variety of physiological responses. With the advent of peptide antigens, it is quite simple to elicit antibodies against precursor and mature segments of proteins. Many proteins are processed as they are translocated through the cell. Using processing status-specific antibodies, it should be possible to assay the localization of protein precursors as compared with the mature protein. The genes encoding proteins are often gene families that express isoforms of proteins that can be differentially accumulated in response to many forms of regulation. Antipeptide antibodies can be used to further characterize the regulation of expression and accumulation of gene family members. Other forms of protein processing, such as attachment and removal of phosphate, regulate protein activity particularly of many important signaling pathways could be localized. Phosphorylated peptides can be produced that will recognize the modified protein that can be used to localize the spatial distribution of an activated protein. When compared with the distribution of the protein using antibodies against the whole protein, it should be feasible to ask important questions of the events that occur following the activation of pathways. As with any technique the only technical limitation in immunocytochemistry is the creativity of investigators in designing new and important uses for immunogold assays.

REFERENCES

1. Herman EM. Immunocytochemical localization of macromolecules with the electron microscope. *Annu Rev Plant Physiol Mol Biol* 1988; 39: 139–155.
2. Bendayan M, Zollinger M. Ultrastructural localization of antigenic sites on osmium-fixed tissues applying the protein A–gold technique. *J Histochem Cytochem* 1983; 31: 101–109.
3. Craig S, Goodchild DJ. Periodic acid treatment of sections permits on-grid localization of pea seed vicilin in ER and Golgi. *Protoplasma* 1984; 122: 91–97.
4. Herman EM. Colloidal gold labeling of acrylic resin embedded plant tissues, in *Colloidal Gold: Methods and Applications,* Vol. 2, (Hayat MA, ed.), Academic Press, New York, 1989, pp. 303–321.
5. Herman EM, Melroy EL. Electron microscopic immunocytochemistry in plant molecular biology, in *Plant Molecular Biology Manual*, Vol. B13, Kluwer Academic Publishers, Norwell, MA, 1990, pp. 1–24.
6. Newman GR, Jasani B, Williams ED. A simple post-embedding system for rapid demonstration of tissue antigens under the electron microscope. *Histochem J* 1983; 15: 543–555.
7. Newman GR, Hobot JA. Modern acrylics for postembedding immunostaining techniques. *J Histochem Cytochem* 1987; 32: 971–981.

8. Carlemalm E, Garvito RM, Villiger W. Resin development for electron microscopy and an analysis of embedding at low temperature. *J Microsc (Oxford)* 1982a; 126: 132–143.

9. Carlemalm E, Villiger W, Hobot JA, Acetarin JD, Kellenberger E. Low temperature embedding with Lowicryl resins: two new formulations and some applications. *J Microsc (Oxford)* 1982b; 140: 55–63.

10. Roth J, Bendayan M, Carlemalm E. Villager W, Garavito M. Enhancement of structural preservation and immunocytochemical staining in low temperature. *J Histochem Cytochem* 1981; 29: 663–671.

11. Craig S, Staehelin LA. High pressure freezing of intact plant tissues. Evaluation and characterization of novel features of the endoplasmic reticulum and associated membrane systems. *Eur J Cell Biol* 1988; 46: 81–93.

12. Dahl R, Staehelin LA. High-pressure freezing for the preservation of biological structure: theory and practice. *Electron Microsc Technol* 1989; 13: 165–174.

13. Ding B, Turgeon R, Parthasarathy MV. Routine cryofixation of plant tissue by propane jet freezing for freeze substitution. *J Electron Microsc Technol* 1991; 19: 107–117.

14. Romano EL, Stolinski C, Hughes-Jones NC. An antiglobulin reagent labeled with colloidal gold for use in electron microscopy. *Immunochemistry* 1974; 11: 521–522.

15. Romano EL, Romano M. Staphyloccal protein A bound to colloidal gold: a useful reagent to label antigen-antibody sites for electron microscopy. *Immunochemistry* 1977; 14: 711–715.

16. Frens G. Controlled nucleation for the regulation of particle size in monodispersed gold suspensions. *Nat Phys Sci* 1973; 241: 20–22.

17. Roth J. The preparation of 3 nm and 15 nm gold particles and their use in labeling multiple antigens on ultrathin sections. *Histochem J* 1982; 14: 791–801.

18. Polak JM, Priestly JV. *Electron Microscope Immunocytochemistry An Introduction: Current Techniques and Problems*, Oxford University Press, Oxford, UK, 1992.

19. Herman EM, Larkins BA. Protein storage bodies. *Plant Cell* 1999; 11: 601–613.

20. Melroy DL, Herman EM. TIP, an integral membrane protein of the soybean seed protein storage vacuole, undergoes developmentally regulated membrane insertion and removal. *Planta* 1991; 184: 113–122.

21. Helm RM, Cockrell G, Herman E, Burks AW, Sampson HA, Bannon GA. Cellular and molecular characterization of a major soybean allergen. *Int Arch Allergy Immunol* 1998; 117: 29–37.

19 Microanalysis

Martin J. Hodson

CONTENTS

1. INTRODUCTION

Microanalytical techniques were first pioneered in the 1960s, and the earliest paper using X-ray microanalysis on plant materials is that of Läuchli and Schwander in 1966 *(1)*. It was soon realized that microanalysis could provide a link between anatomical studies and plant physiology. It allowed scientists who were interested in aspects of plant mineral relations to pursue their interests at a cellular or even subcellular level. Microanalysis, in its various forms, is now a well-established technique, and one that is continuing to develop.

With some exceptions *(2–4)*, there have been relatively few recent reviews of microanalysis that have considered applications to plant science. In a previous review of this topic *(5)*, I concentrated almost entirely on methods of specimen preparation for electron probe X-ray microanalysis. Here I highlight further developments in this area, and also broaden the scope of the review to include other microanalytical techniques. This chapter introduces the main types of hardware that are now available for microanalysis, reviews the main techniques used to prepare plant material prior to analysis, and provides protocols for the two major techniques.

From: *Methods in Plant Electron Microscopy and Cytochemistry*
Edited by: W. V. Dashek © Humana Press Inc., Totowa, NJ

2. CHOICE OF HARDWARE

There are now several different types of machines that are all capable of microanalysis. All have advantages and disadvantages, but the choice of which to use is often governed by expense and availability to a particular institution. Electron probe microanalysis is by far the most popular, but here particle-induced X-ray emission (PIXE), the laser microprobe mass analyzer (LAMMA), electron energy loss spectroscopy (EELS), and secondary ion mass spectrometry (SIMS) are also considered.

2.1. Electron Probe Microanalysis

First, let us briefly consider how electron probe microanalysis works. More detailed accounts are available *(2,6)*. Electron microscopes function by firing electrons through the specimen in a transmission electron microscope (TEM). In the scanning electron microscope (SEM), electrons are fired at the specimen, and secondary electrons are produced that are used to form the image. As electrons pass through or hit the specimens they cause some disturbances in the atoms of the material. The high-velocity incident electrons strike the bound, inner shell, electrons of the atoms in the material knocking them out of their orbits or energy levels. This is known as ionization, and creates a vacancy in the energy levels of the atom, which is filled almost immediately by an electron that was at a higher energy level. As the electron drops into the inner shell vacancy, the excess energy is emitted as an X-ray photon. The difference between the two energy levels is the X-ray energy, which is thus characteristic of the element. If the vacancy is created in the innermost shell (the K shell), it is called a K X-ray, and as several different outer shell electrons can fill the vacancy we have K_α, K_β, etc., the subscript identifying the transition involved. A vacancy in the second shell produces an L X-ray, and in the third an M X-ray. Beyond this the energies are too low to be used. The X-rays given off are collected at detectors (crystal or solid state). The earliest type was the wavelength dispersive system, developed and much used by materials scientists. Here, a crystal is tuned for a particular element. These machines are very sensitive, and many of the papers in the 1970s and early 1980s on X-ray microanalysis of plant materials utilized this technology *(7)*. However, wavelength-dispersive systems had two principal disadvantages for biological work: they only allowed one element at a time to be surveyed; and they rapidly damaged the organic matrix of the specimen being analyzed. Energy-dispersive analysis of X-rays (EDAX), which allowed all elements of the atomic number of sodium and above to be analyzed for simultaneously, quickly superseded wavelength-dispersive systems in biological work in the 1980s. In more recent times, both TEMs and SEMs have often been fitted with cold stages that allow analysis of plant material in a frozen state, hopefully preserving its elemental content in the original in vivo locations (*see* Subheadings 3.4.2. and 3.4.5.).

Results are usually obtained in three forms: spectra showing all the elements found at a particular location in the sample; counts, in which the peaks are integrated to produce quantitative results; and X-ray dot maps (also known as X-ray distribution images), which allow low and high concentrations of elements to be assigned to areas within the sample. Electron probe X-ray microanalysis has a number of advantages. More institutions will have access to this technology than any of those outlined shortly, and it is likely to be cheaper. More work has been done using these systems, and it is more probable that the problems of working on a particular plant material have been tackled previously. The TEM machines have a high spatial resolution (20–50 nm), and quantitative results are relatively easy to obtain. Until recently, most machines could only detect elements with the atomic number of sodium and upward, but recent advances in detectors (e.g., ultrathin windows or windowless) have meant that analysis can be extended to lighter elements in more modern machines. Electron probe analysis does lack the sensitivity of some of the other techniques, and cannot be expected to detect low concentrations of some elements in some cell compartments (e.g., Al or Ca in cytoplasm). The analysis also tells us little or nothing about the state that an element is in. For example, phosphorus could be present in a sample as phosphate or a number of organophosphorus compounds, but the technique will not distinguish between these possibilities. Rather than giving examples of the many uses of this machinery in plant science at this point these are considered in Subheading 3.

2.2. PIXE

Microanalysis using PIXE is similar in basic concept to electron probe X-ray microanalysis except that protons are used to excite the atoms in the specimen. There are two types of machines: macro-PIXE that is used for the analysis of bulk specimens, and micro-PIXE that has a spatial resolution that will allow analysis at the cellular level. The main advantage of PIXE over electron probe microanalysis is in sensitivity; PIXE-detection limits in an organic matrix are in the range 0.1–10 μg g^{-1}, whereas 0.1–1 mg g^{-1} are the limits of EDAX (8). This is due to a much lower background Bremsstrahlung radiation in PIXE, and it means that it is very useful for detecting trace elements at low concentrations. The main disadvantage of PIXE is its lack of spatial resolution. Malmqvist, one of the leading exponents of PIXE, remarked, "...it is important to state that although it requires very thin and carefully prepared samples the EMP (electron microprobe) is the only microprobe which, using a high-brightness ion source, can be used for elemental analysis on intracellular (organelle) level down to a lateral resolution well below 50 nm" *(8)*. Malmqvist saw PIXE as complementary to other microanalytical techniques rather than replacing them. This advice has frequently been neglected, and plant scientists have not infrequently used PIXE when EDAX would give better results. For example, EDAX will give much better elemental maps than PIXE when the elements known to be in the sample are at reasonably high concentrations.

Sample preparation is just as important for PIXE as in other microanalytical techniques. The most often used technique for biological material has involved quench freezing, followed by cryosectioning and freeze drying. Unfortunately, cryosectioning is often technically difficult for plant material, and as we shall see (Subheading 3.4.1.) freeze drying has been heavily criticized as a preparative technique, particularly for vacuolate cells. Too often, plant samples have been air dried prior to analysis in PIXE; this will not be a great problem for deposited elements (e.g., silica), but soluble elements such as potassium may be redistributed during drying (*see* Subheading 3.1.)

Among recent examples of the use of PIXE to locate mineral elements in plant materials are: an investigation of aluminum accumulation in beech roots *(9)*, a study of silicon in sedge nuts *(10)*, and mapping of hypersensitive lesions in leaves caused by invasion of powdery mildew *(11)*. An investigation of nickel localization in the leaves of the hyperaccumulator plant *Alyssum lesbiacum* was unusual in that specimens were prepared by molecular distillation drying (*see* Subheading 3.4.4.) *(12)*.

2.3. LAMMA

Of all of the machines used for microanalysis LAMMA seems to be the most problematic. A laser beam is used to disintegrate a spot in the sample, and the material emitted is then analyzed in a mass spectrometer. It has similar lateral resolution to PIXE, and like SIMS can be used to distinguish between isotopes of the same element. It has, however, proved very difficult to quantify, and is destructive to the specimen. One recent investigation *(13)* of the distribution of stable isotopes of calcium, magnesium, and potassium in Norway spruce used three microprobes: EDAX at 0.3 μm lateral resolution; isotope specific point analysis, using LAMMA at 1.5 μm lateral resolution; and isotope specific imaging using SIMS at 1–3 μm lateral resolution.

2.4. EELS

Like X-ray microanalysis, EELS uses the interactions between the incident electron beam and the electrons in the specimen. Here, we are concerned with the incident electrons from a TEM that collide with atomic shell electrons, are scattered inelastically, and lose energy as a result. The amount of energy loss is related to the atomic number of the element concerned, and so the electrons passing through a specimen contain much information on the composition of the specimen. In a conventional TEM the electrons that have passed through the specimen are focused to produce the image, and this information is lost. In EELS, however, a spectrometer is used to separate the inelastically scattered electrons from each other on the basis of their different energies. These data can be used to produce an electron energy loss spectrum for all of the elements in the sample, or an electron spectroscopic image (ESI) showing the distribution of a single element in a sample *(3,4,14)*. Energy loss near edge structures (ELNES) in the spectra can give information concerning crystal chemistry, coordination numbers, and distances *(4)*. Thus, one of the advantages of EELS is that compounds can be determined: zinc silicate in *Minuartia verna (14)*; zinc and tin silicates in *Silene vulgaris (15)*; calcium and tin silicates in some dicotyledons *(16)*; and chelation of copper with phenolic compounds in *Armeria maritima (4)*. Other recent uses of EELS have included a study of two types of granules in mycorrhizal fungi *(17)*, investigations of silicification processes in monocotyledon leaves *(18,19)*, and the characterization of degradation substances in beech roots *(20)*. The main disadvantages of EELS are difficulties of quantification, and the use of ultrathin (30–70 nm) resin-embedded sections in TEM. The sections must be thin to avoid multiple scattering of electrons, but it is technically almost impossible to cut dry sections of this thickness, and cutting onto water is known to lose soluble elements. Thus, present applications of EELS seem to be limited to deposited, immobile elements.

2.5. SIMS

SIMS was first developed in the 1960s, and has a long history of applications in biology *(21,22)*. The bombardment of a specimen with a primary ion beam (often O_2^+ or Cs^+) results in the emission of secondary positive or negative ions. Analysis of the mass of the secondary ions is the principle on which SIMS is based. The major advantages of SIMS are sensitivity to all elements and isotopes, low detection limits, and relatively high spatial resolution. The disadvantages are that it is highly sensitive to the local chemical environment, and is difficult to quantify.

Recent examples of the use of SIMS in plant science include the measurement of metals in tree rings *(23)*, aluminum in soybean roots *(24)*, a variety of elements in soybean leaf *(25)*, and nitrogen in yeast and soybean leaf *(26)*. The latter paper also demonstrated the advantage of SIMS in distinguishing between different isotopes of the same element (^{15}N and ^{14}N).

3. PREPARATIVE TECHNIQUES

We have seen that there are now a number of types of hardware that can be used to locate elements at the cellular or subcellular level. The key problem for all workers interested in microanalysis is now not so much the machinery but how to prepare the plant material prior to analysis. It cannot be stressed enough that a wrong decision here will often totally invalidate the results obtained. The use of very expensive analytical facilities (e.g., PIXE) will not ensure that the data acquired will be good if the wrong preparative technique is used. Surprisingly, this point is often neglected. The first decision (Fig. 1) rests on whether the element(s) of interest are deposited as solids (and are thus relatively immobile during specimen preparation) or are soluble (and are potentially highly mobile). If the former is the case, then relatively conventional preparative procedures for electron microscopy can be adopted, but if there is thought to be any chance of ions moving during tissue preparation then some type of cryotechnique will often be preferred.

3.1. Conventional Preparation

If it is certain that the elements of interest will be deposited and immobile then conventional preparative techniques will be appropriate. Thus, for SEM, freeze or critical point drying, followed by carbon coating and observation at room temperature in an SEM fitted with EDAX, is often used. If the greater resolution of TEM is required, fixation in glutaraldehyde and/or osmium tetroxide, followed by dehydration, embedding in resin and cutting sections onto water would be a suitable protocol. These kinds of technique have been found suitable for analysis in a variety of investigations including protein globoids in seeds *(27)*, silica deposits in a variety of plant organs *(28–30)*, aluminum in accumulator plants *(31)*, calcium oxalate crystals in the seeds of Norway spruce *(32)*, and calcium and silicon in the idioblasts of mulberry leaves *(33)*.

Conventional preparation is, however, nearly useless for microanalysis of soluble elements as it leads to major loss from and redistribution within the tissue. For example, it has been conclusively shown that a very high percentage (over 80%) of rubidium (as an analogue for potassium) is lost from labeled leaf tissues during conventional preparation

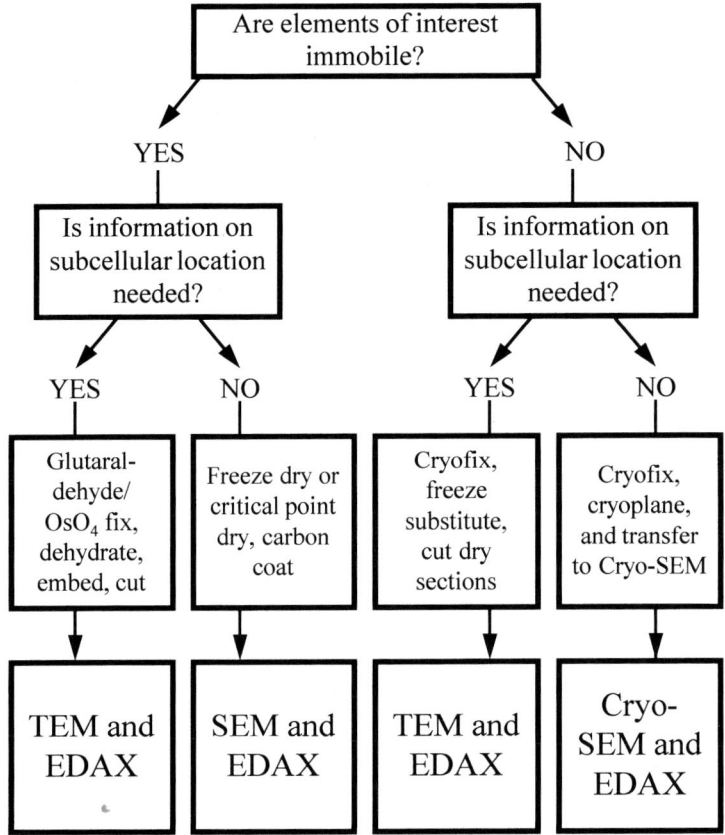

Fig. 1. The main choices in preparing plant tissues for microanalysis.

for TEM *(34)*. If large amounts are lost from the tissue what remains is very unlikely to be in its original in vivo position. So the problem is to keep the soluble components in their original positions. Unfortunately, there are still quite a number of publications appearing each year that pay little attention to these problems.

3.2. Precipitation

In the late 1970s and early 1980s there was a flurry of interest in precipitation techniques as a method of keeping ions in their in vivo locations, and thus avoid the problems mentioned in Subheading 3.1. The idea was simply to add a precipitant to the fixative, which reacted with the ion in question, and hopefully kept it where it was in the tissue. The most popular technique was to use a silver salt to precipitate chloride ions *(35)*. Two problems emerged: loss of ions was not totally prevented *(36)*; and quantification was not possible. This led to the almost complete abandonment of these techniques by the mid 1980s. In one recent example *(37)*, pyroantimonate was used to locate soluble cations in maize roots, but the authors also used freeze substitution for comparison.

3.3. Microdrop

Microdrop analysis has proved to be a very useful technique in the relatively few investigations that have adopted it. This method has been successfully applied to both the analysis of sap collected from the xylem *(38)*, and from cell vacuoles using a modified pressure probe *(39–41)*. Analysis of sap droplets extracted from individual plant epidermal cells and the analysis of cells frozen *in situ* (*see* Subheading 3.5.1.) has been compared *(40)*, and the data suggested that both methods could produce quantitative data accurately reflecting in vivo concentrations in cereal leaf epidermal cells. The advantage of the technique is that it is relatively nondestructive; a single cell from an entire plant is very rapidly extracted for analysis. The main potential errors in this procedure arise from uncertainty over the cell type or compartment being sampled, and the possibility of water movement during sampling due to the drop of turgor pressure in the sampled cell as fluid is extracted. Neither of these problems was seen as insuperable *(40)*, but it is undoubtedly the case that epidermal cells are easier to sample than internal tissues. Recent advances in the quantification of this technique using an internal standard *(42)* suggest that microdrop analysis may be even more used in the future.

3.4. Cryotechniques

There is little doubt that cryotechniques, and particularly cryo-SEM, are now the dominant methods of specimen preparation for electron probe X-ray microanalysis when localization of soluble ions is required. In a previous review *(5)* these techniques were covered in considerable detail and this material is not reiterated here. Instead, protocols for the two major methods are provided and some recent developments and publications in this area are highlighted.

3.4.1. Freeze Drying

Freeze drying can be applied in several different contexts in the preparation of plant material for microanalysis. It can simply be used to extract the last remaining water from samples such as seeds or dry inflorescence bracts prior to analysis in SEM. We have already noted (Subheading 2.2.) that the most common preparative method for PIXE involves freeze drying after quench freezing and cryosectioning. Freeze drying has also been used prior to resin embedding. For example, roots of Norway spruce were prepared according to the following protocol: samples

Table 1
Some Recent Investigations Using Cryo-SEM and EDAX

Tissue	Element(s) analyzed	Topic	Authors
White spruce needle	Various	Biomineralization	(46)
Corn root	K	Xylem embolism	(47)
Tamarix leaf	Ca and others	Ion selectivity in salt glands	(48)
Larch needle	Various	Changes in ion levels over a season	(49)
Larch needle	Ca and Sr	Partitioning of Ca and Sr in the endodermis	(50)
Tea leaf	Al, Mg, Si	Development of low-voltage technique	(51)
Wheat root	Al, Si and others	Aluminum/silicon interactions	(52)
Soybean root nodule	K	Development of X-ray imaging	(53)

were rapidly frozen in a liquid propane–isopentane (3:1, v/v) mixture cooled with liquid nitrogen; freeze dried; infiltrated with ether; embedded in styrol-butylmethacrylate; and 1 μm dry sections were cut and examined in a TEM fitted with EDAX (43). The authors were then able to observe changes in aluminum, calcium, magnesium, and phosphorus in the root cell walls. This type of methodology, however, has been heavily criticized as a method for preparing plant cells for TEM investigations (44). Certainly, it is difficult to see where vacuolar ions go during freeze drying other than onto the nearest membrane or into the neighboring cell walls. Freeze drying seems to be an acceptable technique provided that immobile elements are being studied, nonvacuolate tissues are being analyzed, or relatively low-resolution work is being carried out. The use of cryosectioning in conjunction with freeze drying is further discussed in Subheading 3.4.5.

3.4.2. Cryo-SEM

This technique is simple in basic principle. Material is first rapidly frozen to the temperature of liquid nitrogen. It is then fractured, cryoplaned to produce a flat surface for analysis, and transferred to the cryostage of an SEM. It is analyzed while still frozen, and thus ion movement during tissue preparation should be minimal. A more detailed scheme of a typical procedure (45,46) is given in Subheading 3.4.2.1. This is undoubtedly the most popular microanalytical method with plant scientists at present, and as Table 1 shows it has been applied to a wide range of tissues and research topics (46–53). Recent developments include a method for cryoplaning plant material to produce a flat surface ideal for microanalytical work *(45)*. Up to that point one of the major problems with the cryo-SEM approach had been uneven fracture surfaces, which would result in X-rays being emitted at angles other than the ideal for detection. Hence, uneven surfaces made quantification of the technique very difficult. Now, however, it is becoming routine for quantitative results to be produced using this method *(40,47,49,51,53)*. The problem of whether to etch the frozen hydrated samples before analysis has been previously discussed *(5)*, and there still seems to be little agreement on this point. It is often difficult to visualize material unless it is etched, but etching concentrates ions, and can make quantification difficult. Recent work on soybean root nodules *(53)* has shown that it is possible to obtain quantitative analyses from specimens that have not been etched. It has also proved possible to use low voltages for analysis of light elements *(51)*. Undoubtedly, cryo-SEM has come a long way in recent years, but its key problem is that of SEM itself, lack of spatial resolution. To obtain results at the subcellular level the higher resolution of TEM is needed, and as we shall see the preparation of samples for these machines has caused plant scientists a considerable headache.

3.4.2.1. Protocol for Microanalysis in a Cryo-SEM *(45,46)*. Frozen material is fractured and mounted in brass stubs in a cryo-microtome held at -80°C. The fractured surface is planed with a glass knife to produce a smooth surface. The specimens are transferred under liquid nitrogen and then under vacuum to a cold block in a cryopreparation chamber (at -180°C). They are then moved onto the sample stage (-170°C) of the scanning electron microscope. Slight etching is carried out at -90°C to reveal the cell shapes. The specimens are then rapidly recooled, and transferred back to the preparation chamber, where they are coated with carbon and returned to the specimen stage at -170°C for analysis. Analytical conditions are as follows: raster size 5×5 mm at $\times 2000$ magnification; beam current, 1 nA; accelerating voltage, 15 kV; working distance, 35 mm; takeoff angle, 45°; detector to specimen distance, 50 mm; and live time, 100 s. Windows of 0.16 keV are centered around the various elemental peaks. Beryllium and ultrathin windows are used to collect counts for 100 s with the beam centered on the area at a magnification of $\times 2000$ using a raster size equivalent to a scan area at specimen level of 25 mm^2. Counts are periodically calibrated with a cobalt standard spectrum. The data are transformed to corrected weight percentages using a ZAF-PB program. This is a computer program that uses peak/background ratios (PB), and takes into account differences in mean atomic number (Z), internal absorption of x-rays in the specimen (A), and x-rays generated in the specimen (F).

3.4.3. FREEZE SUBSTITUTION

Freeze substitution was a highly regarded method for specimen preparation in the 1980s, but certainly seems to have declined in use in the 1990s. Again, samples are cryofixed, but then tissue water is gradually substituted by solvent (e.g., acetone or diethyl ether) in a deep freeze. Water in the sample is "mopped up" by a molecular sieve. After returning the samples to room temperature, they are embedded in resin under anhydrous conditions. Sections are cut dry and are mostly examined and analyzed in TEM. A detailed methodology is presented in Subheading 3.4.3.1, and variations have been previously discussed (5,54). The first investigation to produce quantitative results using this technique (55) concerned the subcellular localization of sodium, potassium, and chloride ions in the halophyte *Suaeda maritima*. This spawned a whole series of further papers on related topics, and the results were consistent with evidence produced by other methods. In an investigation of mineral distribution in the wheat root using freeze substitution, soluble Si was observed to be concentrated in the metaxylem (56). In my view, this would be an unlikely result if major redistribution within the tissues had occurred. However, the 1990s have seen a number of critical reviews (3,57), claiming that freeze substitution induces changes in elemental composition in cells. These have certainly dampened enthusiasm for this technique. Despite this resistance, we may be about to see a resurgence of interest in freeze substitution as a preparative method. An excellent investigation of sample preparation procedures for SIMS analysis concluded that freeze substitution was a viable method of sample preparation even for soluble elements such as potassium (25). Soybean leaf material that had been conventionally fixed in glutaraldehyde showed potassium, calcium, and magnesium confined to the parenchyma cell walls. However, freeze-substituted leaf cells showed a very different pattern; for example, potassium was found at the periphery of the cell, in dense spots in the cytoplasm, and evenly distributed in the vacuoles. The authors argued that heterogeneous distribution of soluble elements was a strong indicator that good localization had been obtained. They went on to propose that the K/Ca emission ratio could be used as an index of the degree of natural ion distribution. Other recent examples of the use of freeze substitution include an investigation of Al localization in maize roots (37), Al localization in root tips of two Australian grasses (58), and a series of papers produced by Neumann, Lichtenberger, and their colleagues in Germany (4,14–16,19). In the present author's opinion freeze substitution still deserves to be considered as a valid methodology for preparing plant material for microanalysis, particularly if subcellular localization is to be attempted.

3.4.3.1. Protocol for Freeze Substitution Preparation of Plant Material for Microanalysis *(5,56)*. Small plant segments (1 mm^2) are cryofixed by plunging into supercooled propane at −186°C (note this is hazardous as oxygen/propane mixtures are potentially explosive). The specimens are transferred to anhydrous diethyl ether at −80 °C in a deep freeze. The tissue water is gradually substituted by solvent over a period of 14 d, with three changes of dried ether during that time. Water in the sample is absorbed by a molecular sieve. Samples are then infiltrated with ERL/ether mixtures at −80°C, and gradually warmed to room temperature. After returning to room temperature, samples are handled in an anhydrous dry box, and brought through a Spurr/ERL resin series. Samples are embedded in resin under anhydrous conditions and the resin is polymerized. Sections are cut dry at 250–500 nm thickness, and mounted in hinged grids (thinner wet-cut sections can be cut later for ultrastructural analysis). Specimens are mostly examined and analyzed in TEM, but can also be observed in SEM.

3.4.4. MOLECULAR DISTILLATION DRYING

In a previous review of this topic *(5)*, it was suggested that molecular distillation drying followed by resin embedding needed more work before it could be evaluated as a preparative technique for microanalytical work as there had only been one investigation that had used it *(59)*. Unfortunately, as far as this author is aware, there has only been one more study *(12)* that has used the technique to prepare material for micro-analysis. It is possible that workers have been put off using it by the apparent fall from grace of the related freeze substitution technique or, more likely, by the expense of molecular distillation drying apparatus.

3.4.5. CRYOSECTIONS IN TEM

The ideal solution to microanalysis would be simply to freeze the plant material rapidly to the temperature of liquid nitrogen and then section it while it is still frozen on a cryotome. The frozen sections would then be transferred to a cold stage in a TEM and analyzed. In theory, no ion movement will take place and analysis at the high resolution of TEM should be possible. Indeed, this is a useful technique for liver, kidney, and soft animal tissues, but unfortunately it is almost impossible to cut tough plant material, and maintain the sections in a reasonable state for analysis *(2)*. Even if this problem could be overcome unstained tissues will be difficult to visualize in TEM.

Recently, a number of groups have succeeded in cutting cryosections that were subsequently freeze dried and mounted for observation in SEM or analysis in EDAX *(60)* or SIMS *(24,61)*. This technique is adequate for rough mapping of tissues. Some indication of ionic content differences between the protoplasm and the cell wall is possible, but subcellular localization seems to be beyond the resolution of the method at present.

4. CONCLUSIONS

Writing in 1991, at the beginning of the decade, Van Steveninck and Van Steveninck *(2)* concluded their review of the microanalysis of plant cells as follows: "Thus, at present a reliable means of specimen preparation is the main limitation to microanalysis. Analysis of frozen bulk specimens can provide excellent preliminary results at low resolution, and either cryosectioning or freeze substitution are essential to provide specimen material which can be analyzed with a high degree of resolution. It is difficult to predict which of the cryo-methods will ultimately provide the most favorable results. However, one should keep in mind that no firm conclusion can be reached without comparing several contributing types of microanalytical approach." Where are we now at the end of the decade? The advances in the use of EELS, giving structural details of compounds, and SIMS, opening up the use of isotope work, are very impressive, and show great promise for the future. Microdrop analysis has been developed considerably in the recent years, and is now a useful addition to the plant microanalyst's repertoire. Undoubtedly the major change in plant microanalysis that has occurred during the last 10 years has been the increased application of cryotechniques. The vast majority of investigations now use rapid freezing as the first stage in specimen preparation. Cryo-SEM of frozen bulk samples has become the preferred approach, and freeze substitution has, to some extent, come under a cloud. This author, for one, does not feel the cloud is entirely justifiable. If freeze substitution is killed off as a method of preparation for high resolution work then we are in some trouble, as cryosectioning of plant material for examination of frozen sections in TEM has proven technically very difficult. Should we give up trying to locate soluble ions at the subcellular level in plant tissues? The answer is no.

REFERENCES

1. Läuchli A, Schwander H. X-ray microanalyser study on the location of minerals in native plant tissue sections. *Experientia* 1966; 22: 503–505.
2. Van Steveninck RFM, Van Steveninck ME. Microanalysis, in *Electron Microscopy of Plant Cells* (Hall JL, Hawes C, eds.), Academic Press, London, 1991, pp. 415–455.
3. Stelzer R, Lehmann H. Recent developments in electron microscopical techniques for studying ion localization in plant cells. *Plant Soil* 1993; 155: 33–43.

4. Lichtenberger O, Neumann D. Analytical electron microscopy as a powerful tool in plant cell biology: examples using electron energy loss spectroscopy and x-ray microanalysis. *Eur J Cell Biol* 1997; 73: 378–386.

5. Hodson MJ. Ion localization and x-ray microanalysis, in *Methods in Plant Cell Biology, Part A*, (Galbraith DW, Bohnert HJ, Bourque DP, eds.), Academic Press, London. *Methods Cell Biol* 1995; 49: 21–31.

6. Sigee DC. *X-Ray Microanalysis in Biology: Experimental Techniques and Applications*, Cambridge University Press, Cambridge, UK, 1993.

7. Sangster AG, Hodson MJ, Parry DW, Rees JA. A developmental study of silicification in the trichomes and associated epidermal structures of the grass *Phalaris canariensis* L. *Ann Bot* 1983; 52: 171–197.

8. Malmqvist KG. Proton microprobe analysis in biology. *Scanning Electron Microsc* 1986; III: 821–845.

9. Hult M, Bengtsson B, Larsson NP-O, Yang C. Particle induced x-ray emission microanalysis of root samples from beech (*Fagus sylvatica*). *Scanning Microsc* 1992; 6: 581–590.

10. Ernst WHO, Vis RD, Piccoli F. Silicon in developing nuts of the sedge *Schoenus nigricans*. *J Plant Physiol* 1995; 146: 481–488.

11. Weiersbye-Witkowski IM, Przybylowicz WJ, Straker CJ, Mesjasz-Przybylowicz J. Elemental micro-PIXE mapping of hypersensitive lesions in *Lagenaria sphaerica* (Cucurbitaceae) resistant to *Sphaerotheca fuliginea* (powdery mildew). *Nucl Instr Methods B* 1997; 130: 388–395.

12. Krämer U, Grime GW, Smith JAC, Hawes CR, Baker AJM. Micro-PIXE as a technique for studying nickel localization in leaves of the hyperaccumulator plant *Alyssum lesbiacum*. *Nucl Instr Methods B* 1997; 130: 346–350.

13. Kuhn AJ, Schroder WH, Bauch J. On the distribution and transport of mineral elements in xylem, cambium and phloem of spruce (*Picea abies* [L.] Karst.). *Holzforschung* 1997; 51: 487–496.

14. Neumann D, zur Nieden U, Schwieger W, Leopold I, Lichtenberger O. Heavy metal tolerance of *Minuartia verna*. *J Plant Physiol* 1997; 151: 101–108.

15. Bringezu K, Lichtenberger O, Leopold I, Neumann D. Heavy metal stress of *Silene vulgaris*. *J Plant Physiol* 1999; 154: 536–546.

16. Neumann D, Lichtenberger O, Schwieger W, zur Nieden U. Silicon storage in selected dicotyledons. *Bot Acta* 1997; 110: 282–290.

17. Turnau K, Kottke I, Oberwinkler F. *Paxillus involutus Pinus sylvestris* mycorrhizae from a heavily polluted forest: I Element localization using electron energy loss spectroscopy and imaging. *Bot Acta* 1993; 106: 213–219.

18. Bode E, Kozik S, Kunz U, Lehmann H. Vergleichende elektronenmikroskopische Untersuchungen zur Lokalisation von Silizium in Blättern zweier verschiedener Gräserarten. *Dtsch tierärztl Wschr* 1994; 101: 367–372.

19. Neumann D, Schwieger W, Lichtenberger O. Accumulation of silicon in the monocotyledons *Deschampsia caespitosa*, *Festuca lemanii* and *Schoenus nigricans*. *Plant Biol* 1999; 1: 290–298.

20. Watteau F, Villemin G, Mansot JL, Ghanbaja J, Toutain F. Localization and characterization by electron energy loss spectroscopy (EELS) of the brown cellular substances of beech roots. *Soil Biol Biochem* 1996; 28: 1327–1332.

21. Spurr AR. Applications of SIMS in biology and medicine. *Scanning* 1980; 3: 97–109.

22. Maenhaut W. Applications of ion beam analysis in biology and medicine, a review. *Nucl Instr Methods B* 1988; 35: 388–403.

23. Martin RR, Zanin JP, Bensette MJ, Lee M, Furimsky E. Metals in the annual rings of eastern white pine (*Pinus strobus*) in southwestern Ontario by secondary ion mass spectroscopy (SIMS). *Can J For Res* 1997; 27: 76–79.

24. Lazof DB, Goldsmith JG, Rufty TW, Linton RW. The early entry of Al into cells of intact soybean roots. A comparison of three developmental root regions using secondary ion mass spectrometry imaging. *Plant Physiol* 1996; 112: 1289–1300.

25. Grignon N, Halpern S, Jeusset J, Briancon C, Fragu P. Localization of chemical elements and isotopes in the leaf of soybean (*Glycine max*) by secondary ion mass spectrometry: critical choice of sample preparation procedure. *J Microsc* 1997; 186: 51–66.

26. Gojon A, Grignon N, Tillard P, Massiot P, Lefebvre F, Thellier M, Ripoll C. Imaging and microanalysis of ^{14}N and ^{15}N by SIMS microscopy in yeast and plant samples. *Cell Mol Biol* 1996; 42: 351–360.

27. Egerton-Warburton LM, West M, Lott JNA. Conservative allocation of globoid-held mineral nutrients in *Banksia grandis* (Proteaceae) seeds. *Can J Bot* 1997; 75: 1951–1956.

28. Hodson MJ, Sangster AG. Techniques for the microanalysis of higher plants with particular reference to silicon in cryofixed wheat tissues. *Scanning Microsc* 1990; 4: 407–408.

29. Sangster AG, Hodson MJ. Botanical studies of silicon localization in cereal roots and shoots, including cryotechniques: a survey of work up to 1990, in *The State-of-the-Art of Phytoliths in Soils and Plants* (Pinilla A, Juan-Tresserras J, Machado MJ, eds.), Monografia 4 del Centro de Ciencias Medioambientales, CISC, Madrid, 1997, pp. 113–121.

30. Rafi MM, Epstein E, Falk RH. Silicon deprivation causes physical abnormalities in wheat (*Triticum aestivum* L.). *J Plant Physiol* 1997; 151: 497–501.

31. Cuenca G, Herrera R, Merida T. Distribution of aluminium in accumulator plants by x-ray microanalysis in *Richeris grandis* Vahl leaves from a forest in Venezuela. *Plant Cell Environ* 1991; 14: 437–441.

32. Tillman-Sutela E, Kauppi A. Calcium oxalate crystals in the mature seeds of Norway spruce, *Picea abies* (L.) Karst. *Trees* 1999; 13: 131–137.

33. Sugimura Y, Mori T, Nitta I, Kotani E, Furusawa T, Tatsumi M, Kusakari S-I, Wada M, Morita Y. Calcium deposition in idioblasts of mulberry leaves. *Ann Bot* 1999; 83: 543–550.

34. Hall JL, Yeo AR, Flowers TJ. Uptake and localisation of rubidium in the halophyte *Suaeda maritima*. *Zeit Pflanzenphysiol* 1974; 71: 200–206.

35. Smith MM, Hodson MJ, Öpik H, Wainwright SJ. Salt-induced ultrastructural damage to mitochondria in root tips of a salt-sensitive ecotype of *Agrostis stolonifera*. *J Exp Bot* 1982; 33: 886–895.

36. Harvey DMR, Flowers TJ, Hall JL. Localisation of chloride in leaf cells of the halophyte *Suaeda maritima* by silver precipitation. *New Phytol* 1976; 77: 319–323.

37. Vasquez MD, Poschenreider C, Corrales I, Barceló J. Change in apoplastic aluminum during the initial growth response to aluminum by roots of a tolerant maize variety. *Plant Physiol* 1999; 119: 435–444.

38. Gartner S, Le Faucheur L, Roinel N, Paris-Pireyre N. Preliminary studies on the elemental composition of xylem exudate from two varieties of wheat by electron probe analysis. *Scanning Electron Microsc* 1984; IV: 1739–1744.

39. Malone M, Leigh RA, Tomos AD. Concentrations of vacuolar inorganic ions in individual cells of intact wheat leaf epidermis. *J Exper Bot* 1991; 42: 305–309.

40. Hinde P, Richardson P, Koyro HW, Tomos AD. Quantitative x-ray microanalysis of solutes in individual plant cells: a comparison of microdroplet and *in situ* frozen-hydrated data. *J Microsc* 1998; 191: 303–310.

41. Küpper H, Zhao FJ, McGrath SP. Cellular compartmentation of zinc in leaves of the hyperaccumulator *Thlaspi caerulescens*. *Plant Physiol* 1999; 119: 305–311.

42. Xu W, Marshall AT. A simple method of using an internal standard for x-ray micro-analysis of microdroplets. *J Microsc* 1998; 189: 108–113.
43. Godbold DL, Jentschke G. Aluminium accumulation in root cell walls coincides with inhibition of root growth but not with inhibition of magnesium uptake in Norway spruce. *Physiol Plant* 1998; 102: 553–560.
44. Harvey DMR. Applications of x-ray microanalysis in botanical research. *Scanning Electron Microsc* 1986; III: 953–973.
45. Huang CX, Canny MJ, Oates K, McCully ME. Planning frozen hydrated plant specimens for SEM observation and EDX microanalysis. *Microsc Res Technol* 1994; 28: 67–74.
46. Hodson MJ, Sangster AG. Mineral deposition in the needles of white spruce [*Picea glauca* (Moench.) Voss]. *Ann Bot* 1998; 82: 375–385.
47. McCully ME. Root xylem embolisms and refilling. Relation to water potentials of soil, roots, and leaves, and osmotic potentials of root xylem sap. *Plant Physiol* 1999; 119: 1001–1008.
48. Storey R, Thomson WW. An x-ray microanalysis study of the salt glands and intra-cellular crystals of *Tamarix*. *Ann Bot* 1994; 73: 307–313.
49. Stelzer R, Holste R, Groth M, Schmidt A. X-ray microanalytical studies on mineral concentrations in vacuoles of needle tissue from *Larix decidua* (L.) Mill. *Bot Acta* 1993; 106: 325–330.
50. Gierth M, Stelzer R, Lehmann H. Endodermal Ca and Sr partitioning in needles of the European larch (*Larix decidua* (L.) Mill.). *J. Plant Physiol* 1998; 152: 25–30.
51. Echlin P. Low-voltage energy-dispersive x-ray microanalysis of bulk biological materials. *Microsc Microanal* 1999; 4: 577–584.
52. Cocker KM, Hodson MJ, Evans DE, Sangster AG. Interaction between silicon and aluminum in *Triticum aestivum* L. (cv. Celtic). *Israel J Plant Sci* 1997; 45: 285–292.
53. Marshall AT, Xu W. Quantitative elemental x-ray imaging of frozen-hydrated biological samples. *J Microsc* 1998; 190: 305–316.
54. Harvey DMR. Freeze-substitution. *J Microsc* 1982; 127: 209–221.
55. Harvey DMR, Hall JL, Flowers TJ, Kent B. Quantitative ion localisation within *Suaeda maritima* leaf mesophyll cells. *Planta* 1981; 151: 555–560.
56. Hodson MJ, Sangster AG. Subcellular localization of mineral deposits in the roots of wheat (*Triticum aestivum* L.). *Protoplasma* 1989; 151: 19–32.
57. Lazof DB, Bernstein N. The NaCl induced inhibition of shoot growth: the case for disturbed nutrition with special consideration of calcium. *Adv Bot Res* 1999; 29: 113–189.
58. Crawford SA, Marshall AT, Wilkens S. Localisation of aluminium in root apex cells of two Australian perennial grasses by x-ray microanalysis. *Aust J Plant Physiol* 1998; 25: 427–435.
59. Hajibagheri MA, Flowers TJ. Use of freeze-substitution and molecular distillation drying in the preparation of *Dunaliella parva* for ion localization studies by x-ray microanalysis. *Microsc Res Technol* 1993; 24: 395–399.
60. Frey B, Brunner I, Walther P, Scheidegger C, Zierold K. Element localization in ultrathin cryosections of high-pressure frozen ectomycorrhizal spruce roots. *Plant Cell Environ* 1997; 20: 929–937.
61. Lazof DB, Goldsmith JKG, Rufty TW, Suggs C, Linton RW. The preparation of cryosections from plant tissue: an alternative method appropriate for secondary ion mass spectrometry studies of nutrient tracers and trace metals. *J Microsc* 1994; 176: 99–109.

20 The Use of Electron Microscopy in Molecular Biology

Roslyn A. March-Amegadzie

Contents

1.INTRODUCTION

1.1. Molecular Biology

Molecular biology is considered by some to be a hybrid of biochemistry and cell biology. It is the study of biology in terms of the interactions and interconversions of molecular structures. The goal is to understand biology in chemical terms. In 1945, William Astbury gave us the term "molecular biology," referring to the study of the chemical and physical structure of biological macromolecules. Still, molecular biology encompasses more than biochemistry and cell biology. Molecular biology actually crosses many boundaries including those of biochemistry, cell biology, genetics, organic chemistry, and physics. Molecular biology as it stands today is a new science. One may not be able to pinpoint the time of conception of this new science, but many would include the ground breaking contributions of McLeod, Avery, Watson, and Crick.

Basic to molecular biology is the concept that DNA and RNA are macromolecules that convey information. The sequence of purine and pyrimidine bases in DNA encodes all the information needed to form and direct the chemical reactions within cells. This information thus encodes all catalytic, regulatory, and structural proteins contained within an organism. The flow of information from DNA to RNA to proteins is

From: *Methods in Plant Electron Microscopy and Cytochemistry*
Edited by: W. V. Dashek © Humana Press Inc., Totowa, NJ

universal with a few exceptions being known in some viruses and phages. In these cases, the flow of information begins with RNA. Molecular genetics studies the flow and regulation of the information among DNA, RNA, and protein. The terms "molecular biology" and "molecular genetics" are almost synonymous.

1.2. The Foundation of Immunochemical Techniques

An immunochemical method, whether it is called immunocytochemistry or immunoelectron microscopy, is based on the binding of antigen and antibody. This represents a good distinction between immunochemical methods versus other generalized ligand binding systems (i.e., biotin and avidin) in use. An antigen is a foreign molecule that has the ability to elicit antibody formation. Antigens are mostly proteins although complex polysaccharides and nucleic acids are effective antigens. Antibodies are also called immunoglobulins since they are globulins produced by the immune system. Immunoglobulins (Ig) are tetramers composed of two heavy chains and two light chains. The antigen-binding capacity is found in the amino-terminal portions of the light and heavy chains. The portion of the Ig molecule which has this antigen-binding ability is known as Fab and the remainder of the molecule is known as Fc.

Immunochemical methods are valuable because of their sensitivity and specificity. The sensitivity depends on the method used to determine an end point. One of the reaction components may be tagged with radioactivity, or tagged by covalent binding of an enzyme capable of being detected, or by covalent binding of a totally unrelated species (i.e., fluorescein).

The high specificity lies within the very large number of epitopes. Epitopes, or immunochemical determinants, may be natural immunologic determinants or consist of deliberately added haptenic groups. Haptens are small and nonantigenic. They must be coupled to a weakly antigenic or nonantigenic protein in order to become an effective antigen. Antibodies formed in response to hapten-containing antigens will bind with the antigen and/or the free hapten. In addition, the immunoglobulins themselves are antigens, making double-antibody methods possible where an antibody from one species reacts with an antibody (in this case also the antigen) from another species, for example, rabbit anti-mouse IgG.

1.3. Using the Electron Microscope in Molecular Biology

The electron microscope (EM) has been used extensively in molecular biological research *(1)*. Among its many applications, EM has been used to visualize polyribosomal structure *(2)*, to visualize ribosome substructure *(3)*, and to visualize the elongation factor Tu on the *Escherichia coli* ribosome *(4)*. Conformational changes in DNA topoisomerase II *(5)* and mitochondrial DNA *(6)* have been visualized through EM. EM aided in determining that a specific DNA repeat excludes nucleosomes *(7)* and also aided in the analysis of transcription units *(8)*. This chapter contains a sampling of protocols recently used which have aided investigations of a molecular biological nature. Protocols outlined are not necessarily the first published report of the methods employed but merely represent applications.

The last section of this chapter includes in brief a procedure of McFadden *(9)* for *in situ* hybridization. *In situ* hybridization relies on the complementarity of the bases contained within DNA and RNA. In addition to the hybridization (reassociation) of complementary DNA strands, hybridization is possible between DNA and RNA strands that are complementary. Also, hybridization is possible between a synthetic sequence and a sequence of biological origin. *In situ* hybridization may be used to determine the location of a specific nucleic acid sequence within a cell. The procedure requires the use of a probe for the sequence of interest. The probe, in turn, must be complementary to the sequence of interest. The probe may be either single stranded or double stranded, DNA or RNA. There must exist a method by which to detect the probe.

At the EM level, detection usually involves using a probe (oligonucleotide) in which a hapten has been incorporated. Incorporation of the hapten does not interfere with the hybridization of the complementary sequences. The next step is the binding of a reporter (may be an antibody) to the hapten. The reporter is then subjected to a binding molecule (may be a secondary antibody) that is coupled with an electron-dense material such as colloidal gold for visualization. Nonetheless, the many affinity-detection and immunodetection systems developed for immunocytochemistry may now with ingenuity be applied to molecular biology at the EM level.

2. SELECTED PROTOCOLS

2.1. DNA and Chromatin Visualization

2.1.1. Visualization of Plasmid DNA

For the direct visualization of Plasmid DNA, Chaudhuri et al. *(10)* used the formamide-spreading technique *(11)*. They induced the formation of the various forms of the plasmid and were able to observe the forms using EM. The forms visualized were the supercoiled (control), linear, nicked circular, single-stranded loops, and multimers. After using hydrogen peroxide to induce DNA damage of the plasmid pSV2neo (5.6 kb), the scientists performed the following steps:

1. Dilute DNA solution to 1 μg/mL in TE buffer (10 mM Tris, pH. 8.0, 1 mM EDTA).
2. Add cytochrome C to a final concentration of 100 μg/mL.
3. Add formamide (redistilled) to 60% v/v with 0.1 M Tris (pH 8.5), 0.01 M EDTA.
4. Spread DNA on a hypophase of 30% formamide in 0.01 M Tris (pH 8.5), 0.001 M EDTA.
5. Pick up samples on 400-mesh copper grids with carbon film pretreated with alcian blue.
6. Immerse grids in uranyl acetate (5×10^{-5} M in ethanol) for 30 s.
7. Wash in ethanol and then air dry.
8. Rotary shadow with Pt:Pd (80:20) in a vacuum evaporator at about an angle of 8°C.
9. View with a transmission electron microscope (TEM).

2.1.2. Scanning EM Observation of Chromosomes

Outlined in this section will be the method of Maruyama et al. *(12)*, who were successful in observing metaphase chromosomes in meristematic cells of *Vicia faba* when the fixation of freeze-fractured roots was performed after incubation in buffer solutions. The method allows for experimental studies on chromosome morphology. In fact, the published report also gives the effects of divalent cations on chromosome morphology.

The steps taken by Maruyama et al. *(12)* are as follows:

1. Treat roots with 0.02% colchicine for 5–7 h.
2. Freeze root tips 0.3–0.5 cm long by first putting them in Freon 22 —cooled with liquid nitrogen—and then transferring them to liquid nitrogen.
3. Fracture tips in half longitudinally with a razor blade and a hammer (using a binocular microscope).
4. Thaw roots for 2 min by dipping into 50% dimethyl sulfoxide containing 0.1 M Na-cacodylate, pH 7.4, 2 mM CaCl$_2$, 5 mM MgCl$_2$, and 1 mM phenylmethyl-sulfonyl fluoride (PMSF).
5. Samples are ready for treatment, if any. In this case, samples were washed in a buffer solution of 0.1 M Na-cacodylate, pH 7.4, 4% sucrose, 1 mM PMSF and various concentrations of divalent cations. Incubation is for 10 min at 4°C with three rapid changes.
6. Fix with 4% glutaraldehyde buffered with 0.1 M Na-cacodylate, pH 7.2 for 24 h in a refrigerator (the solution in this case also contained the appropriate concentration of divalent cations).
7. Wash 1X with 0.1 M Na-cacodylate buffer.
8. Postfix with 1% osmium tetroxide in the same buffer for 2 h at 4°C.
9. Make tissues electrically conductive by means of conductive staining of tannic acid–osmium ligation *(13)*.

10. Dehydrate through acetone–water mixtures and pure acetone.
11. Critical point dry with liquid carbon dioxide.
12. Lightly sputter coat the fractured surfaces with platinum-palladium.
13. View specimens with a scanning electron microscope (SEM).

2.1.3. SEM Visualization of the DNA Helix and Nucleosomes

The poor resolution of the fine structures of DNA and nucleosomes by conventional SEM was solved by the development of an ultrahigh-resolution SEM *(14–15)*. Inaga et al. *(16)* observed the DNA double helix and nucleosomes of chicken erythrocytes by using an ultrahigh-resolution SEM. They modified the microspreading technique of Seki et al. *(17)* and combined it with the carbon plate method devised by Tanaka et al. *(18)*. Briefly, they (procedures were performed at 0–4°C before fixation with the formalin solution):

1. Washed cells with CKM buffer (0.05 M Na-cacodylate, 0.025 M KCl, 0.005 M MgCl$_2$ and 0.25 M sucrose, pH 7.5).
2. Suspended cells in 0.2 M KCl at a concentration of 4×10^7 cells/mL.
3. Added 10 volumes of 0.08% Joy detergent (Proctor and Gamble, pH 8.7) and incubated for 1 min.
4. Added 1/10 volume of 10% formalin in 0.1 M sucrose and incubated for 30 min.
5. Centrifuged the suspension at 7000g for 5 min in an Eppendorf tube containing a carbon plate prepared according to the method of Tanaka et al. *(18)*.
6. Rinsed specimens briefly in distilled water.
7. Stained specimens with 2% aqueous uranyl acetate for 1 min.
8. Dehydrated specimens through an ethanol series.
9. Critical-point dried specimens with dry ice *(19)*.
10. Viewed specimens with the ultrahigh-resolution SEM without any metal coating at accelerating voltages of 15 kV.

2.1.4. Determination of Chromatin Loop Size

Tohno et al. *(20)* used similar steps to determine the chromatin loop size in human leukemia cells (an ultrahigh-resolution EM was not needed). To determine chromatin loop size, the lengths of 102 chromatin fibers protruding from four nuclei were measured and their average lengths were estimated. The steps they used (essentially those of Miller and Bakken *(21)* are summarized as follows:

1. Suspend cells in 0.2 M KCl at a concentration of 108 cells/mL.
2. Add 1 mL of 0.11% and 0.09% detergent Joy to 0.1 mL of the cell suspension.
3. Mix gently and incubate for 1 min.

4. Add 0.2 mL of 10% formalin in 0.1 M sucrose to the cell suspension to fix.
5. Treat a carbon-covered and collodion-coated grid (400 mesh) with glow discharge (glow discharge is addressed in Subheading 2.2.).
6. Place the grid on the bottom of a cylindrical vessel (5 mm in radius, 15 mm in height) containing 0.35 mL of 10% formalin in 0.1 M sucrose (pH 7.0).
7. Layer about 10 µL of the formalin-fixed cells on the solution in the vessel.
8. Centrifuge vessel at 10,000 rpm in a Kubota RS-150A rotor for 10 min.
9. Remove grid and rinse in 0.4% Kodak Photo-Flo.
10. Dry and then stain with 2% uranyl acetate for 5 min.
11. Rotary-shadow at an angle of 7° with platinum-palladium (80:20).
12. View with a JEM 100SX EM operated at 100 kV.

2.2. Immunoelectron Microscopy

2.2.1. IMMUNOCYTOCHEMISTRY IN RICE COLEOPTILES

When the light microscope (LM) alone is not sufficient to study a system, the binding of an electron-dense ligand to an antibody facilitates the use of an EM. There are times when, in addition to EM study, immunoelectron microscopic analysis is necessary. Observed anatomical changes often lead to inquiries on the molecular events. This proved true in the case of the report by Inada et al. *(22)* in which they described a three-dimensional analysis of the senescence program in rice (*Oryza sativa* L.) coleoptiles. Immunoelectron microscopy was used to determine the behavior of cellular DNA during senescence. The procedure employed by the investigators is detailed below.

1. Fix tissue (in their case it was coleoptiles) in 2% glutaraldehyde in 20 mM cacodylate buffer, pH. 7.0 for 4 h at 4°C.
2. Dehydrate tissue in a graded ethanol series and propylene oxide.
3. Embed tissue in L.R. White resin.
4. Cut ultrathin sections using an ultramicrotome.
5. Mount on nickel grids.
6. Treat sections on the grids with phosphate-buffered saline (PBS) at pH 7.4 containing 0.05% Triton X-100 for 15 min at room temperature.
7. Incubate in blocking buffer (5% BSA in PBS) for 30 min at room temperature.
8. Incubate with mouse antibodies (in the case of Inada et al. *(22)* it was mouse monoclonal antibodies raised against human single- and double-stranded DNA) at a dilution of 1:10 in blocking buffer for 1 h at 37°C.
9. Wash 5X with PBS at pH 7.4.
10. Incubate with goat anti-mouse IgM conjugated with 15-nm colloidal gold at a dilution of 1:160 in blocking buffer for 2 h at 37° C.

11. Wash 5X with PBS at pH of 8.2.
12. Rinse 1X with distilled water.
13. Stain with 3% uranyl acetate for 20 min at 37°C.
14. View with an EM.

With micrographs from an immunoelectron microscopic procedure, one may simply note the distribution of the tag or one may quantify the results as in the cases of Yamashita *(23)* and Hosaka et al. *(24)*. Yamashita reported his data as a comparison of labeling densities (particles/μm^2). Hosaka et al. compared the ratios of binding under control and experimental conditions in different cell lines. They also compared the ratios of gold labelling efficiencies using the following formula: labeling efficiency (untreated cells) − control cell efficiency/labeling efficiency (treated cells) − control cell efficiency. The labeling efficiency = number of gold/cell surface length.

2.2.2. DOUBLE-LABELING IMMUNOELECTRON MICROSCOPY

Double-labeling immunoelectron microscopy is a technique useful for the marking of multiple sites on a specimen. Two different probes are employed. The probes must have differing distinguishable detection methods. One system of detection uses probes that are distinguishable by the difference in sizes of the colloidal gold particles attached to protein A (*Staphylococcus aureus* coat protein). Romano and Romano *(25)* were the first to report the use of gold particles complexed with protein A. Protein A binds the Fc portions of IgG molecules *(26)*. In 1981, Geuze et al. *(27)* described the use of colloidal gold particles in double-labeling immunoelectron microscopy. Their study showed that protein A/gold probes are suitable for use with frozen sections. They compared the localization of amylase and that of a glycoprotein in the rat pancreas.

For an example of double-labeling immunoelectron microscopy, the procedure of Yamaguchi and Kondo *(28)* is outlined. This serves a triple purpose. In addition to double labeling, the specimen used is a bacterium. Because bacteria are sometimes used in molecular biological techniques, the need may arise to subject the bacteria to immunoelectron microscopic study. Also, the plasma polymerization replica method was used *(29)*. In this method, a specimen has polymerized molecules attached in such a manner as to copy its three-dimensional surface ultrastructure. The specimen is digested and the replica film is viewed by EM. Ionized gas molecules from a glow discharge make up the polymerizing molecules. Yamaguchi and Kondo *(28)* performed differential staining of flagellar and somatic antigens in *P. vulgaris*. The summary of their procedure follows.

2.2.3. Yamaguchi and Kondo's Immunoelectron Microscopy of *P. vulgaris* (Performed at Room Temperature)

1. Fix bacteria in 5% formaldehyde for 30 min.
2. Put fixed bacteria on a 400-mesh grid coated with a collodion support film.
3. Treat grid with blocking solution (1% bovine serum albumin [BSA], 3% gelatin, 0.9% NaCl in 0.01 M phosphate buffer, pH 7.4).
4. Incubate with rabbit anti-H factor serum (first antibody) without dilution for 10 min.
5. Wash with PBS (0.9% NaCl in 0.01 M phosphate buffer, pH 7.4).
6. Incubate with protein A/5 nm colloidal gold without dilution for 5 min.
7. Refix bacteria on grid with 2.5% glutaraldehyde to destroy antigenicity of first antigens and immunoglobulins bound to first antigen, preventing crossover binding.
8. Wash with PBS.
9. Treat with blocking solution.
10. Incubate with rabbit anti-O factor serum (second antibody) without dilution for 10 min.
11. Wash grid with PBS.
12. Incubate with goat anti-rabbit IgG/10 nm colloidal gold without dilution for 5 min.
13. Wash with PBS and water.
14. Place grids with labeled bacteria on the cathode plate in a bell jar of the plasma polymerization replica apparatus.
15. Evacuate the air.
16. Fill jar with naphthalene gas to about 6 Pa by subliming solid naphthalene with heat.
17. Apply a high voltage (2 kV) across the electrodes for 10 s to generate a glow discharge for generation of the replica film.
18. Dip grids with replica film in acetone (dissolves the collodion support film).
19. Transfer to 1% hypochlorite solution for 5 min (digests samples).
20. Wash with water and examine in an electron microscope without any metal coating.

2.2.4. Correlative LM Immunocytochemistry and EM Immunocytochemistry

In many studies it is desirable to gain information from both LM and EM. Methods have been devised for such correlative studies *(30–32)*. Sawada and Esaki's method *(33)* was designed to address factors that they believed to be important for accurate comparison. These factors are the transparency and thinness of embedded blocks, the flatness of tissue sections, and ease of removal of embedding molds from polymerized Epon blocks. Sawada and Esaki's method is summarized below.

SLIDE   FILTER PAPER   TISSUE SECTION

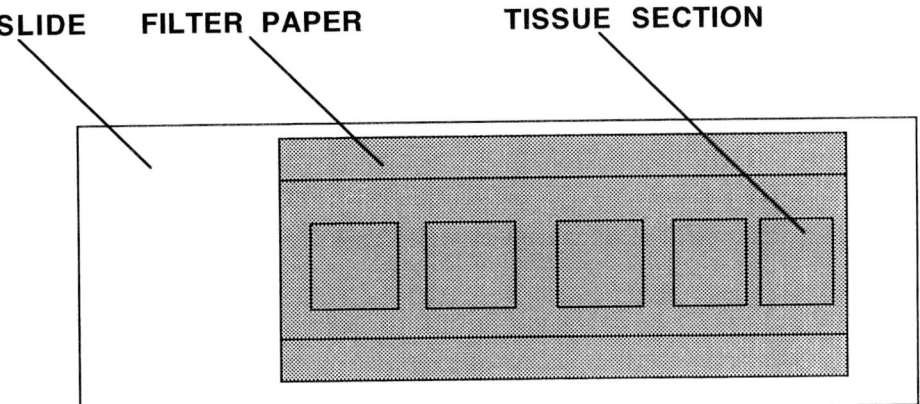

Fig. 1. Slide with tissue sections. Gray area represents applied Epon. The aclar film (cut to fit) is applied on top of this setup.

Correlative Microscopic Procedure Suitable for Immunocytochemistry

1. Fix tissue with periodate-lysine-paraformaldehyde (PLP) fixative.
2. Immerse in 30% sucrose in 0.1 M phosphate buffer, pH 7.4.
3. Embed in octanol (OCT) compound.
4. Freeze in liquid nitrogen.
5. Cut 10–20 μm cryostat sections.
6. Align sections on center of slides coated with 3-aminopropytriethoxy-silane (silane)—*see* Fig. 1.
7. Immunolabel sections (in the case of Sawada and Esaki, nanogold was conjugated to the secondary antibodies and silver enhancement was employed *[34]*.)
8. Postfix with 1% osmium tetroxide.
9. Stain with uranyl acetate.
10. Dehydrate through an ethanol series substituted with propylene oxide and add several drops of Epon before sections dry.
11. Put a spacer made of a thin ribbon of filter paper along both sides of the aligned sections.
12. Put aclar film or a polyester overhead projector film on the Epon.
13. Wipe off excess Epon with filter paper.
14. Polymerize resin in a 60°C oven for 2 d.
15. Peel off the plastic film with forceps or razor blades.
16. View under an LM.
17. Heat slides to 70–80°C with a hot plate.
18. Trim and peel the blocks from the slides with razor blades and forceps.
19. Glue the sections, tissue side outward, to support stubs for an ultramicrotome.
20. Cut sections and stain with uranyl acetate and lead citrate.
21. View with a TEM.

2.2.5. *IN SITU* HYBRIDIZATION AND DETECTION OF PROBE

Armed with the common techniques of molecular biology and immunocytochemistry, an investigator is in a good position to apply *in situ* hybridization to EM for localization of nucleic acids at the ultrastructural level. McFadden *(9)* has a review on such use of *in situ* hybridization techniques. In the review, McFadden has included details of some of his laboratory protocols needed in *in situ* hybridization from fixation and labeling to probe labeling to the hybridization steps for localization of specific RNAs at the EM level. His protocol for hybridization is outlined below:

1. Digest sections with Proteinase K (1 µg/mL in 50 mM EDTA, 100 mM Tris-HCl, pH 8.0) for 15 min at 20°C.
2. Remember to heat denature the probe if using a double-stranded probe.
3 Put 3 µL aliquots of probe into a sterile 4 cm polycarbonate Petri dish.
4. Place grids (pioloform-coated gold or nickel) on probe droplets.
5. Place Petri dish into a small resealable container.
6. Put approximately 0.25 mL of hybridization buffer in a small cap and put into the container.
7. Seal the container using autoclave tape.
8. Hybridize for at least 2 h.
9. Dip grids in 4X standard saline citrate (SSC): 0.15 M NaCl$_2$, 0.015 M Na-citrate, sodium phosphate buffer, 7.0.
10. Dip grids in 2X SSC.
11. Incubate for 2 h in the vapor-tight container in a droplet of 1X SSC at the hybridization temperature.
12. Transfer grids to a drop of SC buffer (50 mM PIPES, 0.5 M NaCl, 0.5% Tween 20).
13. Block grids in 1% BSA in SC buffer.
14. Incubate in antihapten antibody (5 µg/mL) in SC buffer with 1% BSA for 30 min at room temperature. Antibody may be goat anti-biotin, rabbit anti-biotin, or sheep anti-digoxygenin.
15. Rinse with SC buffer from squirt bottle.
16. Incubate for 30 min at room temperature in secondary antibody conjugated to gold marker. Secondary antibody may be goat anti-rabbit/5 or 15 nm gold, protein A-gold, goat anti-rabbit/10 nm gold, rabbit anti-sheep/15 nm gold, or protein G conjugated to colloidal gold.

For double labeling, McFadden suggests labeling one probe with digoxygenin and the other with biotin:

1. Verify that each system works well separately.
2. Combine equal quantities of the two probes.
3. Hybridize for 90 min.
4. Wash in SSC for 90 min.

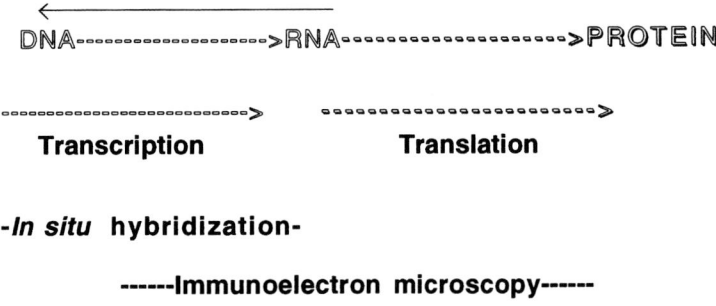

Fig, 2. Representation of the central dogma of molecular biology. Immunoelectron microscopy accompanies many other techniques used to study gene expression including *in situ* hybridization.

5. Detect with sheep anti-digoxygenin.
6. Mark with rabbit anti-sheep/15 nm gold.
7. Use unconjugated protein A (36 µg/mL in SC) to block protein A binding sites on the Fc portion of rabbit anti-sheep.
8. Detect biotinylated probe with rabbit anti-biotin.
9. Mark with protein A-gold/10 nm.

3. CONCLUSIONS

Molecular biology involves the study of the major macromolecules, DNA, RNA, and protein. The central dogma of molecular biology is illustrated in Fig. 2. The central dogma shows the relationship among the macromolecules in the processes of transcription and translation. Figure 2 also gives the relationship between immunoelectron microscopy and *in situ* hybridization. *In situ* hybridization allows one to localize a specific nucleic acid sequence. Immunoelectron microscopy is an essential component to the technique of *in situ* hybridization when applied at the EM level.

Immunoelectron microscopy is not limited to nucleic acid localization but is also an essential component in the localization of a specific protein, polysaccharide, or theoretically any hapten under study. Therefore, immunoelectron microscopy is a valuable tool when it comes to the study of gene expression. Electron microscopy is a valuable tool in molecular biology and is even more powerful when combined with immunochemical techniques.

REFERENCES

1. Sommerville J, Scheer U, eds. *Electron Microscopy in Molecular Biology, A Practical Approach*, IRL Press, Oxford, Washington, DC, 1987.
2. Slayter H, Warner J, Rich A, Hall C. The visualization of polyribosomal structure. *J Mol Biol* 1963; 7: 652–657.

3. Shelton E, Kuff E. Substructure and configuration of ribosomes isolated from mammalian cells. *J Mol Biol* 1966; 22: 23–31.

4. Stark H, Rodnina M, Rinke-Appel J, Brimacombe R, Wintermeyer W, van Heel M. Visualization of elongation factor Tu on the *Escherichia coli* ribosome. *Nature* 1997; 389: 403–405.

5. Schultz P, Olland S, Oudet P, Hancock R. Structure and conformational changes of DNA topoisomerase II visualized by electron microscopy. *Proc Natl Acad Sci USA* 1996; 93: 5936–5940.

6. Yaffee M, Walter P, Richter C, Muller M. Direct observation of iron-induced conformational changes of mitochondrial DNA by high-resolution field-emission in-lens scanning electron microscopy. *Proc Natl Acad Sci USA* 1996; 93:5341– 5346.

7. Wang Y, Griffith J. The $((G/C)_3NN)_n$ motif: a common DNA repeat that excludes nucleosomes. *Proc Natl Acad Sci USA* 1996; 93: 8863–8867.

8. Osheim Y, Mougey E, Windle J, Anderson M, O'Reilly M, Miller O Jr, Beyer A, Sollner-Webb B. Metazoan rDNA enhancer acts by making more genes transcriptionally active. *J Cell Biol* 1996; 133: 943–954.

9. McFadden G. *In situ* hybridization techniques: molecular cytology goes ultrastructural, in *Electron Microscopy of Plant Cells* (Hall J, Hawes C, eds.), Academic Press, London, 1991, pp. 219–255.

10. Chaudhuri S, Bhattacharyya N, Bhattacharjee B. Hydrogen peroxide-induced DNA damages *in vitro* revealed by electron microscopy. *J Electron Microsc* 1992; 41: 57–59.

11. Coggins L. Preparation of nucleic acids for electron microscopy, in *Electron Microscopy in Molecular Biology-A Practical Approach* (Sommerville J, Scheer U, eds.), IRL Press, Oxford, Washington, DC, 1987, pp. 1–29.

12. Maruyama K, Kume N, Okuda M. A scanning electron microscopic observation of chromosomes in freeze-fractured cells of *Vicia faba*. *J Electron Microsc* 1985; 34: 162–168.

13. Murakami T. A metal impregnation method of biological specimens for scanning electron microscopy. *Arch Histol Jpn* 1973; 35: 323–326.

14. Tanaka K, Matsui I, Kuroda K, Mitsushima A. A new ultra-high resolution scanning electron microscope (UHS-TI). *Biomed SEM* 1985; 14: 23–25. (in Japanese)

15. Tanaka K, Mitsushima A, Kashima Y, Osatake H. A new high resolution scanning electron microscope and its application to biological materials, in *Proc Eleventh Intl Cong Electron Microsc, vol III* (Imura T, Maruse S, Suzuki T, eds.), Publication Committee of the XIth International Congress on Electron Microscopy, Kyoto, Japan, 1986, pp. 2097–2100.

16. Inaga S, Osatake H, Tanaka, K. SEM images of DNA double helix and nucleosomes observed by ultrahigh-resolution scanning electron microscopy. *J Electron Microsc* 1991; 40: 181–186.

17. Seki S, Nakamura T, Oda T. Supranucleosomal fiber loops of chicken erythrocyte chromatin. *J Electron Microsc* 1984; 33: 178–181.

18. Tanaka K, Mitsushima A, Yamagata N, Kashima Y, Takayama H. Direct visualization of colloidal gold-bound molecules and a cell surface receptor by ultrahigh resolution scanning electron microscopy. *J Microsc* 1990;161: 455–461.

19. Tanaka K, Iino A. Critical point drying method using dry ice. *Stain Technol* 1974; 49: 203–206.

20. Tohno Y, Tohno S, Tanaka Y. Chromatin loop size in human leukemia (HL-60) cells. *J Electron Microsc* 1995; 44: 35–38.

21. Miller O, Bakken A. Morphological studies of transcription. *Acta Endocrinol Suppl* 1972; 168: 155–177.

22. Inada N, Sakai A, Kuroiwa H, Kuroiwa T. Three-dimensional analysis of the senescence program in rice (*Oryza sativa L.*) coleoptiles. *Planta* 1998; 206: 585–597.

23. Yamashita S. Intranuclear localization of hormone-occupied and -unoccupied estrogen receptors in the mouse uterus: application 1 nm immunogold-silver enhancement procedure to ultrathin frozen sections. *J Electron Microsc* 1995; 44: 22–29.

24. Hosaka Y, Taguchi K, Iwamoto T, Kuroda K, Tsuruoka H, Xu H, Hamaoka T. Ultrastructure of murine tumour cell lines defective in MHC class I expression before and after interferon-γ treatment. *J Electron Microsc* 1998; 47: 495–503.

25. Romano E, Romano M. Staphylococcal protein A bound to colloidal gold: a useful reagent to label antigen-antibody sites in electron microscopy. *Immunochemistry* 1977; 14: 711–715.

26. Forsgren A, Sjoquist X. "Protein A" from S. aureus. I. pseudo immunoreaction with human g-globulin. *J Immunol* 1966; 97: 822–827.

27. Geuze H, Slot J, Van Der Ley P, Scheffer R. Use of colloidal gold particles in double-labeling immunoelectron microscopy of ultrathin frozen tissue sections. *J Cell Biol* 1981; 89: 653–665.

28. Yamaguchi M, Kondo I. Immunoelectron microscopy of *Proteus vulgaris* by the plasma polymerization metal-extraction replica method: differential staining of flagellar (H) and somatic (0) antigens by colloidal golds. *J Electron Microsc* 1989; 38: 382–388.

29. Tanaka A, Sekiguchi Y, Kuroda S. A new replica method for electron microscopic studies of a plasma polymerization film with a glow discharge. *Seikagaku* 1983; 55: 1212–1219. (in Japanese)

30. Aldes L, Boone T. A combined flat-embedding, HRP histochemical method for correlative light and electron microscopic study of single neurons. *J Neurosci Res* 1984; 11: 27–34.

31. Reymond O, Pickett-Heaps J. A routine flat embedding method for electron microscopy of microorganisms allowing selection and precisely oriented sectioning of single cells by light microscopy. *J Microsc* 1983; 130: 79–84.

32. De Felipe J, Fairen A. A simple and reliable method for correlative light and electron microscopic studies. *J Histochem Cytochem* 1993; 41: 769–772.

33. Sawada H, Esaki M. A simple flat embedding method for the correlative light and electron microscopic immunocytochemistry. *J Electron Microsc* 1998; 47: 535–537.

34. Sawada H, Esaki M. Use of nanogold followed by silver enhancement and gold toning for postembedding immunolocalization in osmium-fixed, Epon-embedded tissues. *J Electron Microsc* 1994; 43: 361–366.

21 Summation

William V. Dashek

Transmission (Chapter 14) and scanning (Chapter 13) electron microscopies remain the mainstays of the functioning EM laboratory. It is
likely that the field of EM will witness new fixation and tissue processing procedures, especially in the development of novel embedding media
(Chapter 14). It is anticipated that cryofixation and microwave fixation
(Chapter 14) will become more commonplace, thereby overcoming chemical fixation artefacts. In addition, marked improvements in instrumentation, i.e., computerized (Chapter 9) and multifunctional EMs are a
continuing process. Some examples of recent improvements are atomic
force and scanning tunneling microscopies (Chapter 15).

Some of the well-established ancillary EM techniques have found new
life in their applications to molecular biology (Chapter 20). Negative
staining has been very useful to certain biochemists interested in tentatively identifying subcellular organelles isolated from cellular and tissue homogenates (Chapter 11) for biochemical studies. It is anticipated
that more molecular biologists and biochemists will become cognizant
of the merit of certain EM techniques, e.g., *in situ* hybridization in visualizing macromolecular complexes (Chapter 20). Tissue printing on nitrocellulose membranes (Chapter 7) offers the molecular biologists and
biochemists a rapid means of imaging macromolecules within cells and
tissues.

Other recently developed techniques, such as immunoelectron microscopy (Chapter 18), electron systems imaging, and X-ray microanalysis
(Chapter 19), should become routine practice in most EM laboratories
and, thus, mainstays rather than ancillary EM methods.

Many electron microscopists appreciate the value of the array of
sophisticated, light microscopic techniques that are currently available
(Chapter 1). In fact, volumes regarding correlative LM and EM are making

From: *Plant Electron Microscopy and Cytochemistry*
Edited by: W. V. Dashek © Humana Press Inc., Totowa, NJ

their appearances. Combined LM and EM autoradiography on alternative thick and thin sections of resin-embedded specimens is routine in certain laboratories (Chapters 3 and 17). It is anticipated that correlative light (Chapter 6) and electron (Chapter 18) immunocytochemistry will become routine. Although the development of specific cytochemical stains, especially fluorochromes (Chapters 4 and 5) for LM, continues at a faster rate than that for EM, there are electron microscopists (Chapter 16) who have committed themselves to developing stains with enhanced specificity for the localization of chemicals at the ultrastructural level. Indeed, the coupling of certain LM cytochemical reagents with heavy metals, e.g., PAS-silver or zinc iodide osmium tetroxide (Chapter 16), offers the promise for improving stain specificity at the EM level.

Finally, the array of modern LMs, including nuclear magnetic resonance, confocal laser, dark-field, phase-contrast fluorescence (Chapter 1), continues to be extended. The array offers the electron microscopist many opportunities for correlative LM and EM possibilities.

INDEX